D1206961

ROBERT E. KRIEGER
PUBLISHING COMPANY INC.
MALABAR
FLORIDA 32950

INSECT THERMOREGULATION

INSECT THERMOREGULATION

Edited by

BERND HEINRICH
UNIVERSITY OF VERMONT, BURLINGTON

A Wiley-Interscience Publication

JOHN WILEY & SONS, New York • Chichester • Brisbane • Toronto

Library of Congress Cataloging in Publication Data:

Main entry under title:
Insect thermoregulation.

Revised papers originally presented as a symposium
sponsored by the American Society of Zoologists at the annual meeting, Dec. 27-30, 1978.
"A Wiley-Interscience publication."
Includes index.
1. Insects—Physiology—Congresses. 2. Body temperature—Regulation—Congresses. I. Heinrich, Bernd,
1940- II. American Society of Zoologists.

QL495.I52 595.7'0188 80-19452
ISBN 0-471-05144-6

Printed in the United States of America

10 9 8 7 6 5 4 3 2 1

Contributors

Dr. George A. Bartholomew, Department of Biology, University of California, Los Angeles, California 90024

Dr. Timothy M. Casey, Department of Environmental Physiology, Cook College of Rutgers University, New Brunswick, New Jersey 08903

Dr. Bernd Heinrich, Department of Zoology, University of Vermont, Burlington, Vermont 05405

Dr. Robert K. Josephson, Department of Developmental and Cell Biology, University of California, Irvine, California 92664

Dr. Ann E. Kammer, Division of Biology, Kansas State University, Manhattan, Kansas 66506

Dr. Thomas D. Seeley, Museum of Comparative Zoology, Harvard University, Cambridge, Massachusetts 02138

Preface

Both as editor of this volume and as organizer of the symposium from which it is spawned, I faced the difficult responsibility of selecting the contributors. Several criteria went into my choices. The number of contributors was limited in order to encourage synthesis of ideas. In addition, an opportunity was provided to present a wide diversity of expertise and opinion.

This volume is not intended to be a collection of specialized research reports. It is, rather, meant to be an up-to-date summary and review of the field of insect thermoregulation from a diversity of perspectives. The discerning reader will perceive that the lacunae of knowledge represent real gaps in information rather than lack of coverage of published material.

These papers were originally presented as a symposium sponsored by the American Society of Zoologists at the Annual Meeting held in December 27–30, 1978, at Richmond, Virginia. They have subsequently been revised and in some cases greatly expanded. James E. Heath, a pioneer who did much to stimulate the recent revival of interest in the field, was unfortunately unable to attend the meeting and felt that it would be inappropriate to provide a manuscript after not having been present for the formal presentations and discussions. His anticipated contributions in this synthesis of insect temperature regulation are missed.

BERND HEINRICH

Vermont, Burlington
October 1980

Preface

Contents

INSECT THERMOREGULATION

INTRODUCTION

BERND HEINRICH

Body temperature measurements of insects date back nearly 2½ centuries, and measurements of the temperature in beehives are at least as old. Most of the earlier measurements (as well as many recent ones) have shown little or no difference between body and ambient air temperature of individual insects. Insects have thus been labeled "poikilothermic" animals. In comparison with the high and endothermically maintained levels of body temperature of homeothermic birds and mammals, the temperature responses of insects were perhaps prematurely considered either primitive or uninteresting. Occasional reports of endothermy in insects were apparently either treated with skepticism or considered to represent isolated aberrations. Meanwhile, a rich literature on the intricacies of body temperature regulation of vertebrate animals developed.

A surge of research in the last decade has indicated that the thermal responses of insects constitute a field of investigation rivaling in intellectual content that long associated with vertebrates. Rapid progress has been made in part by borrowing from the already existing concepts developed in vertebrates.

Insects are an ideal group of animals for the study of evolutionary problems of temperature adaptation. Some have evolved to survive repeated freezing and thawing and can be revived after having been cooled almost to −273°C. Some are regularly active with tissue temperatures at 10°C or less, and still others maintain a high stable body temperature for extended periods of time over wide fluctuations of ambient temperature.

There is not only a broad spectrum of temperature-control phenomena, but there is also a spectrum of different ways of examining them. One can look at temperature control in terms of biochemical, cytological, morphological, physiological, behavioral, ecological, and evolutionary perspectives. Physiological aspects can in turn be further dissected and examined with reference to the muscles that generate the heat, the neurological mechanisms that control them, or the possible sensors and presumed neurological set points that drive the behavior. It is our aim to cover all these points of view, insofar as information is available. In addition we will, for the sake of context, discuss in a more superficial way some of the biochemical strategies related to temperature adaptation, even though they are examined in greater detail in other sources.

DEFINITIONS AND THERMOREGULATORY TAXONOMY

BERND HEINRICH

Throughout this volume we use a number of terms that may have different meanings for different people. For the sake of uniformity we here briefly review some of the common terms, indicating how we have used them to signify different aspects of temperature regulation in insects.

The variety of temperature-control phenomena defies concise classification. Most of the terms are borrowed from vertebrate physiology, and even the ones most commonly used by workers (see Table 1) have a range of meanings. Potential misunderstandings arise because not all the terms are mutually exclusive, and some of the terms refer to mechanisms in vertebrates that may or may not occur in insects.

Which of the terms best suit insects? The answer is that many terms fit, depending on the species and on what aspect of body temperature one wishes to emphasize. First, by *temperature regulation*, as opposed to *poikilothermy*, we imply that the animal maintains a stable temperature in at least a portion of the body, either above or below ambient temperature, by behavioral or physiological means. The limits within which body temperature is regulated may be narrow or wide. (The maintenance of a relatively stable body temperature, as such, because of passive processes such as thermal inertia, reflectance properties, or evaporative cooling resulting from normally moist surfaces, does not constitute temperature regulation.) *Implicit in temperature regulation is the assumption that the animal maintains its body temperature below some maximum set point, above some minimum set point, or both, for at least a portion of its activity period, even though thermal conditions in the environment may vary.*

The terms *ectothermy* and *endothermy* refer to the source of heat used to maintain body temperature above (or below) ambient. They do not imply the existence of temperature regulation. Ectothermy refers to heat gained from the environment, while endothermy refers to heat gained from biochemical processes inside the animal (produced as a necessary by-product of activity metabolism and/or produced specifically to prevent or retard cooling). Neither term has implicit assumptions regarding the mechanisms of heat gain or heat production, or of temperature regulation. In addition, an animal may be an ectotherm under one set of circumstances and an endotherm under a different set of circumstances. In either case it may or may not regulate its body temperature.

The term endothermy does not specify time or duration, but *heterothermy* refers more specifically to periodic endothermy. Finally, *homeothermy* refers to continual regulation of body temperature between both a lower and a higher set point, by any of a variety of mechanisms, either continuously or during some defined period. For example, a marmot is homeothermic in summer, but its body temperature decreases

Table 1 Some definitions

Homeothermy	Body temperature is relatively constant and independent of ambient temperature. The term is usually applied to birds and mammals.
Poikilothermy	Body temperature is variable and dependent on ambient temperature.
Endothermy	Heat that determines body temperature is produced by the animal's own energy metabolism. The only continuously endothermic terrestrial animals are birds and mammals.
Ectothermy	Heat that determines body temperature is acquired from the environment by radiation, convection, or conduction.
Heterothermy	Endothermic part of the time and ectothermic part of the time.

during hibernation. Depending on the season it may be classified as a homeotherm or a heterotherm. (Although the marmot maintains its body temperature above some minimum set point in winter, it does not attempt to cool down to this set point if ambient temperature increases.) A moth that is endothermic for an hour per day and ectothermic the rest of the time is clearly a heterotherm. When not active, it behaves like a poikilotherm, with body temperature passively following ambient temperature. When in flight, it regulates its body temperature and shows many of the attributes of homeothermy.

In summary, there is leeway in the use of the terms because they refer to different arbitrary points along a continuum. Many insects are primarily or exclusively poikilothermic. Others are periodic ectotherms and/or periodic endotherms, thus qualifying as heterotherms. Some endothermic insects regulate their body temperature, and all endothermic insects are heterothermic (i.e., either periodic poikilotherms or periodic homeotherms). A number of different terms may apply to the same animal, each emphasizing a different phenomenon.

Throughout the text we also use the following terms and abbreviations that refer to body temperature regulation. By *shivering* we mean contractions of the flight muscles, largely against each other rather than on the wings, that result in elevation of thoracic temperature. The rise and/or maintenance of an elevated thoracic temperature in a stationary insect (whether achieved by basking or shivering) is called *warm-up*. Shivering is, in some species, accompanied by low-amplitude wing vibrations which are sometimes also accompanied by buzzing. (Some early authors referred to endothermic warm-up as *wing whirring* before they were aware of its functional significance.) Shivering and endothermic warm-up are accomplished in some species without wing movements and without sound.

Another term in common use is *wing loading*, and this refers to the weight the animal holds aloft during flight divided by the area of its wings. In general, the smaller the wing area relative to body weight, the more rapidly the insect must beat its wings to stay aloft and the greater its rate of energy expenditure and heat production. However, most insects also vary the pitch of the wings in flight at different loads and speeds, hence the resistance of the wing through the air is not constant. As a result, the body weight/wing area ratio provides only a first-order approximation of the actual load or force generated during any one wingbeat. Ideally wing loading should be measured in terms of force per area, and while this has been examined mathematically in terms of aerodynamic theory, it has so far not been adequately linked with physiological studies of insect thermoregulation.

The following abbreviations are in common use: T_A, T_B, T_{Th}, and T_{Ab}. They refer to ambient, "body," thoracic, and abdominal temperature, respectively. Ambient temperature refers to air temperature (in shade) within several centimeters of the insect. Any one thoracic or abdominal temperature generally refers to the highest measured near the center of the body (unless specified otherwise).

Since many insects can cool several degrees Celsius within a few seconds, and since the temperature probe can itself withdraw heat and significantly reduce the T_B of the insect measured, the highest temperatures measured are usually underestimates of the actual body temperature.

Temperature measurements, though deceptively easy to take, are not always accurate unless the proper cautions and corrections are made. Discrepancies between different studies can be expected from a variety of sources of error; body temperatures cannot always be accepted at face value when the characteristics of the probe used are not evaluated, when the skill and speed of probe insertion vary, and when the instants at which the readings are made relative to the insect's cooling rate are not compared.

1
A Brief Historical Survey

BERND HEINRICH

As a springboard to the contributions that follow, I here provide a brief, simplified chronology of observations and ideas on insect temperature regulation to trace the major routes of intellectual evolution leading to our current concepts. However, this summary is not meant to provide a trajectory for future research and exploration, since, like random mutations, discoveries are mostly unplanned. Progress is inevitable so long as intellectual pressure is applied and the routes of development remain free.

In 1899 the Russian physicist Porfirij I. Bachmetjev (also spelled Bachmetjew), then living in Switzerland, wrote a historical review of insect body temperature measurements: In his publication on the body temperature of insects observed in Bulgaria, he said: "Die Frage über die Temperatur der Insekten interressiert seit langem die Gelehrten." (The question of the temperature of insects has long been of interest to scholars.) According to his review, body temperature measurements of insects date back nearly 2½ centuries, and regulation of the nest temperature in social insects was well known. As early as 1810, Pierre Huber commented on the thermal significance of ant mounds, while the great French naturalist and physicist René A. F. Réaumur measured the temperatures in honeybee hives in the early 1700s. However, despite the great value of the earlier observations, most of the seminal experimental work that provides the basis of our modern understanding of the physiology of insect thermoregulation is less than 10 years old.

The first significant event that ultimately led to our present concepts of insect thermoregulation was the simple step of taking body temperature measurements, even though these observations were at first crude and inadequate. Patterns began to emerge gradually as more and better measurements were taken in many species under many conditions. The patterns in turn generated hypotheses that ultimately yielded experimental work and insights into the how, the why, and the where of insect thermoregulation.

The first person credited with making measurements was Réaumur. Aside from measuring hive temperatures, he also observed that the caterpillars of the Painted Lady butterfly, *Vanessa cardui*, either beheaded or intact, did not freeze solid when cooled down to temperatures below the freezing point of water, and that pupae survived cooling to the same temperatures. Thus the concepts of heat production in beehives, and supercooling in lepidopterans, date back at least 240 years. Henri Dutrochet, who reviewed body temperature measurements of invertebrate animals in 1840, credits the noted geologist Johann F. Hausmann with taking the first temperature measurements showing that individual insects produce heat. Hausmann in 1803 put sphinx moths (*Sphinx convolvuli*) and beetles (*Carabus hortensis*) into small glass vials, along with a small mercury thermometer. The temperature in the vials rose about 2°C above air temperature after ½ hr and then declined. Dutrochet presumed they asphyxiated.

In 1826 John Davy made the first internal temperature measurements. He inserted a small mercury thermometer directly into the body. Although he made no distinction between thorax and abdomen, he recorded body temperature to within 0.1°C in a cockroach, a cricket, a wasp, and a lantern beetle. He found that the cockroach (which we now know to be poikilothermic) was 0.5°C above ambient temperature. However, he found the same small temperature difference in wasps (which we now know to be capable of considerable endothermy). There were of course three main problems with these measurements. First, an animal skewered on a thermometer is something less than a functioning organism. Second, Davy took into account neither the difference between thorax and abdomen nor the distinction between activity and rest. Last, a mercury thermometer is a huge heat sink into which a large amount of body heat can flow from the object being measured.

To avoid most of the above problems it was necessary to reduce the size of the thermoprobes. A step in the right direction was taken in 1831 by Leopoldo Nobili and Macedonio Melloni, who were the first to use thermocouples to measure body temperatures of insects. They measured the temperatures of caterpillars, pupae, and adult butterflies. Like Davy, they were concerned primarily with the *precision* of the temperature measurements, rather than with biologically relevant accuracy. They found that the body temperatures were several tenths of degree Celsius above ambient, but they probably measured only resting butterflies, because they found that the temperature excess of the caterpillars exceeded that of the adults (and the pupae).

The first extensive measurements of insect body temperature that had biological significance were published in 1837 by George Newport, a member of the Entomological Society of London and the Royal College of Surgeons. The introduction to his paper entitled, "On the Temperature of Insects, and Its Connexion with the Functions of Respiration and Circulation in This Class of Invertebrate Animals," has a remarkably modern ring to it. He wrote: "Every naturalist is aware that many species of insects, particularly hymenopterous insects, which live in society, maintain a degree of heat in their dwellings considerably above that of the external atmosphere, but no one, I believe, has hitherto demonstrated the interesting fact that every individual insect when in a state of activity maintains a separate temperature of body considerably above that of the surrounding atmosphere, or medium in which it is living, and that the amount of temperature varies in different species of insects, and in different states of those species.''

Newport's major contribution was the demonstration of a correlation between activity and elevated body temperature, a condition that contrasted with the available data on vertebrate animals at that time. He also found that flying insects, even when not in flight, tended sometimes to be hotter than crawling ones. An adult *Sphinx ligustri* measured 5.5°F above air temperature, and a *Bombus terrestris* heated up 9.3°F. Running beetles, such as the mainly terrestrial predaceous *Carabus* spp., had a temperature excess of 0.3°F or less. He used mercury thermometers ''scarcely larger than crow quills,'' which he generally laid against the side of the abdomen; thus the temperatures he reported underestimated the actual body temperatures.

Newport's paper was a landmark for its time. It stimulated later research, possibly as much for its shortcomings as for its contributions. However, Newport appeared to have been carried away by his enthusiasm and vertebrate biases. In the same publication he wrote on the blood circulation in caterpillars and on the pulsations of the blood vessels in the dura mater of a human patient that had been trepanned. In both he noted a large difference in pulsation rate between rest and during activity. In addition, he made other comparisons, trying to show that insects were like vertebrates. For example, he described bumblebees incubating their brood like birds. (This, though it must have seemed fantastic at the time, turned out to be true.) He made additional observations which probably did *not* appear to be extraordinary at the time but which we now know to be wrong. He said, ''If the excited state of the insect be excessive, and the consequent evolution of heat greatly exceed its usual amount, nature has resorted to an expedient of cooling down the animal body, through means of profuse perspiration, which is carried on in insects perhaps to a greater extent than in other animals.'' (He probably observed the glistening intersegmental membranes of bumblebees when they had their abdomens extended while incubating and mistook the lipid covering for sweat.)

In defense of Newport, and others who subsequently made similar errors in applying analogies to the wrong species, under the wrong conditions, it might be worth noting that for almost every mechanism and idea described in insects there is an analogous situation in vertebrates that provides an often useful (but sometimes misleading) intellectual precedent. Indeed, when it comes to social insects, one wonders if it might not be possible to reverse the direction of transfer and profit from their example in architecture by applying some of the techniques of temperature control that they have evolved to buildings constructed by humans.

Unfortunately specifics cannot always be predicted from statistical generalities or from generalities applying to both vertebrates and insects. There are different ways of achieving the same results, and each case must be treated as unique. The fascinating thing is that comparisons can be made between insects and vertebrates at all, for insects are morphologically and anatomically so different from vertebrates that they could have evolved in another world. Thus, because of the different body plans, the details of the mechanisms of both control of heat loss and heat production are quite different in insects and vertebrates. To observe how insects have evolved similar (analogous) solutions to similar thermal problems is to gain greater insights into the evolution of temperature regulation. But the study of insect body temperatures was left fallow for about 60 years after Newport.

We can obtain additional insights from the publication of Bachmetjev in 1899. Bachmetjev enlarged on Newport's findings and refined the measurements. He inserted thermocouples into the thorax, thus identifying the source of the heat. He measured thoracic temperature continuously, showing the extreme lability of body temperature with time. He plotted, for example, the thoracic temperature of a *Saturnia pyri* silk moth, showing how it fluctuated between 18 and 25°C for an hour. He observed that, when the moth began wingbeats, thoracic temperature immediately rose and, when wingbeats stopped, it immediately declined. These observations clearly associated the elevation of body temperature with muscle activity, a fact already alluded to by Maurice Girard in his paper on the body temperature of invertebrate animals in 1869.

After Bachmetjev there followed a period of about 30 years in which there were no new significant observations on insect thermoregulation. But in 1928 Heinz Dotterweich from the Zoological Institute at Kiel published a landmark paper first scratching the surface of the "why" —the biological significance—of insect thermoregulation. Dotterweich worked mainly with sphinx moths and established that the wing vibrations and associated rise in thoracic temperature were related to achieving flight readiness; a high thoracic temperature was necessary for flight. Warm-up lasted several minutes, the exact duration depending on the temperature difference between the initial body (and ambient) temperature and the 34–36°C at takeoff. M. J. Oosthuizen from the College of Agriculture in Potchefstroom, South Africa, in 1939 made similar observations on the saturniid moth *Samia* (now *Hyalophora*) *cecropia* while on leave at the laboratory of H. H. Shepard at the University of Minnesota, and in addition he measured abdominal temperature simultaneously with thoracic temperature and showed that the abdomen remained relatively cool throughout warm-up.

The next landmark paper was that by August Krogh, then the "dean" of respiratory physiology, and Eric Zeuthen in 1941. This study was sparked in part by the work of Marius Nielsen (1938) from the same laboratory in Denmark, who showed that body temperature in humans rose during strenuous activity and that it was regulated at higher levels corresponding to work output. Krogh and Zeuthen were interested in warm-up and flight exercise not only in Lepidoptera but also in beetles and bumblebees. They found that preflight warm-up in beetles and bees was *not* accompanied by externally visible wing vibrations. But by making recordings from electrodes placed within the thorax, they found that the thoracic muscles were nevertheless actively exercising. Meanwhile, a decade before R. B. Cowles launched the study of behavioral thermoregulation in reptiles, Gottfried Fraenkel inferred that some insects, specifically migrating locusts, *Schistocerca gregaria*, were able to elevate their body temperature significantly without shivering by complex postural adjustments to sunshine. These observations on behavioral temperature control were later greatly extended by many others in work primarily on butterflies but also with many other insects (see Casey, this volume).

Norman S. Church, a student of V. B. Wigglesworth at the University of Cambridge, worked on the biophysics of heat exchange in flying insects, which set the stage for subsequent physiological work. The first indication that an insect might be capable of regulating its body temperature in flight—rather than merely raising it to initiate activity—was indicated by some data reported by D. A. Dorsett from University College, Ibadan, Nigeria, in 1962. Dorsett focused on sphinx moths and he confirmed and extended Dotterweich's findings by showing that the wingbeat frequency rose during warm-up until it achieved that characteristic of flight just before takeoff. He also showed that different individuals may maintain different thoracic temperatures in flight and (45 specimens of *Deilephila nerii* had different thoracic temperatures ranging from 34 to 45°C) that thoracic temperature was correlated with wing loading. It appeared as though the moths were regulating their thoracic temperature, but he did not prove it because he did not make the measurements necessary to see if the same thoracic temperatures were maintained in a *variety* of ambient temperatures.

Three years later, in 1965, James E. Heath, until then primarily a physiologist working on behavioral thermoregulation in reptiles, and Phillip A. Adams published a short paper in *Nature* entitled "Temperature Regulation of a Hawk Moth in Flight." This paper provided the first evidence for thoracic temperature regulation by individual insects in flight. These workers had flown *Celerio* (now *Hyles*) *lineata* on a tether at *different* ambient temperatures and found that these moths, which weigh ½ g or less, maintained similar thoracic temperatures at different ambient temperatures. This first clear evidence for the *regulation* of thoracic temperature in flight was surprising and exciting news which ushered in a lot of enthusiasm and ultimately a sudden flurry of work on insect thermoregulation.

Since there has been so much work done in the last 10 years or so, there is no way for me to do justice to everyone's contribution in this brief historical sketch. However, these contributions will be discussed in detail by the various contributors to this volume. Before indulging myself by presenting my personal involvement and perspective, I would like to make brief reference to those workers who contributed greatly to the understanding of insect body temperature, though they did not specifically focus on the area of thermoregulation. They contributed by providing the physiological basis upon which much of the subsequent work rested.

In the 1960s a number of workers were involved in a complex cross-pollination that ultimately provided the basic physiology of insect flight muscle and its control by the nervous system. This had bearing on insect thermoregulation, for the muscles are the "furnace" that provides almost all the endothermic heat. In Denmark in August Krogh's laboratory, Torkel Weis-Fogh (1964, 1967) and A. C. Neville (1963) examined the contraction kinetics of locust muscles as a function of temperature, as well as the functional design of the tracheal system necessary for muscle functioning. Weis-Fogh later went to Cambridge University in England, while J. W. S. Pringle (1968) and co-workers (Machin et al., 1962) at Cambridge and Oxford were then also working on temperature-related aspects of muscle contraction. Donald M. Wilson, after taking his Ph.D. in neurophysiology in T. H. Bullock's laboratory at the University of California at Los Angeles, took a postdoctoral position with Weis-Fogh in Copenhagen and concentrated on studying the neural aspects of muscle contraction in the locust (Wilson, 1962; Wilson and Weis-Fogh, 1962). Heath and Josephson (1970) later extended this work to an examination of the singing muscles of katydids whose temperature is regulated during singing.

When Wilson was in the Zoology Department at the University of California at Berkeley, his student Ann E. Kammer (1968) provided the first detailed neurophysiological picture of the control of heat generation by shivering in large Lepidoptera. In addition, A. Himmer had previously (1925) shown that honeybees were capable of considerable endothermy, and Harald Esch (1964), a student of K. von Frisch, and his student Joseph Bastian (1970) at the University of Notre Dame, were beginning detailed examinations of the physiology of honeybee muscles during shivering and flight. Others among the numerous contributors to our understanding of insect muscle physiology who later aided our comprehension of insect thermoregulation included Kenneth Roeder from Tufts University and Kazuo Ikeda and Edward Boettiger from the University of Connecticut. Both provided important data on the relationship between electrical activity and contraction kinetics.

Within this milieu there emerged another paper by Heath and Adams in 1967, entitled "Regulation of Heat Production by Large Moths." In it they attempted to show that the temperature regulation they had previously reported for the white-lined sphinx moth, *C. lineata*, was due entirely to an increase in heat production to offset cooling. Although this was subsequently found not to apply to sphinx moths in flight, several years later it turned out that the *idea* was sound (for bumblebees while they were *not* in flight).

Heath and Adams' paper left an open question of how the moths regulated their thoracic temperature in flight. Indeed, I was not convinced that they did so at all, and George Bartholomew at the University of California in Los Angeles wisely encouraged me to pursue the question as a thesis problem.

It seemed that more controlled experiments needed to be made, and I designed a flight mill on which to fly moths and measure their thoracic temperature continuously. These measurements with the large 1.5 to 3.0-g sphinx moth, *Manduca sexta*, gave little or no indication of temperature regulation (Heinrich, 1971). I was almost ready to go to press with this information, when I recalled Dorsett's paper on wing loading and decided that maybe the moths were doing something different in free flight than in partially supported flight. Measurements of body temperatures of *Hyles lineata* captured in the Mojave Desert at different ambient temperatures and measured within 2–3 sec convinced me that sphinx moths were indeed capable of regulating their thoracic temperatures. In fact, in the field they maintained an even higher and more stable thoracic temperature than Adams and Heath had reported. I had almost fallen into the trap of many of the earlier workers—of presuming that greater precision of measurement substituted for relevant conditions at the time the observations were being made. The key was that there was a large difference between wing beating, as such, and free, unsupported flight.

When it became clear later that the metabolic rate during flight was the same at different air temperatures, there was no alternate conclusion except that *all* of the temperature regulation during flight was due to regulation of heat loss. I had wished it were some of each, in order to seem less dogmatic, but the data allowed no other interpretation. I had no idea of what the mechanism might be. But an old paper by Franck Brocher (1920) showing the loop of the aorta through the flight muscles in sphinx moths immediately rang a bell—it looked like a perfectly designed cooling coil—and this idea ultimately led to the solving of an interesting riddle (Heinrich, 1970). Similarly, the detailed anatomical studies by Freudenstein (1928) on the honeybee led to stimulating comparative studies between bumblebees and honeybees.

Meanwhile, a fellow student down the hall at the University of California in Los Angeles, F. Gary Stiles, was studying the energetics of hummingbirds. We took several field trips to the Anza Borego Desert, and from him I learned with fascination about taking nectar from flowers to measure the energy uptake of nectivores. When I returned home to Maine after receiving my Ph.D., I idly measured temperatures of bumblebees. Noting differences at different flowers I at once thought of the work of Vance Tucker (1966) and that of other of Bartholomew's co-workers long interested in torpor and small rodents and birds. This started a new line of research on energetics in relation to thermoregulation, and since this involved bees it also led to an interest in pollination.

More recent work, a lot of it done by Bartholomew, Timothy M. Casey, and others, has emphasized comparative aspects. We are now beginning to understand not only the mechanisms of thermoregulation in insects, but we are also beginning to see general patterns that promise to provide insights into the ecology and the evolution of thermoregulation as a general phenomenon. Although there is reason to feel optimistic about the progress that has been made, I also feel that because of the great variety of insects there is much to be done, and that we can expect many more surprises.

REFERENCES

Bachmetjer, P. (1899). Über die Temperaturen der Insekten nach Beobachtungen in Bulgarien. *Z. Wiss. Zool.* **66,** 521–604.

Bartholomew, G. A. (1963). Behavioral adaptations of mammals to the desert environment. *Proc. XVI Int. Congr. Zool. Wash., D.C.* **3,** 49–52.

Bastian, J. and Esch, H. (1970). The nervous control of the indirect flight muscles of the honey bee. *Z. Vergl. Physiol.* **67,** 307–324.

Brocher, F. (1920). Étude expérimentale sur le functionnement du vaisseau dorsal et sur la circulation du sang chez les insectes. *Arch. Zool. Exp. Gen.* **60,** 1–45.

Church, N. S. (1960). Heat loss and body temperature of flying insects. *J. Exp. Biol.* **37,** 171–212.

Davy, J. (1826). Observations sur la témperature de l'homme et des animaux des divers genres. *Ann. Chim. Phys. Ser. 2* **33,** 180–197.

Dorsett, D. A. (1962). Preparation for flight by hawk moths. *J. Exp. Biol.* **39,** 579–588.

Dotterweich, H. (1928). Beitrage zur Nervenphysiologie der Insekten. *Zool. Jahrb. Abt. Allg. Zool. Tiere* **44,** 399–450.

Dutrochet, H. (1840). Recherches sur la chaleur de êtres vivants à basse température. *Ann. Sci. Nat. (Zool.) Ser. 2* **13,** 5–58.

Esch, H. (1964). Über den Zusammenhang twischen Temperatur, Aktionpotentialen und Thoraxbewegungen bei der Honigbiene (*Apis mellifica* L.). *Z. Vergl. Physiol.* **48,** 547–551

Fraenkel, G. (1930). Die Orientierung von *Schistocerca gregaria* zu strahlender Wärme. *Z. Vergl. Physiol.* **13,** 300–313.

Freudenstein, K. (1928). Das Herz und das Circulationssystem der Hönigbiene (*Apis mellifica* L.). *Z. Wiss. Zool.* **132,** 404–475.

Girard, M. (1869). Études sur la chaleur libre dégagée par les animaux invertébrés et spécialment les insectes. *Ann. Sci. Nat. (Zool.) Ser. 5* **11,** 135–274.

Heath, J. E. and Adams, P. A. (1965). Temperature regulation in the sphinx moth during flight. *Nature* **205,** 309–310.

Heath, J. E. and Adams, P. A. (1967). Regulation of heat production by large moths. *J. Exp. Biol.* **47,** 21–33.

Heath, J. E. and Josephson, R. K. (1970). Body temperature and singing in the katydid, *Neoconocephalus robustus* (Orthoptera, Tettigoniidae). *Biol. Bull. Woods Hole, Mass.* **138,** 272–285.

Heinrich, B. (1970). Thoracic temperature stabilization in a free-flying moth. *Science* **168,** 580–582.

Heinrich, B. (1971). Temperature regulation of the sphinx moth, *Manduca sexta*. I. Flight energetics and body temperature during free and tethered flight. *J. Exp. Biol.* **54,** 141–152.

Himmer, A. (1925). Körpertemperaturmessungen an Bienen und anderen Insekten. *Erlangen Jahrb. Bienenk.* **3,** 44–115.

Kammer, A. E. (1968). Muscle activity during warm-up in Lepidoptera. *J. Exp. Biol.* **48,** 89–109.

Huber, P. (1810). Recherches sur les moeurs des fourmis indigènes. Paschoud: Paris.

Krogh, A. and Zeuthen, E. (1941). The mechanism of flight preparation in some insects. *J. Exp. Biol.* **18,** 1–10.

Machin, K. E., Pringle, J. W. S., and Tamasige, M. (1962). The physiology of insect fibrillar muscle. IV. The effect of temperature on beetle flight muscle. *Proc. Roy. Soc.* **B 155,** 493–499.

Neville, A. C. and Weis-Fogh, T. (1963). The effect of temperature on locust flight muscle. *J. Exp. Biol.* **40,** 111–121.

Newport, G. (1837). On the temperature of insects, and its connexion with the functions of respiration and circulation in this class of invertebrate animals. *Phil. Trans. R. Soc. Lond.* **127,** 259–338.

Nobili, L. and Meloni, M. (1831). Recherches sur plusiers phénoménes calorifigues entreprises en moyen du thermomultipicateur. *Ann. Chem.* **48,** 198–218.

Oosthuizen, M. J. (1939). The body temperature of *Samia cecropia* Linn. (Lepidoptera, Saturniidae) as influenced by muscular activity. *J. Entomol. Soc. S. Africa* **2,** 63–73.

Pringle, J. W. S. (1967). The contractile mechanism of insect fibrillar muscle. *Prog. Biophys. Mol. Biol.* **17,** 3–60.

Tucker, V. A. (1966). Diurnal torpor and its relation to food consumption and weight changes in the California pocket mouse *Perognathus californicus*. *Ecology* **47,** 245–252.

Weis-Fogh, T. (1964). Functional design of the tracheal system of flying insects as compared to the avian lung. *J. Exp. Biol.* **41,** 207–227.

Weis-Fogh, T. (1967). Respiration and tracheal ventilation in locusts and other flying insects. *J. Exp. Biol.* **47,** 561–587.

Wilson, D. M. (1962). Bifunctional muscles in the thorax of grasshoppers. *J. Exp. Biol.* **39,** 667–677.

Wilson, D. M. and Weis-Fogh, T. (1962). Patterned activity of co-ordinated motor units, studied in flying locusts. *J. Exp. Biol.* **39,** 643–667.

2

Temperature and the Mechanical Performance of Insect Muscle

ROBERT K. JOSEPHSON

1 INTRODUCTION

In view of the current interest in endothermy and temperature regulation by insects as expressed in this volume, it is perhaps useful to remember that most insect behavior involves muscles operating essentially at ambient temperature. This is because most insects are small, less than a few millimeters in length and a few tens of milligrams in weight, and therefore have a high surface/volume ratio which facilitates heat transfer between the animal and its environment (see Chapter 3). Further, excluding flight and a few other activities, the metabolic rates of insects are moderately low, considerably lower than one would predict for birds or mammals of equivalent size by extrapolation of regression curves for oxygen consumption against body weight into the insect size range (Bartholomew and Casey, 1978). A high surface/volume ratio and a moderately low metabolic rate ensure that the body temperature of most insects will be negligibly different from that of their environment.

Some poikilothermic insects live in rather stable thermal environments; for example, soil insects, ectoparasites of birds and mammals, and the water striders of tropical oceans. But many insects are exposed to temperatures that vary considerably on a daily or seasonal basis. The effects of temperature on muscle performance will determine in part the temperature range over which these insects can operate.

Some activities of a fair number of larger insects do occur at body temperatures significantly greater than ambient (see reviews by Heinrich, 1974; Kammer and Heinrich, 1978). These elevated temperatures are achieved behaviorally by basking in the sun (Casey, this volume), or physiologically by enhanced heat production. In almost every instance the elevated heat production of endothermic insects is due to activity of the wing muscles, the exception being cicadas in which the heat source is the singing muscles in the first abdominal segment. The association between flight muscles and endothermy is a reflection of the high metabolic cost of flight. For an animal to fly it must have muscles capable of high mechanical power output which, because of biochemical inefficiencies, is associated with high heat output. For example, hovering is a particularly expensive mode of flight, but one achievable by many insects. Weis-Fogh (1973) estimated the mechanical power required by hovering flight in a wide range of insects to be 13–46 $W(kg\ body\ weight)^{-1}$. If it is assumed that flight muscles make up approximately 20% of an insect's

Original research was supported by NSF Grant BNS 75-90530, NIH Grant NS-14564, and a grant from the Guggenheim Foundation. Part of the research on tettigoniid muscles was done on the Alpha Helix East Asian Bioluminescence Expedition which was supported by NSF grants OFS 74-01830, OFS 74-02888, and BMS 74-23242.

weight (Greenewalt, 1962; Weis-Fogh, 1977) and that the efficiency of the conversion of metabolic fuel to mechanical power is 20% (Weiss-Fogh, 1972), the mechanical power output during hovering is accompanied by a heat output of 260–920 W(kg muscle)$^{-1}$ which would, in the absence of heat loss, warm the muscle at a rate of 4.7–16 deg/min. Many insects take advantage of this high power capacity of flight muscles and use the muscles for furnaces as well as motors.

Insects that can achieve warm body temperatures, and even the subset of these that regulate body temperature at reasonably constant levels (e.g., hawkmoths, Heinrich, 1970; bumblebees, Heinrich, 1972, 1975; dragonflies, May, 1976; dung beetles, Bartholomew and Heinrich, 1978) are not free from the constraints of temperature. Endothermy is expensive in animals as small as insects, and insects are at best intermittent endotherms, warming during periods of intense activity and cooling down nearly to the ambient temperature when quiescent. But unlike birds and mammals that allow their body temperature to fall during periods of relative inactivity, insects generally do not become torpid when their body temperature approaches that of the environment. Many potentially endothermic insects can walk around, feed, mate, and lay eggs when their metabolic rate is relatively low and their body temperature approximately that of the environment. In many insects some muscles are bifunctional and are used both to move the wings during flight and to move the legs during walking (Wilson, 1962). When used in flight, these muscles may operate at elevated temperature, and during walking at much lower temperature. Thus even in endothermic insects the muscles may operate over a temperature range and the capabilities of the animals may be limited by temperature effects on muscle performance.

The following considers the effects of temperature on the contractile performance of insect muscles. It should be pointed out that the data base for evaluating the effects of temperature on insect muscle is limited and rather biased. For reasons of interest and convenience most measurements have been made on the flight muscles of relatively large insects. The flight muscles of insects are of particular interest because of their extraordinary performance which, for both power output and contraction frequency achieved, is unmatched elsewhere in the animal kingdom. And, of course, the muscles of large insects are more convenient experimental preparations than the often minute muscles of small insects. But it is the flight muscles of large insects that warm up during activity. Consequently most of the available data apply to muscles that normally operate at elevated temperatures. There is no equivalent information on muscles able to function adequately at low temperatures, even though this would be most interesting to have.

2 SYNCHRONOUS AND ASYNCHRONOUS MUSCLES

Muscles with two, basically different, modes of neuronal control must be considered when discussing insect flight muscles: synchronous or neurogenic muscles, and asynchronous or myogenic muscles. In synchronous muscles each contraction is initiated by a muscle action potential or burst of action potentials. The muscle action potentials in turn are initiated by arriving impulses in motoneurons. Muscles are termed synchronous because there is a one-to-one correspondence between electrical activity and contraction, or neurogenic because the pattern of contraction is determined by the activity pattern in motoneurons from the central nervous system. Muscle action potentials are also necessary for contraction in asynchronous or myogenic muscles, but for oscillatory activity such as that in flight the action potentials are permissive rather than directive. Action potentials activate the muscle, and the muscle, when active, can oscillate if connected to a resonant load. The action potential frequency required to maintain an asynchronous muscle in an active state can be quite low. For example, in the flies *Calliphora* and *Lucilia* the action potential frequency recorded from an electrode in the flight muscles may be only about 10 Hz, although the wings are beating, and therefore the flight muscles contracting, at nearly 150 Hz (Roeder, 1951). Muscles like the flight muscles of flies are called asynchronous because there is no direct correspondence between electrical activity of the muscle fiber membrane and contraction, or myogenic because the frequency of contraction is determined by the resonant properties of the muscle and its load and not by the output pattern from the central nervous system (for reviews of the physiology of asynchronous muscle see Pringle, 1967, 1978).

Asynchronous muscle is found in several orders of insects and has probably evolved independently at least 9 or 10 times (Cullen, 1974). The ability of asynchronous muscle to oscillate against a resonant load is a feature of the contractile proteins themselves, for it occurs in glycerinated fibers as well as in intact muscles (Jewell and Rüegg, 1966). Glycerinated fibers are prepared by teasing fibers or fiber bundles from muscles that have been soaked in glycerol and treated with detergent so as to remove most of the cytoplasmic components except the contractile filaments.

Most investigations on the physiology of asynchronous muscle have utilized either "free oscillation" or "driven oscillation" (reviewed by Pringle, 1978). In free oscillation the muscle is attached to a resonant load, that is, a load with mass and stiffness that has a natural period of vibration. In this case the oscillatory frequency of the muscle is the resonant frequency of the load (Machin and Pringle, 1959). In driven oscillation the muscle is attached to a device that can rapidly alter the length of the muscle. Typically muscle tension is measured during imposed sinusoidal length changes over a range of frequencies. Figure 1 gives examples of driven oscillations in glycerinated fibers of the water bug *Lethocerus*. At some frequencies there is a phase shift between the length and tension changes such that the tension change follows or lags the length change. It is this feature of asynchronous muscle that allows it to do positive work. Because of the delay in the tension change, the apparatus does less work on the muscle in stretching it than the muscle does on the apparatus during the subsequent shortening, so over a full cycle there is positive work output by the muscle. The closed loops in Fig. 1 are plots of muscle tension (ordinate) against length (abscissa). In the frequency range in which the muscle does positive work, these loops are traversed in a counterclockwise direction. The area between the zero tension axis and the lower limb of the loop from maximum shortening to maximum lengthening is equal to the work required to stretch the muscle from its shortest to its longest length. The area between the upper limb of the loop and the zero tension axis is the work done by the muscle during shortening. The area of the loop itself is the difference between the work put in and the work obtained, and therefore the net work done by the muscle per cycle. As discussed below, both the maximum work per cycle and the frequency at which work output is maximum are dependent on muscle temperature. The work from oscillating muscle is obtained at the expense of ATP which is hydrolyzed during these oscillatory contractions.

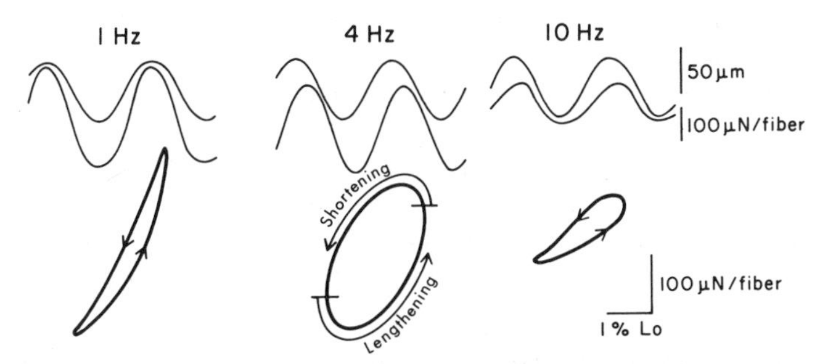

Fig. 1 Work produced by glycerinated flight muscle fibers of the bug **Lethocerus** during imposed sinusoidal length change, 22°C. The fibers were immersed in a solution containing Ca^{2+} and ATP. The upper curves are the length changes, and the curves below are the resulting tension changes. The closed loops are plots of tension against length; the arrows indicate the direction in which these loops are traversed. The area enclosed within the loops is equal to the work produced per oscillatory cycle. L_0 is the rest length of the fibers. (Redrawn from Steiger and Ruegg, 1969.)

Oscillatory contraction by asynchronous muscle requires a resonant load; when tension is measured isometrically, the responses of asynchronous muscle are like those of synchronous muscle in that there is a direct correspondence between muscle action potentials and contraction. There is considerable variability in the twitch responses of asynchronous muscle. In beetle flight muscle single stimuli evoke no measurable tension and the twitch/tetanus ratio is zero (Machin and Pringle, 1959), while in asynchronous tymbal muscles of cicadas single twitches are large and the twitch/tetanus ratio is about 0.3 (Pringle, 1954; Hagiwara et al., 1954; Josephson and Young, 1980). Bumblebee flight muscle is intermediate with small but detectable single twitches (Boettiger, 1957). It is seemingly paradoxical that asynchronous muscles achieve the highest repetition frequencies found in the animal kingdom, up to 1000 Hz in some small insects (Sotavalta, 1953), yet in nonoscillatory contraction the muscle is inherently slow. Tension rises and falls slowly during isometric twitches or tetanic contraction (Pringle, 1954; Boettiger, 1957; Fig. 6), indicating that the duration of the contractile activity evoked by each muscle action potential is long. This is advantageous for oscillatory contractions, because only a low neural input frequency is needed to maintain the muscle fully activated. The muscles known to be bifunctional and used both to move the wings during flight and the legs during walking are synchronous muscles (Wilson, 1962). Because their nonoscillatory contractions are quite slow, asynchronous muscles would be useless for even moderately rapid walking movements. If there are bifunctional asynchronous muscles, the nonflight function must involve slow tonic contraction rather than phasic responses.

Synchronous and asynchronous muscles differ in ultrastructure as well as in physiology. This is nicely seen in cicada tymbal muscles (Fig. 2), since synchronous and asynchronous tymbal muscles occur in different species (Pringle, 1954; Hagiwara et al., 1954; Young, 1972). The ultrastructure of tymbal muscles is strikingly similar to that of synchronous and asynchronous flight muscles (e.g., Cullen, 1974). The obvious ultrastructural difference between synchronous and asynchronous muscles is in the

---

Fig. 2 Structure of synchronous and asynchronous tymbal muscles of cicadas. (A and B) Longitudinal and transverse sections of the synchronous muscle of **Cyclochila australasiae.** (C and D) Longitudinal and transverse sections of the asynchronous muscle of **Platypleura capitata.** The transverse section in (D) is slightly oblique and runs from the Z line (lower right) through the center of the sarcomere (upper left). F, Myofibril; M, mitochondria; SR, sarcoplasmic reticulum; T, transverse tubule; Z, Z line. The sarcoplasmic reticulum and T tubules are abundant in the synchronous muscle, but in the asynchronous muscle they occur as isolated tubules surrounding the center of the sarcomere and at the level of the Z line [arrows in (C) and (D)]. Scale bar = 1 μm.

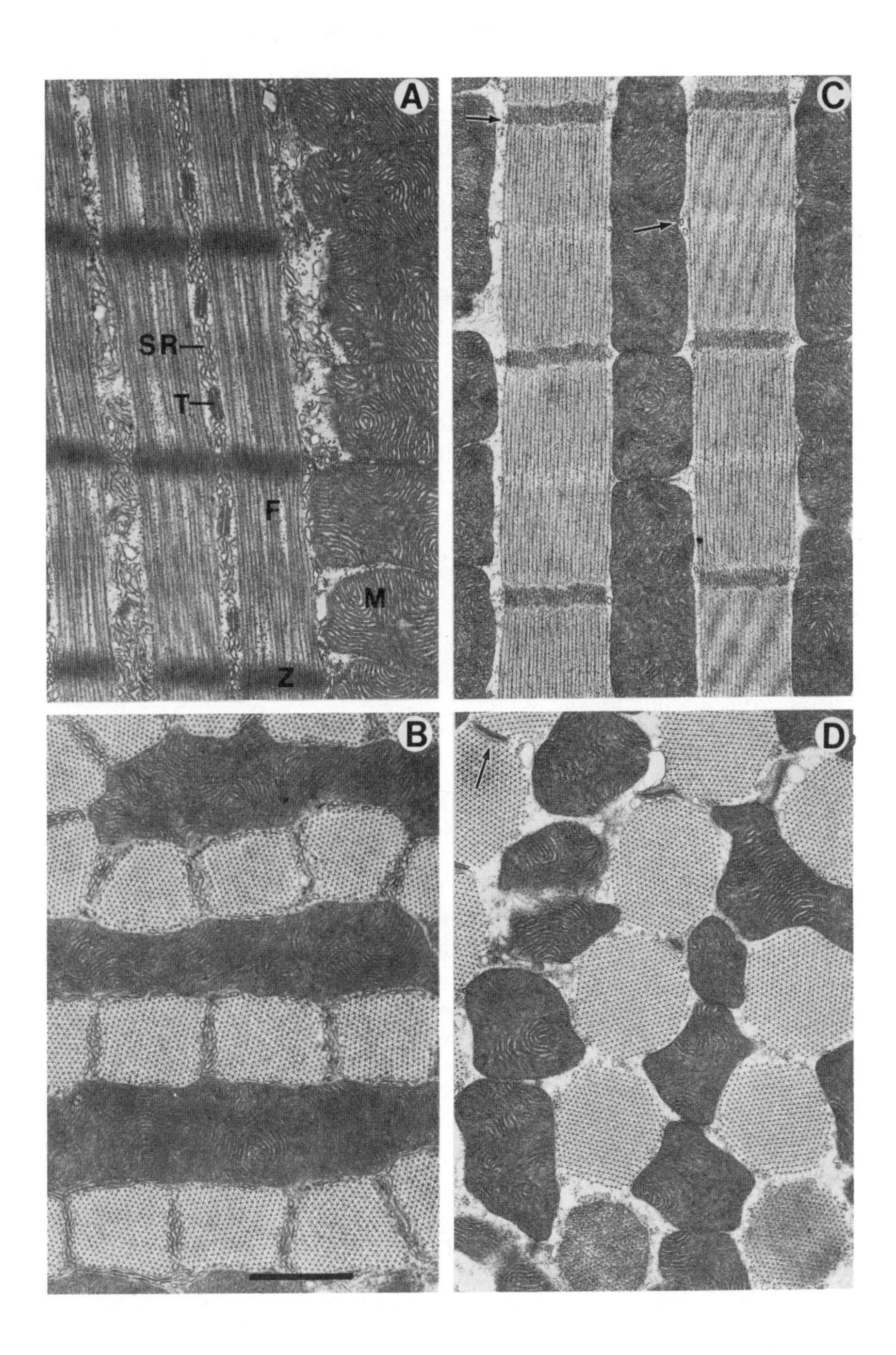
A
SR
T
F
M
Z
B
C
D

development of the sarcoplasmic reticulum which is abundant in synchronous flight and tymbal muscles but exceedingly sparse in asynchronous ones. This is related to the long duration of twitches in asynchronous muscles. Contractile activity in skeletal muscle is controlled by cytoplasmic calcium. Contraction is initiated by the release of calcium from the sarcoplasmic reticulum and terminated by the resequestration of the calcium by the sarcoplasmic reticulum. Throughout the animal kingdom twitch duration in striated muscle is inversely related to the relative abundance of the sarcoplasmic reticulum (Josephson, 1975). Yet another name given to asynchronous muscle is fibrillar muscle, based on the relatively large size of the myofibrils in most examples and the ease with which individual myofibrils can be teased apart in living material. Presumably it is because the myofibrils are not bound together by an investiture of sarcoplasmic reticulum that they are readily separable in asynchronous muscle.

3 TWITCH TIME COURSE

One of the most obvious effects of temperature is on the time course of muscle contraction; as the temperature increases, the rate of tension rise and decay in an isometric twitch increases and the twitch duration shortens (Fig. 3).

The change in twitch duration with temperature in wing muscles of two orthopterans, the locust *Schistocera gregaria* and the tettigoniid *Euconocephalus nasutus*, is illustrated in Fig. 4. In *E. nasutus* there are marked differences in performance and ultrastructure between wing muscles in the mesothorax and the metathorax, differences associated with different functions of the muscles in the two segments. Metathoracic wing muscles are used only for flight, and the wing stroke frequency during flight is 19 Hz at an ambient temperature of 25°C. Mesothoracic wing

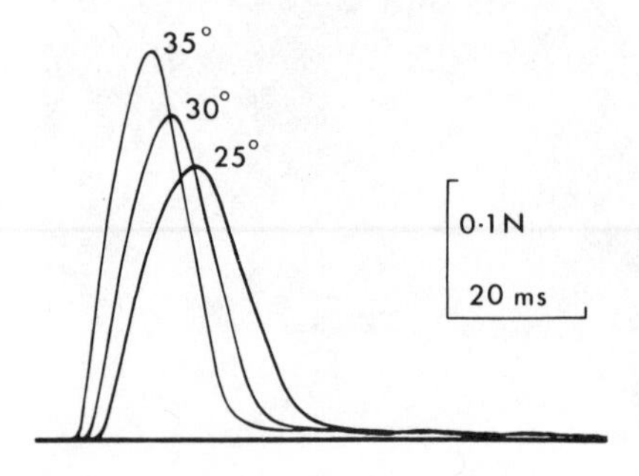

Fig. 3 Twitch tension from the tymbal muscle of the cicada **Cystosoma saundersii** during contraction at constant muscle length (isometric contraction). (From Josephson and Young, 1979.)

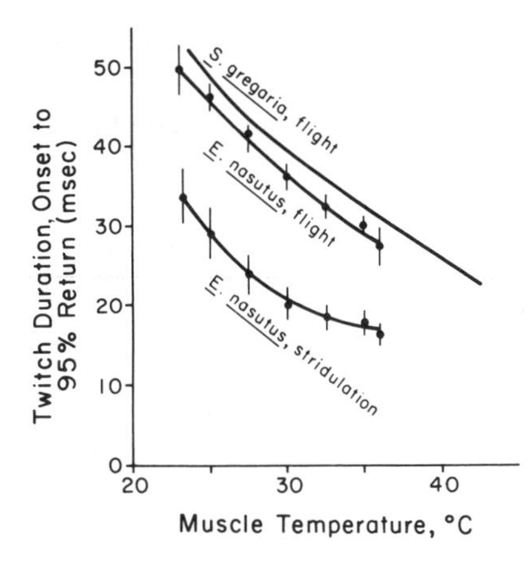

Fig. 4 Twitch duration (onset to 95% relaxation) in three orthopteran wing muscles as a function of muscle temperature. The flight muscle from **S. gregaria** is the metathoracic dorsal longitudinal. (Curve redrawn from Neville and Weis-Fogh, 1963.) The flight muscle from **E. nasutus** is the male metathoracic first tergocoxal; the stridulation muscle is the male mesothoracic first tergocoxal. The data points for the muscles of **E. nasutus** are means (n = 6 and 9 preparations); vertical lines indicate standard errors.

muscles, on the other hand, are used for both flight and stridulation, and the wing stroke frequency during stridulation is 150 Hz at a thoracic temperature of 35°C (Josephson, 1973). Reflecting their higher operating frequency, the mesothoracic muscles of *E. nasutus* produce faster twitches than the metathoracic muscles and have a correspondingly better developed sarcoplasmic reticulum (Elder and Josephson, 1980). The wing stroke frequency during flight for *S. gregaria* is 17 Hz at ambient temperatures of 25–35°C (Weis-Fogh, 1956a), similar to the flight frequency of *E. nasutus*, and the flight muscles of *S. gregaria* and the metathoracic wing muscles of *E. nasutus* have similar twitch durations.

The wing muscles of insects are organized as two antagonistic sets: elevators and depressors for flight, and wing openers and closers for stridulation. For maximum efficiency the tension generated by each set should be confined to its half-cycle, otherwise part of the work done by one set of muscles will be used to stretch resisting antagonists and not be available to move the wings. The muscle illustrated for *S. gregaria* in Fig. 4 is the dorsal longitudinal, a wing depressor. In *S. gregaria* the downstroke takes up 60% of the full wing cycle or about 35 msec (Weis-Fogh, 1956a). From Fig. 4 it is seen that relaxation of the dorsal longitudinal muscle is more than 95% complete in 35 msec at muscle temperatures of 33°C and above. During sustained flight the muscle temperature of *S. gregaria* warms to approximately 6°C above the environmental temperature, and this temperature excess is independent of the ambient temperature in the range for which sustained flight is possible (Church, 1960). Thus the dorsal longitudinal muscle has time to contract and nearly fully relax during the depression phase of the wing cycle at ambient temperatures greater than 27°C, which approximates the minimum environmental temperature at which sustained flight is possible (22–25°C, Weis-Fogh, 1956a).

Figure 4 is based on contractions initiated by single action potentials. During vigorous flight the wing muscles of *S. gregaria* may be activated by pairs of action potentials separated by 4–8 msec (Wilson and Weis-Fogh, 1962; Wilson, 1964). Such double firing greatly increases the muscle tension and work output as compared to that in a twitch, but it also increases the duration of the contraction by an amount equal to the interval between the action potentials (Neville and Weis-Fogh, 1963). Especially at the lower end of the temperature range some fraction of the work resulting from paired action potentials is wasted, because the duration of the contraction exceeds that of the depression phase of the wing stroke. Further, in *S. gregaria* the elevation phase of the wing stroke is approximately 24 msec. If the contraction kinetics of the elevator muscles are like those of the depressors, twitches will be 95% or more complete during the elevation phase only at thoracic temperatures greater than 43°C or, assuming a 6°C temperature excess, ambient temperatures greater than 37°C. If the elevator muscles are not appreciably faster than the depressors, some of the work done by the depressors must be used to stretch resisting elevator muscles and therefore will be unavailable to move the wings.

The inefficiency caused by overlapping activity in antagonistic sets of muscles is much more severe for stridulation in *E. nasutus*. If it is assumed that the opening and closing strokes of the wing are of equal duration, at a stridulation frequency of 150 Hz the duration of each half-cycle is but 3.3 msec. This is much shorter than the twitch duration of the mesothoracic muscles at their operating temperature of 35°C. *Euconocephalus nasutus* warms its thorax by muscular activity before the onset of stridulation (Stevens and Josephson, 1977), and warm-up is absolutely necessary for stridulation. During the early evening when *E. nasutus* sings, the environmental temperature is approximately 25°C. At 25°C the wing muscles contract in a nearly smooth tetanus when activated at the stridulation frequency (Fig. 5). At 35°C, on the other hand, the muscles are able to relax about halfway between tension peaks, so there is oscillating tension available to drive the wings.

The relative force on the wings of *E. nasutus* during stridulation is illustrated in Fig. 6. The peak force available to drive the wings is about one-half the peak tension due to the openers or closers alone at the stridulation frequency, and much less than the isometric tension of which these muscles are capable.

Given that much of the work output of the stridulation muscles must be wasted internally in stretching resisting antagonists, it is surprising that

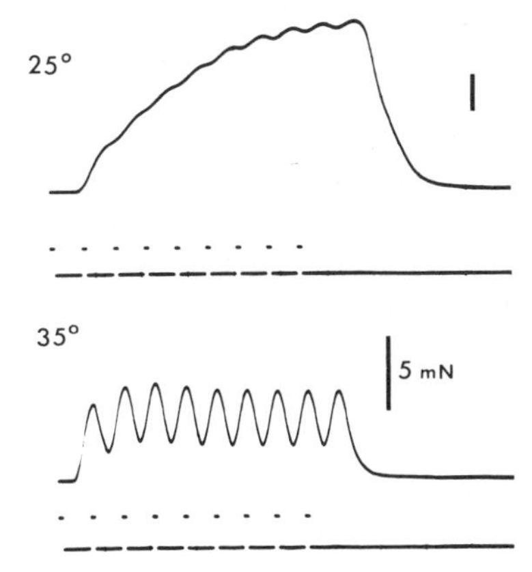

Fig. 5 Isometric tension from the first tergocoxal muscle of the tettigoniid **E. nasutus** when stimulated at the stridulation frequency of 150 Hz. The lower trace marks the stimuli.

the power output as sound is as high as it is. The song of *E. nasutus* is a loud, continuous buzz. The intensity of the sound produced by *E. nasutus* has not been measured, but that of a related katydid, *Neoconocephalus robustus,* whose muscle performance and song are quite similar to that of *E. nasutus* (Josephson and Halverson, 1971; Josephson et al. 1975), is 110 dB at 10 cm, equivalent to an energy output of 1.22×10^{-2} W (Counter, 1977). The mesothoracic muscles of *N. robustus* are approximately 78 mg (Stevens and Josephson, 1977), so the energy output as sound is 156

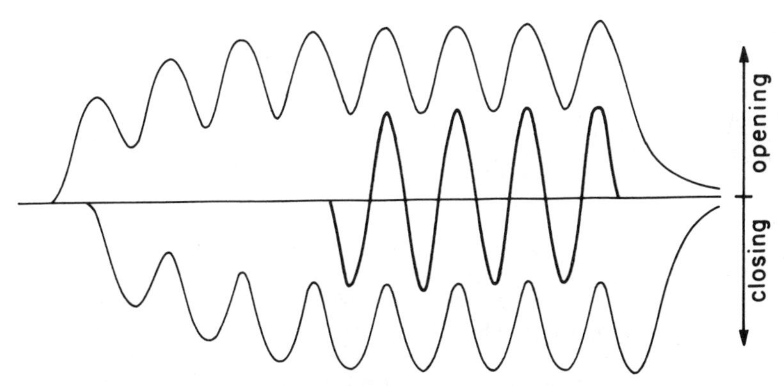

Fig. 6 Estimated force available to move the wings during stridulation in **E. nasutus**, 35°C. It is assumed that forces on the wings from the opener and closer muscles, acting through their respective level arms, are equal; that the contraction kinetics of the opener and closer muscles are all like that of the first tergocoxal muscle illustrated in Fig. 5; and that the openers and closers are activated in exact antiphase. The upper curve is the force generated by the opener muscles; the lower curve is the force generated by the closer muscles. The central curve, which begins on the fourth closing cycle, is the net force on the wings, obtained as the difference between the upper and lower curves.

W/kg muscle. This output level is almost identical to the mechanical power output of wing muscles in *Schistocerca* and *Drosophila* during strenuous flight (170 and 160 W/kg, Weis-Fogh and Alexander, 1977).

In a group of 23 orthopteran flight and cicada tymbal muscles, increasing the temperature from 30 to 35°C resulted in a 10–20% decrease in the duration of isometric twitches, equivalent to a Q_{10} for the reciprocal of duration of 1.2–1.6 (Fig. 7). These were all tropical animals or temperate animals active during the warmest part of the year, and 30–35°C should lie within their normal operating range. There is no obvious correlation in Fig. 7 between the relative decrease in duration and twitch brevity; the proportional decrease in duration is nearly the same for very fast muscles as for slower ones. The asynchronous muscle in the group, a cicada tymbal muscle, is exceptional in that not only was the twitch duration much longer than for any of the synchronous muscles tested but also the effect of temperature was anomalous in that, on average, twitches became slightly longer as the temperature was raised from 30 to 35°C.

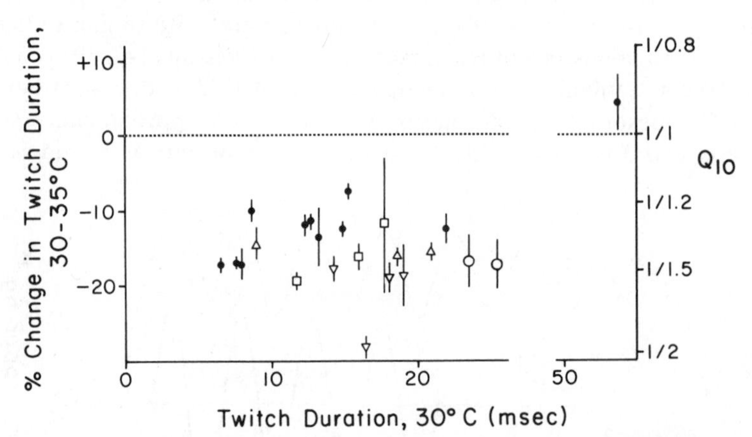

Fig. 7 Change in isometric twitch duration as insect muscles are warmed from 30 to 35°C. Solid circles are cicada tymbal muscles, each circle representing a different species. Open symbols are first tergocoxal muscles from tettigoniids. The several entries with the same symbol are different muscle types from the same species. Vertical bars are standard errors, n = 5–9 separate examples of each muscle type. See Josephson (1973) for recording methods. The cicada tymbal muscles, in order of increasing twitch duration, are those from: **Psaltoda claripennis, P. harrissii, P. argentata, Tomasa tristigma, Cyclochila australasiae, Abricta curvicosta, Thopha saccata, Arunta perulata, Chlorocysta viridis, Cystosoma saundersii, Platypleura capitata.** The tettigoniid muscles, listed in order of increasing duration for each species, are: Δ, **E. nasutus,** ♂ mesothoracic first tergocoxal muscle ($t.cx_1$), ♀ mesothoracic $t.cx_1$, ♂ metathoracic $t.cx_1$; □, **E. cornutus,** ♂ mesothoracic $t.cx_1$, ♀ mesothoracic $t.cx_1$, ♂ metathoracic $t.cx_1$; ▽, **Mecapoda elongata,** ♂ mesothoracic $t.cx_1$, ♀ mesothoracic $t.cx_1$, ♂ mesothoracic tergotrochanteral (t.t.), ♂ metathoracic $t.cx_1$; ○, **Sexava coriacea,** ♂ metathoracic $t.cx_1$, ♂ mesothoracic $t.cx_1$.

The temperature range considered in Fig. 7 is limited because many of the measurements were made under field conditions in the tropics and temperatures less than 30°C often were not available. In seven of the cicada muscle types and three of the orthopteran muscles, twitches were measured at 25°C as well as at the two higher temperatures. In these muscles the average decrease in twitch duration was 21.4% (SE = 1.2%) from 25 to 30°C and 14.4% (SE = 0.8%) from 30 to 35°C. As is also apparent in Fig. 4, the effects of temperature on the twitch time course are greater at the lower than at the upper end of the operating temperature range.

4 TENSION

4.1 Tetanic Tension

The maximum tension generated by an insect muscle generally increases with temperature, but the change is not very great, especially at the upper end of the temperature range a muscle is likely to experience. In a set of 20 orthopteran flight and cicada tymbal muscle types, the isometric tetanic tension averaged 5% greater at 35°C than at 30°C (SE = 2%). There was significant variation between the temperature responses of different muscle types, and in several muscles the tetanic tension declined rather than increased as the temperature was raised (Fig. 8). The effects of temperature on tetanic tension appear to be greater at lower than at higher

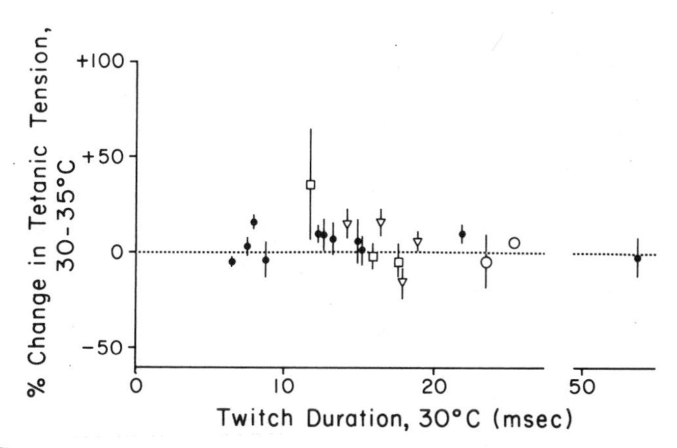

Fig. 8 Change in isometric tetanic tension as muscles are warmed from 30 to 35°C. Symbols as in Fig. 7.

temperatures. In tymbal muscles of 7 of the cicada species, the average increase in tetanic tension was 22% (SE = 8%) from 25 to 30°C and 9% (SE = 2%) from 30 to 35°C, for a total change of 33% from 25 to 35°C. In locust flight muscle the tetanic tension doubles from 11 to 25°C but increases by only another 33% from 25 to 35°C (Weis-Fogh, 1956b). The frog sartorius muscle, because of the many studies done on it, forms a benchmark with which mechanical properties of other muscles can be compared. In the frog sartorius tetanic tension increases with temperature, but the increase is even less than in most insect muscles, amounting to only about 20% for the temperature range 0–24°C (Hill, 1951). To the best of my knowledge the temperature effects on tetanic tension have yet to be explained in terms of the actin–myosin interactions leading to contraction.

4.2 Twitch Tension

The influence of an increase in temperature on twitch tension can be expected to be the resultant of two opposing effects: (1) an increase in the rate of biochemical activity of the contractile proteins, in particular an increase in the rate at which thick and thin filaments can slide by one another; and (2) a decrease in the duration of muscle activation, manifest by the decrease in twitch duration with increasing temperature.

Muscle contraction is usually measured experimentally as length change under a constant load (isotonic contraction) or as tension change at a constant length (isometric contraction). In isotonic contraction there is an inverse relation between the force on a muscle and its shortening velocity. This relation is usually expressed as a force–velocity curve, and an example from a locust flight muscle is seen in Fig. 9. The force–velocity relation obtained from an isotonically contracting muscle reflects basic properties of the contractile component of the muscle, the contractile component being the set of thick and thin filaments whose movement relative to one another produces movement and tension. By definition an isometric contraction is one in which there is no muscle shortening. In practice, in most isometric tension measurements there is shortening, especially of the contractile component, because of compliance in the recording apparatus, in the tendons and apodemes of the muscle, and in the muscle structure itself. These compliances can be treated as elastic elements in series with the contractile component. The stiffness of these series elastic elements and the force–velocity relation for the contractile component together determine the time course and extent of the tension generated.

At the onset of an isometric contraction the force on the contractile

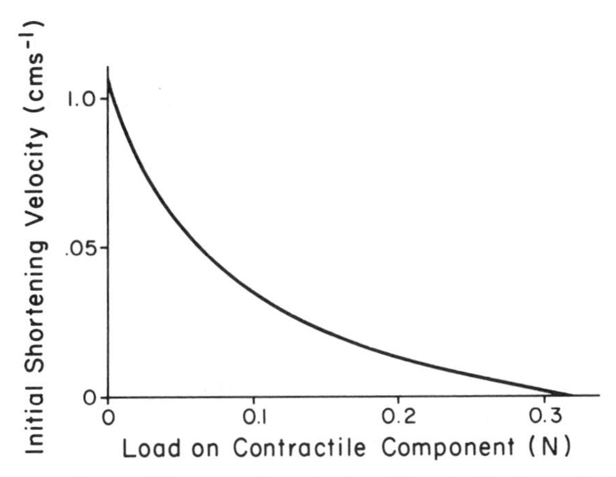

Fig. 9 Shortening velocity of the metathoracic dorsal longitudinal muscle of *S. gregaria* as a function of load, 11°C. (Redrawn from Buchthal et al., 1957.)

component is low, because the series elastic elements are unstretched, and the shortening velocity of the contractile component is high. As the contractile component shortens, the series elastic elements become stretched and the tension on the contractile elements increases, leading to a decrease in the shortening velocity of the contractile component. In effect the contractile component's response shifts along the force–velocity curve, beginning at the upper left corner (zero force) and moving toward the lower right corner (zero velocity). When muscle activation is continued, as in a tetanic contraction, the shortening velocity of the contractile component eventually becomes zero as the load reaches the maximum value the contractile component can support, given by the intercept of the force–velocity curve and the force axis. In a twitch, relaxation begins before this force is reached, and the twitch tension achieved is less than the maximum tetanic tension. The situation is actually more complex than this indicates, because the force–velocity curve itself varies with time (e.g., as activation wanes in the later part of a twitch, the curve collapses back toward the origin) and because the time course of contractile component activation varies with load and the amount of muscle shortening allowed (Pringle, 1960). The point to be made is that in this model isometric twitch tension is increased either by increasing the shortening velocity of the contractile component at any given load (shifting the force–velocity curve upward) so that series elastic elements become more rapidly stretched, or by increasing the period of full activation so that the contractile component has more time to stretch the series elastic elements. Increasing temperature increases the shortening velocity of the contractile component, as evidenced by the more rapid rise in tension (e.g., Fig. 3), but it also decreases the twitch duration. The overall effect of temperature on isometric twitch tension will then depend on which has the greater temperature coefficient, the shortening velocity as a function of load or the set of processes that determine twitch duration.

In locust flight muscle the isometric twitch tension increases by a factor of 2 between 11 and 25°C (Buchthal, Weis-Fogh and Rosenfalck, 1957) but does not change significantly from 25 to 42°C (Neville and Weis-Fogh, 1963). In the higher temperature range the demonstrated decrease in twitch duration must be just balanced by an increase in intrinsic shortening velocity. Figure 10 summarizes changes in twitch tension when the temperature is increased from 30 to 35°C in 23 tettingoniid wing muscles and cicada tymbal muscles. The temperature effects are much more variable for twitch tension than for tetanic tension, both for individual samples of a single muscle type and between muscle types (compare standard error bars and the dispersion of averages in Figs. 8 and 10). This greater variability probably arises from the interacting, competing effects of temperature on twitch tension. The average change in tension in Fig. 10 is an increase of 16% (SE = 6%), but many of the individual muscles lie close to the zero line; for these the temperature increase resulted in little

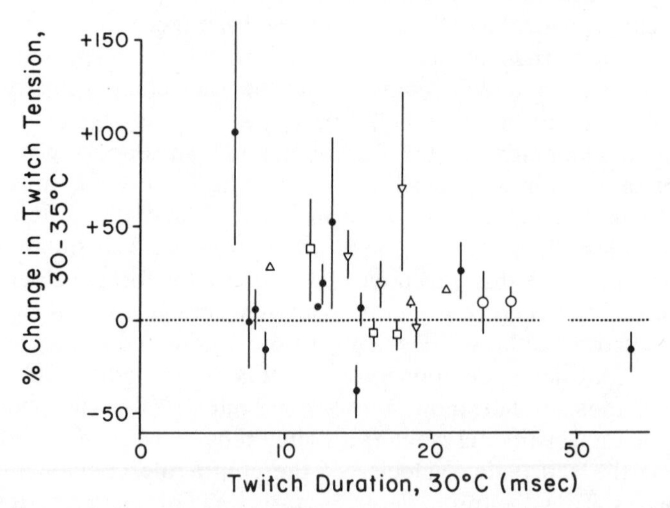

Fig. 10 Change in isometric twitch tension as muscles are warmed from 30 to 35°C. Symbols as in Fig. 7. Standard error bars are not shown when these are smaller than the symbol.

change in twitch tension. For the 7 cicada muscles tested at 25°C as well as at the higher temperatures, the increase in twitch tension averaged 15% (SE = 14%) from 25 to 30°C and 10% (SE = 11%) from 30 to 35°C.

5 POWER

The work done by a contracting muscle is the product of the force on the muscle and the distance of shortening. The rate at which work is done is the mechanical power output of the muscle. The power available from the muscles is of obvious importance in determining the behavioral capabilities of an animal. This is particularly true for flight. Terrestrial or aquatic locomotion can proceed, albeit at low velocities, when muscles deliver low power. Flight, on the other hand, has a power threshold, the power required to keep the organism airborne. Temperature greatly affects the power output of muscles, and adequate power for flight is available for only part of the temperature range experienced by many insects.

Unfortunately the mechanical power output of insect muscles during normal behavior is not readily determinable. In principle the power output of an active muscle could be obtained by simultaneously measuring the force developed and the velocity of shortening. The product of these two, integrated over a full cycle, would give the work output per cycle. In practice, measuring either the force on or the shortening velocity of a muscle in a behaving insect without greatly altering the behavior would indeed be difficult, and direct measurement of the *in situ* power output of an insect muscle has yet to be achieved. Measurements of activity metabolism in insects from oxygen consumption, heat production, or fuel utilization are principally measures of the mechanical power output of the muscles and the chemical metabolism needed to support this power output. Correlations between metabolic rate and realizable mechanical power output by the muscles are complicated, however, by inefficiencies in the mechanical system (part of the muscle work may be used to overcome skeletal inertia, part may be used to overcome tension in antagonistic muscles and, in flying insects, part of the work output is dissipated as nonuseful air vortices) and by inefficiencies in the metabolic production of ATP and in the conversion of ATP to mechanical power. Further, through much of a full cycle a muscle is not shortening but rather being stretched, and during lengthening work is done on the muscle rather than by the muscle. Work done on rather than by a muscle is conventionally described as negative work. The appropriate way to handle negative work on energy balance sheets is still uncertain; it is not clear whether negative work is simply converted to heat or if part of it can be sequestered for subsequent positive work output (e.g., Hill and Howarth, 1959; Mannherz, 1970; Ulbrich and Rüegg, 1977).

Although there are no direct measurements of mechanical power output from insect muscles *in vivo,* a few determinations of work output from isolated insect muscles have been made. Figure 11, which gives the work produced during isotonic twitches, is from a detailed study of Buchthal et al. (1957) on the mechanical properties of the dorsal longitudinal flight muscle of the locust *S. gregaria*. The work plotted ("active work") is that ascribable to the contractile component and is obtained by subtracting from the total work the work in shortening due to passive elasticity in an unstimulated muscle. As in other muscles examined, the work per twitch is maximal at a load about one-third of the maximal isometric tension the muscle can develop (maximum isometric tension in Fig. 11 is given by the intersection of the work curve with the load axis). At larger loads the force is greater, but the distance of twitch shortening is disproportionally small; at smaller loads the shortening distance increases, but not enough to make up for the reduced force.

It can be seen in Fig. 11 that the twitch work is a strong function of temperature. The maximum tension increases with temperature, but the effect is not large; in these experiments tension increased by 20% from 10 to 20°C and 30% from 20 to 30°C. The major part of the increase in twitch work is due to an increase in the shortening velocity and the distance of

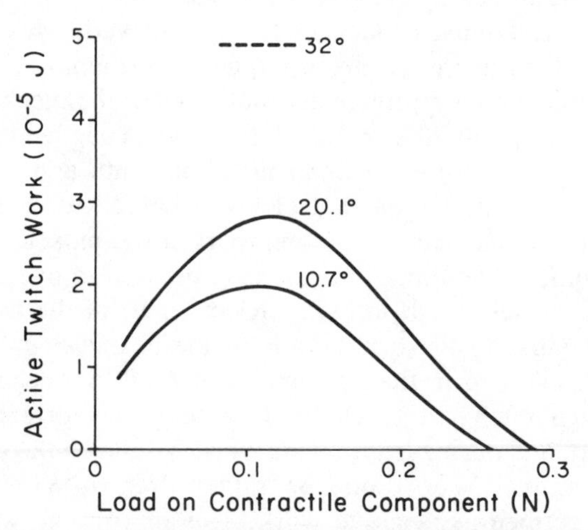

Fig. 11 Mechanical work produced during isotonic twitches of the dorsal longitudinal muscle of *S. gregaria* as a function of load and temperature. (Redrawn from Buchthal et al., 1957.)

shortening. The maximum twitch work measured in a few experiments at 32°C approached 0.05 mJ. Neville (1963) found the twitch work of the same muscle contracting against a spring to be 0.065 mJ at 36°C. Since the dorsal longitudinal muscle weighs about 11 mg (Buchthal et al., 1957), the twitch work is 4.5 J/kg at 32°C and 5.9 J/kg at 36°C. At a wing stroke frequency of 17 Hz (Weis-Fogh, 1956a), the twitch work is equivalent to a power output of 77–100 W/kg muscle, a range that overlaps the calculated continuous power output of the muscles in a flying locust (60–90 W/kg muscle, Jensen, 1956; Weis-Fogh, 1977) but which is less than the maximum power output in strenous flight.

In asynchronous muscles, too, the maximum power output increases with temperature both in living muscle (Machin et al., 1962) and in glycerated fibers (Pringle and Tregear, 1969; Steiger and Rüegg, 1969). The total work output during oscillatory contraction is the product of work per cycle and cycle frequency. With increasing temperature both the maximum work per cycle and the frequency at which the work per cycle is maximum increase. In glycerinated fibers of the bug *Lethocerus*, the increase in frequency at which maximum work per cycle is performed has a Q_{10} of 2.0–2.8 in the temperature range 12–32°C (Pringle and Tregear, 1969). Because work per cycle does not drop rapidly at frequencies above the optimum, the frequency at which power output is maximum is 1.3–3 times greater than the frequency at which work per cycle is maximum (Jewell and Rüegg, 1966, Pringle and Tregear, 1969; Steiger and Rüegg, 1965).

6 MATCHING CONTRACTION FREQUENCY AND TEMPERATURE

As indicated above, there is an optimum oscillation frequency for power output by asynchronous muscle, and this frequency varies with temperature. There is an optimum contraction frequency for maximum power output in synchronous muscles also. In synchronous muscles capable of fast twitch contractions the work per twitch is essentially constant at low activation frequencies (Neville and Weis-Fogh, 1963), and therefore at low repetition frequencies the power output is linearly related to frequency. As frequency is increased, the work per twitch declines as individual twitches begin to fuse. At the optimum frequency the increase in power output with frequency due to more frequent contraction is balanced by a decrease in power output due to declining work per twitch. Since the twitches become shorter with increasing temperature, the temperature at which twitch fusion limits power output increases as the muscle is warmed. The effective power output from systems of antagonistic muscles such as the elevators and depressors of the flight system should be maximum at activation frequencies considerably lower than the frequency for maximum power in individual muscles. Even before there is fusion of tension in the individual muscles, the contractions of the antagonistic muscles begin to overlap and part of the work done by one set of muscles is expended against its antagonists. Again because the twitches become shorter as the temperature rises, the frequency at which power output is maximal from the muscle system should increase as the muscles are warmed.

The total power output, including both heat and work, of insect wing muscles during flight is extraordinarily high (values summarized by Weis-Fogh and Alexander, 1977; Kammer and Heinrich, 1978), which suggests that the muscles must operate at or near conditions that maximize power output. One of the factors determining power output is repetition frequency. Since the frequency at which mechanical power output is greatest is a function of temperature, if an insect is to maximize power output, the contraction frequency of its muscles must increase with temperature. For activities involving synchronous muscles, the firing frequency of motoneurons from central ganglia should increase with temperature; in asynchronous muscles, the resonant frequency of the muscle load, which determines the contraction frequency, should vary with temperature.

The contraction frequency during flight of synchronous wing muscles is apparently generated in the thoracic ganglia. In the locust, *Schistocerca,* flight can be elicited following section of the ventral nerve cord either anterior to or posterior to the thorax, indicating that the thoracic ganglia contain the necessary neuronal circuitry to generate the flight pattern (Wilson, 1961; reviewed by Mullony, 1975). Most of the motoneurons supplying the flight muscles originate in the ganglia of the two wing-bearing segments, the mesothoracic and the metathoracic ganglia (some flight muscle bundles of the mesothoracic segment are innervated from the prothoracic ganglia; see, for example, Neville, 1963; Bentley, 1973; Stokes, Josephson, and Price, 1975). The mesothoracic and methathoracic ganglia are surrounded by flight muscles, and the temperature of the ganglia must be nearly the same as that of the muscles. If the neural oscillators controlling flight frequency are also in the mesothoracic or metathoracic ganglia, all that is required for effective matching of muscle temperature and activation frequency is that the neural oscillators have appropriate temperature coefficients. Directly warming the thoracic ganglia of the moth *Hyalaphora cecropia* initiates the transition from the motor pattern of warm-up to that of flight (Hanegan and Heath, 1970), indicating that the neuronal circuitry is directly sensitive to temperature or is supplied by thermoreceptors monitoring the temperature of its surroundings.

Many authors have noted a correlation between wing stroke frequency and ambient temperature, or the lack thereof, in insects with synchronous flight muscles. Since wing muscles may warm up during flight, ambient temperature may only inaccurately reflect muscle temperature. More relevant are the demonstrations that wing stroke frequency increases with thoracic temperature in several insects with synchronous muscles (e.g., Sotavalta, 1954; Kammer, 1970; Heinrich and Bartholomew, 1971). There are some interesting exceptions. In the locust, *Schistocerca,* the wing stroke frequency is independent of ambient temperature in the range 25–35°C (Weis-Fogh, 1956a), even though the thoracic temperature of flying animals increases through this range (Church, 1960). In the cockroach, *Periplaneta americana,* the steady-state wing stroke frequency increases with the temperature of the flight muscles, but during the first few minutes of flight the thoracic temperature may increase by several degrees without a correlated increase in frequency, suggesting that the oscillators determining frequency do not experience the same temperature changes as the muscles and therefore in this species may not lie in the pterothoracic ganglia (Farnworth, 1972).

An interesting case in which there is warming of a synchronous muscle during activity without a concomitant temperature change in the controlling ganglion is found in cicadas. The tymbal muscles of cicadas lie in the first abdominal segment surrounded by an air sac and are thermally isolated from the composite thoracic ganglion that supplies the motoneurons to the muscle. Because of this thermal isolation, the temperatures of the tymbal muscles and the ganglion can be quite different. During singing, the tymbal muscles of the Australian bladder cicada, *Cystosoma saundersii,* warm by 10–15°C, while the thoracic temperature, and presumably that of the thoracic ganglion, increases by only 2°C (Josephson and Young, 1979). The differential warming has interesting consequences for the song produced. When the tymbal muscle temperature is increased, the rate of tension rise in twitch contractions increases markedly as the twitches become shorter and the twitch tension increases (Fig. 3). Each tymbal muscle contraction produces a pair of sound pulses corresponding to the sequential buckling of ribs on the tymbal membrane to which the muscle is attached (Simmons and Young, 1978). The individual pulses are each an envelope of sound at 850 Hz (Young, 1972). As the tymbal muscle warms during singing and its twitches become shorter and stronger, the interval between the two sound pulses produced by a single muscle contraction decreases by up to 50%. At the same time the interval between the pulse pairs, determined by the output from the cooler ganglion, decreases by only 20%, while the carrier frequency (850 Hz) of the individual pulses, determined by the mechanical resonance of the abdomen (Fletcher and Hill, 1978), scarcely changes at all (Josephson and Young, 1979).

The oscillatory frequency of an isolated asynchronous muscle depends on the mechanical resonance of its load and is essentially independent of the firing frequency in motoneurons supplying the muscle so long as the firing frequency is high enough to keep the muscle fully activated (Machin and Pringle, 1959). *In vivo* the resonance controlling wing stroke frequency is determined by the mass and stiffness of the wing–thorax–flight muscle system. The passive mechanical resonance of this load should not vary appreciably with temperature. It might therefore be anticipated that the wing stroke frequency of insects with asynchronous flight muscles will be independent of temperature. If this were true, the physiological properties of the muscle and the mechanical resonance of the load would be correctly matched at only a single temperature, since the frequency at which potential power output by the muscle is maximum, unlike the passive mechanical resonance of the load, varies with temperature (Machin et al., 1962). The anticipation that wing stroke frequency should be independent of temperature is in general not realized. In Diptera and Hymenoptera, at least, wing stroke frequency increases significantly with the temperature of the thoracic muscles (Sotavalta, 1954; Esch, 1976). In addition, in intact bees and flies the wing stroke frequency is correlated with the frequency of muscle action potentials to an extent not predicted from studies on isolated asynchronous muscles (Wilson and Wyman, 1963; Esch and Bastian, 1968). The change in frequency may be achieved through the activity of small, synchronous muscles in the thorax, muscles that change the stiffness and therefore the resonant frequency of the thorax (Heide, 1971; Pringle, 1974). If the activity in these muscles were temperature-dependent, the mechanical properties of the thorax and the physiological properties of the flight muscles could in principle be correctly matched over a wide temperature range.

7 SUMMARY

The mechanical performance of insect muscles is strongly dependent on temperature. In the muscles that have been examined, as temperature is increased the rate of tension rise and fall increases, the twitch duration decreases, the potential work per twitch (or per oscillatory cycle in asynchronous muscles) increases, and the repetition frequency at which mechanical power output is maximal increases. Isometric tension is less dependent on temperature than are temporal parameters of contraction. Both twitch and tetanic tension tend to increase with temperature, but there is much variability in temperature sensitivity and, for many muscle types, changing temperature has little effect on either twitch or tetanic tension.

It should be noted that these conclusions are based on the performance of flight and tymbal muscles of large insects, most of which are tropical or subtropical. These muscles warm up during activity and are adapted to operate most effectively at relatively high temperatures and over a relatively narrow temperature range. The asynchronous muscles of larger beetles and the synchronous muscles of larger moths may produce enough power for flight only at thoracic temperatures approaching 40°C (Machin et al., 1962; Bartholomew and Heinrich, 1978; reviewed by Heinrich, 1974, and by Kammer and Heinrich, 1978). This temperature limitation does not apply to many smaller insects such as the housefly, *Musca,* in which flight can occur at thoracic temperatures ranging from at least 16 to 45°C (Heinrich, this volume) or to mosquitoes which, as every camper knows, can be annoyingly active even on cool evenings and which are so small that their thoracic temperature must nearly match the environmental temperature. Some insects can operate at remarkably low temperatures. Alpine grylloblattids and arctic beetles are active at temperatures below 0°C and go into a heat coma if warmed to 15–20°C (Henson, 1957; Baust and Miller, 1970; Morrissey and Edwards, 1979). It will be interesting to see how concepts about how temperature effects on insect muscle change when information becomes available on muscles of cold-tolerant insects and of insects that operate over a wide range of muscle temperatures.

REFERENCES

Bartholomew, G. A. and Casey, T. M. (1978). Oxygen consumption of moths during rest, pre-flight warm-up, and flight in relation to body size and wing morphology. *J. Exp. Biol.* **76,** 11–25.

Bartholomew, G. A. and Heinrich, B. (1978). Endothermy in African dung beetles during flight, ball making, and ball rolling. *J. Exp. Biol.* **73,** 65–83.

Baust, J. G. and Miller, L. K. (1970). Variations in glycerol content and its influence on cold hardiness in the Alaskan carabid beetle, *Pterostichus brevicornis. J. Insect Physiol.* **16,** 979–990.

Bentley, D. R. (1973). Postembryonic development of insect motor systems. In *Developmental Neurobiology of Arthropods*, D. Young, Ed., pp. 147–177. Cambridge University Press: Cambridge.

Boettiger, E. G. (1957). The machinery of insect flight. In *Recent Advances in Invertebrate Physiology*, B. T. Scheer, Ed., pp. 117–142. University of Oregon: Eugene.

Buchthal, F., Weis-Fogh, T., and Rosenfalck, P. (1957). Twitch contractions of isolated flight muscle of locusts. *Acta Physiol. Scand.* **39,** 246–276.

Church, N. S. (1960). Heat loss and body temperatures of flying insects. I. Heat loss by evaporation of water from the body. *J. Exp. Biol.* **37,** 171–185.

Counter, S. A. (1977). Bioacoustics and neurobiology of communication in the tettigoniid *Neoconocephalus robustus*. *J. Insect Physiol.* **23,** 993–1008.

Cullen, M. J. (1974). The distribution of asynchronous muscle in insects with particular reference to the Hemiptera: an electron microscope study. *J. Entomol.* **A49,** 17–41.

Elder, H. Y. and Josephson, R. K. (1980). Unpublished observations.

Esch, H. (1976). Body temperature and flight performance of honey bees in a servo-mechanically controlled wind tunnel. *J. Comp. Physiol.* **109,** 265–277.

Esch, H. and Bastian, J. (1968). Mechanical and electrical activity in the indirect flight muscles of the honey bee. *Z. Vergl. Physiol.* **58,** 429–440.

Farnworth, E. G. (1972). Effects of ambient temperature and humidity on internal temperature and wing beat frequency of *Periplaneta americana*. *J. Insect Physiol.* **18,** 359–371.

Fletcher, N. H. and Hill, G. K. (1978). Acoustics of sound production and of hearing in the bladder cicada *Cystosoma saundersii* (Westwood). *J. Exp. Biol.* **72,** 43–55.

Greenewalt, C. H. (1962). Dimensional relationships for flying animals. *Smithson. Misc. Coll.* **144**(2), 1–46.

Hagiwara, S., Uchiyama, H., and Watanabe, A. (1954). The mechanism of sound production in certain cicadas with special reference to the myogenic rhythm in insect muscles. *Bull. Tokyo Med. Dent. Univ.* **1,** 113–124.

Hanegan, J. L. and Heath, J. E. (1970). Temperature dependence of the neural control of the moth flight system. *J. Exp. Biol.* **53,** 629–639.

Heide, G. (1971). Die Function der nicht-fibrillaren Flugmuskeln von *Calliphora*. Teil II. Muskulare Mechanismen der Flugsteuerung und ihre nervöse Kontrolle. *Zool. Jahrb. Physiol.* **76,** 99–137.

Heinrich, B. (1970). Thoracic temperature stabilization by blood circulation in a free-flying moth. *Science* **168,** 580–582.

Heinrich, B. (1972). Energetics of temperature regulation and foraging in a bumblebee, *Bombus terricola* Kirby. *J. Comp. Physiol.* **77,** 49–64.

Heinrich, B. (1974). Thermoregulation in endothermic insects. *Science* **185,** 747–756.

Heinrich, B. (1975). Thermoregulation in bumblebees. II. Energetics of warm-up and free flight. *J. Comp. Physiol.* **96,** 155–166.

Heinrich, B. and Bartholomew, G. A. (1971). An analysis of preflight warmup in the sphinx moth, *Manduca sexta*. *J. Exp. Biol.* **55,** 223–239.

Henson, W. R. (1957). Temperature preference of *Grylloblatta campodeiformis*. *Nature* **179,** 637.

Hill, A. V. (1951). The influence of temperature on the tension developed in an isometric twitch. *Proc. R. Soc. Lond.* **B138,** 349–354.

Hill, A. V. and Howarth, J. V. (1959). The reversal of chemical reactions in contracting muscle during an applied stretch. *Proc. R. Soc. Lond.* **B151,** 169–193.

Jensen, M. (1956). Biology and physics of locust flight. III. The aerodynamics of locust flight. *Phil. Trans. R. Soc. Lond.* **B239,** 511–552.

Jewell, B. R. and Rüegg, J. C. (1966). oscillatory contraction of insect fibrillar muscle after glycerol extraction. *Proc. R. Soc. Lond.* **B164,** 428–459.

Josephson, R. K. (1973). Contraction kinetics of the fast muscles used in singing by a katydid. *J. Exp. Biol.* **59,** 781–801.

Josephson, R. K. (1975). Extensive and intensive factors determining the performance of striated muscle. *J. Exp. Zool.* **194,** 135–153.

Josephson, R. K. and Halverson, R. C. (1971). High frequency muscles used in sound production by a katydid. I. Organization of the motor system. *Biol. Bull. Woods Hole, Mass.* **141,** 411–433.

Josephson, R. K., Stokes, D. R., and Chen, V. (1975). The neural control of contraction in a fast insect muscle. *J. Exp. Zool.* **193,** 281–300.

Josephson, R. K. and Young, D. (1979). Body temperature and singing in the bladder cicada, *Cystosoma saundersii. J. Exp. Biol.* **80,** 69–81.

Josephson, R. K. and Young, D. (1980). Unpublished observations.

Kammer, A. E. (1970). Thoracic temperature, shivering and flight in the monarch butterfly, *Danaus plexippus* (L.). *Z. Vergl. Physiol.* **68,** 334–344.

Kammer, A. E. and Heinrich, B. (1978). Insect flight metabolism. *Adv. Insect Physiol.* **13,** 133–228.

Machin, K. E. and Pringle, J. W. S. (1959). The physiology of insect fibrillar muscle. II. Mechanical properties of a beetle flight muscle. *Proc. R. Soc. Lond.* **B151,** 204–225.

Machin, K. E., Pringle, J. W. S., and Tamasige, M. (1962). The physiology of insect fibrillar muscle. IV. The effect of temperature on beetle flight muscle. *Proc. R. Soc. Lond.* **B155,** 493–499.

Mannherz, H. G. (1970). On the reversibility of the biochemical reactions of muscular contraction during the absorption of negative work. *FEBS Lett.* **10,** 233–236.

May, M. L. (1976). Thermoregulation and adaptation to temperature in dragonflies (Odonata: Anisoptera). *Ecol. Monogr.* **46,** 1–32.

Morrissey, R. and Edwards, J. S. (1979). Neural function in an alpine grylloblattid: a comparison with the house cricket, *Acheta domesticus. Physiol. Entomol.* **4,** 241–250.

Mullony, B. (1975). Control of flight and related behaviour by the central nervous systems of insects. In *Insect Flight*, R. C. Rainey, Ed., Symposium of the Royal Entomological Society of London, Vol. 7, pp. 16–30.

Neville, A. C. (1963). Motor unit distribution of the dorsal longitudinal flight muscles in locusts. *J. Exp. Biol.* **40,** 123–136.

Neville, A. C. and Weis-Fogh, T. (1963). The effect of temperature on locust flight muscle. *J. Exp. Biol.* **40,** 111–121.

Pringle, J. W. S. (1954). The mechanism of the myogenic rhythm of certain insect striated muscles. *J. Physiol.* **124,** 269–291.

Pringle, J. W. S. (1960). Models of muscle. *Symp. Soc. Exp. Biol.* **14,** 41–68.

Pringle, J. W. S. (1967). The contractile mechanism of insect fibrillar muscle. *Prog. Biophys. Mol. Biol.* **17,** 1–60.

Pringle, J. W. S. (1974). Locomotion: Flight. In *The Physiology of Insects,* Vol. III, M. Rockstein, Ed., pp. 433–476. Academic: New York.

Pringle, J. W. S. (1978). Stretch activation of muscle: Function and mechanism. *Proc. R. Soc. Lond.* **B201,** 107–130.

Pringle, J. W. S. and Tregear, R. T. (1969). Mechanical properties of insect fibrillar muscle at large amplitudes of oscillation. *Proc. R. Soc. Lond.* **B174,** 33–50.

Roeder, K. D. (1951). Movements of the thorax and potential changes in the thoracic muscles of insects during flight. *Biol. Bull. Woods Hole, Mass.* **100,** 95–106.

Simmons, P., and Young, D. (1978). The tymbal mechanism and song patterns of the bladder cicada, *Cystosoma saundersii. J. Exp. Biol.* **76,** 27–45.

Sotavalta, O. (1953). Recordings of high wing-stroke and thoracic vibration frequency in some midges. *Biol. Bull. Woods Hole, Mass.* **104,** 439–444.

Sotavalta, O. (1954). On the thoracic temperature of insects in flight. *Ann. Zool. Soc. Zool. Bot. Fenn. Vanamo* **16,** 1–22.

Steiger, G. J. and Ruëgg, J. C. (1969). Energetics and "efficiency" in the isolated contractile machinery of an insect fibrillar muscle at various frequencies of oscillation. *Pfluegers Arch. Gesamte Physiol. Menschen Tiere* **307,** 1–21.

Stevens, E. D. and Josephson, R. K. (1977). Metabolic rate and body temperature in singing katydids. *Physiol. Zool.* **50,** 31–42.

Stokes, D. R., Josephson, R. K., and Price, R. B. (1975). Structural and functional heterogeneity in an insect muscle. *J. Exp. Zool.* **194,** 379–408.

Ulbrich, M. and Rüegg, J. C. (1977). Mechanical factors affecting the ATP-phosphate exchange reaction of glycerinated insect fibrillar muscle. In *Insect Flight Muscle*, R. T. Tregear, Ed., pp. 317–333. North-Holland: Amsterdam.

Weis-Fogh, T. (1956a). Biology and physics of locust flight. II. Flight performance of the desert locust (*Schistocerca gregaria*). *Phil. Trans. R. Soc. Lond.* **B239,** 459–510.

Weis-Fogh, T. (1956b). Tetanic force and shortening in locust flight muscle. *J. Exp. Biol.* **33,** 668–684.

Weis-Fogh, T. (1972). Energetics of hovering flight in hummingbirds and in *Drosophila. J. Exp. Biol.* **56,** 79–104.

Weis-Fogh, T. (1973). Quick estimates of flight fitness in hovering animals, including novel mechanisms for lift production. *J. Exp. Biol.* **59,** 169–230.

Weis-Fogh, T. (1977). Dimensional analysis of hovering flight. In *Scale Effects in Animal Locomotion*, T. J. Pedley, Ed., pp. 405–420. Academic: New York.

Weis-Fogh, T. and Alexander, R. McN. (1977). The sustained power output from striated muscle. In *Scale Effects in Animal Locomotion*. T. J. Pedley, Ed., pp. 511–525. Academic: New York.

Wilson, D. M. (1961). The central nervous control of flight in a locust. *J. Exp. Biol.* **38,** 471–490.

Wilson, D. M. (1962). Bifunctional muscles in the thorax of grasshoppers. *J. Exp. Biol.* **39,** 669–677.

Wilson, D. M. (1964). Relative refractoriness and patterned discharge of locust flight motor neurons. *J. Exp. Biol.* **41,** 191–205.

Wilson, D. M. and Weis-Fogh, T. (1962). Patterned activity of co-ordinated motor units, studied in flying locusts. *J. Exp. Biol.* **39,** 643–667.

Wilson, D. M. and Wyman, R. J. (1963). Phasically unpatterned nervous control of dipteran flight. *J. Insect Physiol.* **9,** 859–865.

Young, D. (1972). Neuromuscular mechanisms of sound production in Australian cicadas. *J. Comp. Physiol.* **79,** 343–362.

3

A Matter of Size: An Examination of Endothermy in Insects and Terrestrial Vertebrates

GEORGE A. BARTHOLOMEW

1 SOME BIOLOGICAL CORRELATES OF BODY SIZE

It is only a slight overstatement to say that the most important attribute of an animal, both physiologically and ecologically, is its size. Size constrains virtually every aspect of structure and function and strongly influences the nature of most inter- and intraspecific interactions. Body mass, which in any given taxon is a close correlate of size, is the most widely useful predictor of physiological rates. The range of size among living multicellular animals covers about 13 orders of magnitude, from whales with a mass of more than 100 metric tons to rotifers with a mass of less than 0.01 mg. Even among animals within a single class, and therefore sharing the same basic anatomical and physiological organization, the range of body size can be enormous. For example, among mammals body mass spans approximately 8 orders of magnitude from a 10^8-g blue whale (*Baelaenoptera musculus*) to a 2-g masked shrew (*Sorex cinereus*); among birds body mass ranges from the 100-kg (10^5-g) ostrich (*Struthio camelus*) to the 2-g bee hummingbird (*Calypte helense*); among insects it ranges from beetles with a mass of almost 30 g to midges with a mass of less than 0.1 mg. Even within a single order of insects (Coleoptera) the range of body mass spans 4 orders of magnitude, 1.0 mg to 30 g.

In every major taxon the problems of physiological control vary with differences in body size. The problems of maintaining physiological homeostasis—that is, the regulation of dynamic but stable internal chemical and physiological conditions in the face of a varying external environment—differ markedly in very large and very small animals. At the smaller limit there is a minimum size necessary for reducing the randomizing effect of Brownian movement to the level of tolerable noise. At the upper limit there are inevitable problems of transport, physical support, and inertia.

2 THE TAXONOMIC DISTRIBUTION OF ENDOTHERMY

It is intriguing to a student of ecologically relevant physiology to examine the mechanisms by which taxa that differ greatly in size and pattern of morphological organization exert control over a physiological function that plays a central role in their performance. Control over body temperature is a case in point. To keep the discussion within manageable dimensions, I shall limit the topic to the phenomenon of endothermy, the maintenance of body temperature by means of endogenous, that is, physiological or metabolic, heat production (see p. 5 for definitions).

Dependence on endogenous heat production for the maintenance of body temperature is of course most familiar in the two classes of homeothermic vertebrates, birds and mammals. However, endothermy has evolved independently in a number of different groups of animals, including sharks of the family Lamnidae (Carey, 1973), bony fish of the family Scombridae, tuna and related forms (Carey and Teal, 1966; Graham, 1975), at least one snake, *Python molurus* (Hutchinson et al., 1966), and a number of different kinds of insects.

Some of the salient features of endothermy in animals are thrown into particularly sharp focus at the lower limits of size at which the phenomenon occurs, namely, in the smallest birds and mammals and the endothermic insects. The physiological convergence in control of body temperature shown by these representatives of different phyla despite their widely divergent patterns of morphological organization is fascinating. On the one hand, heterothermy in birds and mammals represents a relaxation of homeothermy and the temporary acquisition of limited ectothermy, while on the other, heterothermy in insects is an endothermic supplement to the normal ectothermic condition of the class. Despite this major contrast, some of the shared physiological attributes (and presumably the ecological consequences) are strikingly similar and offer an almost unique example of convergent evolutionary responses of physiological functions to a common set of physical and physiological parameters operating in vastly different morphological structures.

3 MAIN FEATURES OF INSECT ENDOTHERMY

The general nature of insect endothermy now seems clear. Although the details remain to be explored and the variations in pattern found in the various orders have been only roughly delineated, the major features of the phenomenon are readily summarized. For almost a century and a half, it has been known that some insects during flight sustain body temperatures significantly above air temperatures and that the heat responsible for this difference is a by-product of the activity of the thoracic muscles. However, the mechanisms of endothermic temperature control of insects have been examined in detail only during the last decade.

Flapping flight is the most energetically demanding mode of animal locomotion. The flight muscles of insects are the most metabolically active of tissues. Because the efficiency of muscle is approximately 20%, about four-fifths of the large energy expenditure during insect flight appears as heat, some of which is retained in the thorax and can result in body temperatures as high or higher than those of birds and mammals.

Although the heat production that results in elevated body temperature is an obligatory consequence of the activity of flight muscle, in some moths, bees, beetles, and dragonflies, elevated body temperatures must be attained before flight is possible. Before these insects can become airborne they must undergo a preflight warm-up during which the elevator and depressor muscles contract nearly simultaneously rather than alternately, generating much heat and little wing movement. Preflight warm-up typically results in thoracic temperatures between 35 and 40°C at the time of takeoff. Thoracic temperatures remain elevated during flight. When the animals land, they cool down and thoracic temperatures usually fall to, or nearly to, ambient. The endothermically elevated temperatures of heterothermic insects generally represent a regulated state. Regulation is usually by means of controlled heat transfer to the abdomen (and thence heat loss to the environment) and sometimes by variations in rates of metabolic heat production. The primary functional result of elevated thoracic temperature is an increase in the power output and frequency of wingbeat. However, the heat produced by the flight muscles can also be important during activity in processes other than flight; for example, brooding in bees (Heinrich, 1974b), singing in katydids (Heath and Josephson, 1970), terrestrial locomotion in beetles (Bartholomew and Casey, 1977), and ball making and ball rolling in large dung beetles (Bartholomew and Heinrich, 1978). Thus we find that heterothermy (intermittent endothermy) in insects is of wide taxonomic occurrence and subserves a variety of functions.

4 BODY SIZE AND ENDOTHERMY

4.1 Body size, Heat Loss, and Heat Production

For the student of adaptive physiology the small size of endothermic insects as compared with the size of even the smallest birds and mammals is a matter of particular interest. This is because the role of size is so important in heat retention, and heat retention and heat production are the two most critical features of endothermic temperature control.

Effective endothermic control of body temperature depends on the capacity to match rates of heat production and heat loss. Other things being equal, the rate of heat loss of a body is proportional to its surface area, and objects of similar geometry have surface areas that are proportional to the two-thirds power of their volumes (see Section 4.3). It is therefore not surprising that, in both birds and small mammals, rates of mass-specific heat loss are inversely proportional to body weight (Morrison, 1960; Herreid and Kessel, 1967; Lasiewski et al., 1967).

The rate of heat production depends primarily on biochemical rather than dimensional factors. Most aspects of the cellular physiology of birds and mammals are fundamentally similar. Therefore the members of these taxa may reasonably be presumed to share an upper limit for mass-specific rate of heat production. If the upper limit of mass-specific rate of heat production is fixed while the mass-specific rate of heat loss continues to increase as size decreases, there should be a minimum size below which a bird or mammal with a body temperature of 35–40°C cannot produce heat rapidly enough to balance its heat loss even at moderate (~20°C) environmental temperatures. At or near this critical size, maintenance of stable and endothermically elevated body temperatures should become difficult or impossible. In any event, natural selection has produced no endothermic vertebrate weighing less than 2–3g—shrews of the genus *Sorex* and hummingbirds of the genus *Calypte* are familiar examples. There are of course many vertebrates weighing less than 1 g, but these are all ectotherms. Moreover, except for some shrews, most of the birds and mammals that weigh less than 5 g are heterothermic and maintain high body temperatures only when they are active or when energy is available in excess (see Bartholomew, 1972, and Wolfe and Hainsworth, 1979, for review).

In view of the cutoff at 2–3 g for vertebrate endothermy, it is of special interest that many insects, some weighing less than 0.1 g, generate and maintain high body temperatures during flight. For example, moths of the genus *Hylesia* (family Saturniidae) weighing as little as 40 mg elevate body temperature during preflight warm-up as much as 12°C above ambient by means of their own metabolism. How is it possible for these tiny creatures, weighing only $\frac{1}{50}$ as much as the smallest birds and mammals, to achieve endothermic control of body temperature similar to that of mammals which, with anthropocentric arrogance, we usually consider to be the pinnacle of evolution?

From a physical point of view the two basic elements in the determination of an animal's body temperature are the rate of heat gain and the rate of heat loss. In an endothermic animal that does not bask in the sun and that maintains a body temperature substantially above ambient, heat gain can be equated with metabolic heat production. Thus, when heat production H_p exceeds heat loss H_l, body temperature T_b increases, and vice versa:

$$H_p > H_l \rightarrow T_b \text{ increases}$$
$$H_p < H_l \rightarrow T_b \text{ decreases}$$
$$H_p = H_l \rightarrow T_b \text{ constant}$$

4.2 Two Central Questions

These relationships allow us to approach the general question we previously asked, How is it possible for insects to be endothermic despite their small size? by asking a pair of more specific questions: (1) To what extent does insect endothermy depend on high levels of heat production? (2) To what extent does insect endothermy depend on low levels of heat loss?

Operationally these two questions can be answered by comparing the rates of heat loss and the rates of heat production in endothermic insects with those of birds or mammals of the same size. However, birds and mammals are generally much larger than insects. Indeed, except for a few beetles and a few moths there are no endothermic insects as large as the smallest birds and mammals. This dilemma can be resolved in a reasonably satisfactory manner by scaling procedures.

4.3 Scaling Physiological Variables—A Brief Statement

As every biologist knows, if the size of an organism changes but its shape remains the same, certain of its physical attributes change in a predictable manner. This situation can be quantified in simple geometric terms. In planar figures, area increases as the square of length ($a \propto l^2$), and in a three-dimensional object volume increases as the cube of length ($v = l^3$). In a given taxon mass is directly proportional to volume ($m \propto v$). Therefore the length and surface area of an organism can be expressed as fractional powers of its mass, $l \propto m^{1/3}$ and $a \propto m^{2/3}$. These scaling procedures can be applied to physiological variables such as rate of heat loss and rate of heat production as well as to morphology. Relationships based on scaling are usually expressed in terms of the allometric equation $Y = aX^b$, where Y is the physiological variable, a is the proportionality constant that characterizes the variable in the group of animals being considered, X is the body mass, and the exponent b is the power function that specifies the effect of mass on the physiological variable. When Y and X are plotted on logarithmic coordinates, the regression of Y on X is linear. The exponent b is the slope of linear regression, and a is the Y intercept at $X = \log 0$ and therefore specifies the magnitude of the physiological variable for an animal with a mass of 1 unit.

To compare the rates of heat loss, or rates of heat production, in endothermic insects and birds and mammals, one plots the log-transformed data for the physiological variable against the log-transformed data for body mass. One then calculates the linear regressions for Y on X for each group and compares the slopes b and the intercepts a of the regressions. If the slopes and intercepts do not differ significantly, the value for the physiological variable in the groups being compared would be the same if size were the same.

4.4 The Questions Rephrased

Now let us return to our questions (see Section 4.2) and rephrase them slightly: (1) Do endothermic insects produce more or less heat relative to their size than endothermic vertebrates? (2) Do endothermic insects have better or poorer heat retention relative to their size than endothermic vertebrates? In operational terms these questions become, Do endothermic insects have higher rates of energy metabolism relative to their body mass than birds and mammals? and Relative to their size, are the cooling constants of endothermic insects larger or smaller than those of birds and mammals?

5 HEAT LOSS AND BODY SIZE

When a heterothermic animal is in an ectothermic state, its rate of heat loss can conveniently be determined by measuring the rate at which it cools passively under constant environmental conditions. When it is in a homeothermic state, its rate of heat loss can conveniently be determined by measuring its heat production (see Section 5.1), because heat loss and heat production are by definition equal as long as body temperature, and therefore heat content, remain constant.

The cooling curve of an animal (i.e., the decline in body temperature during passive cooling) can be described by the cooling constant, k in the equation $dT_b/dt = k\,(T_b - T_a)$, which specifies the temperature change per unit time per unit difference in temperature between the animal and the environment. The constant k has units of °C/$t\cdot$°C or $1/t$ and can be converted to thermal conductance C expressed as calories or joules $(g\cdot °C\cdot t)^{-1}$ by multiplying it by the specific heat of tissue which is usually assumed to be 0.83 cal/g·°C. Thermal conductance can also be calculated from heat production if mean body temperature, hence heat content, remains constant.

$$C = \frac{H_p}{(T_b - T_a)}$$

A more detailed but still simplified discussion of cooling curves and thermal conductance is available in Bartholomew (1977), and an extensive and rigorous statement can be found in Bakken (1976a,b).

5.1 Cooling Constants for Heterothermic Insects

Data on cooling constants and thermal conductance are available for a substantial range of sizes in moths, bees, flies, dragonflies, and beetles (Heinrich and Bartholomew, 1971; Bartholomew and Epting, 1975a,b; May, 1976a,b; Bartholomew and Heinrich, 1978). I shall base my discussion primarily on moths and to a limited extent on beetles, because they overlap in size the smallest birds and mammals.

The cooling constants for heterothermic moths are easily measured. The moth is anesthetized with carbon dioxide. A 40-gauge or finer copper–constantan thermocouple is implanted on one side of the midline in the thorax through a hole punctured in the cuticle. The hemolymph coagulates and holds the thermocouple securely in place. The moth is placed in a screened cage and then stimulated by touch until it arouses and begins the shivering and wing whirring typical of warm-up. After a few minutes it reaches flight temperature and takes off, but as soon as it is airborne its flight is impeded either by the walls of the cage or by the thermocouple leads and it immediately alights and starts to cool. The rate at which the thoracic temperature returns to ambient temperature (i.e., the cooling rate) can be used to determine the cooling constant k.

Typical heating and cooling curves are shown in Fig. 1. If plotted semilogarithmically, the decrease in temperature of a homogeneous object passively cooling in a constant environment describes a straight line. The cooling segments of the curves such as those in Fig. 1, when plotted on semilogarithmic coordinates, are usually straight lines, although

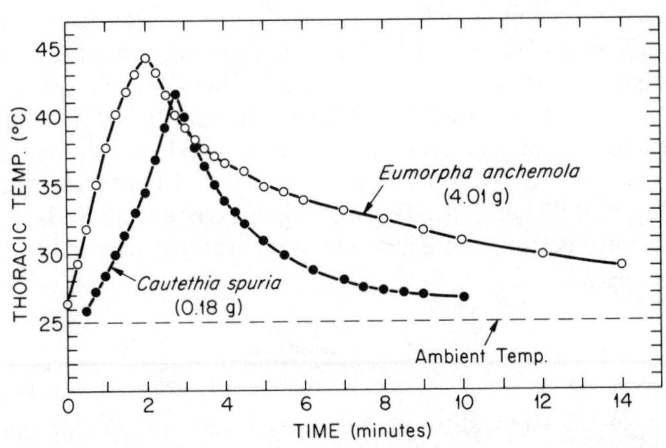

Fig. 1 Typical curves of preflight warm-up and postflight cooling in the sphinx moth (family Sphingidae). (Data from Bartholomew and Epting, 1975a.)

sometimes only the terminal part is linear while the initial part is curvilinear (Fig. 2). The curvilinear segments represent cooling that is physiologically facilitated (Bartholomew and Epting, 1975a), presumably by accelerated flow of blood from thorax to abdomen (Heinrich, 1970). In any event the slope of the linear segment of the cooling curve represents the rate of passive cooling and is equal to k in the equation

$$\frac{dT_{\text{th}}}{dt} = k\,(T_{\text{th}} - T_{\text{a}})$$

where T_{th} is thoracic temperature, T_{a} is ambient temperature, and t is time. For animals that are morphologically similar, the bigger the animal, the slower it cools (Fig. 3*a*). Consequently, when one plots the slopes of the cooling curves or morphologically similar animals of a variety of sizes against body mass, one then has a description of the relation of the cooling constant to body mass (Fig. 3*b*). As previously noted, k is readily converted to units that allow one to plot the relationship of mass-specific thermal conductance C to total body mass. Such plots, through the use of scaling procedures, allow one to compare the rates of heat loss of animals belonging to different taxa and differing greatly in size.

Thus one can compare the magnitude of heat loss in moths, beetles, birds, and mammals by graphing their respective mass-specific thermal conductances as functions of body mass. As pointed out in Section 4.3, if the slopes of the regressions are the same, the effect of size on rates of heat loss in the various groups is the same. If the similar slopes have the

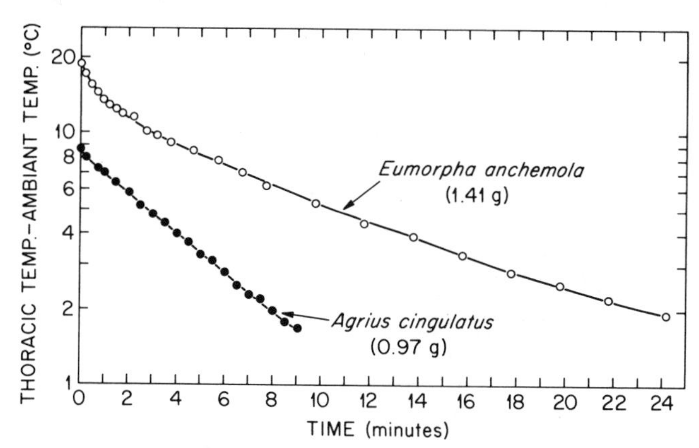

Fig. 2 Semilog plot of cooling curves of two sphinx moths. Note accelerated initial cooling in Eumorpha. (Data from Bartholomew and Epting, 1975a.)

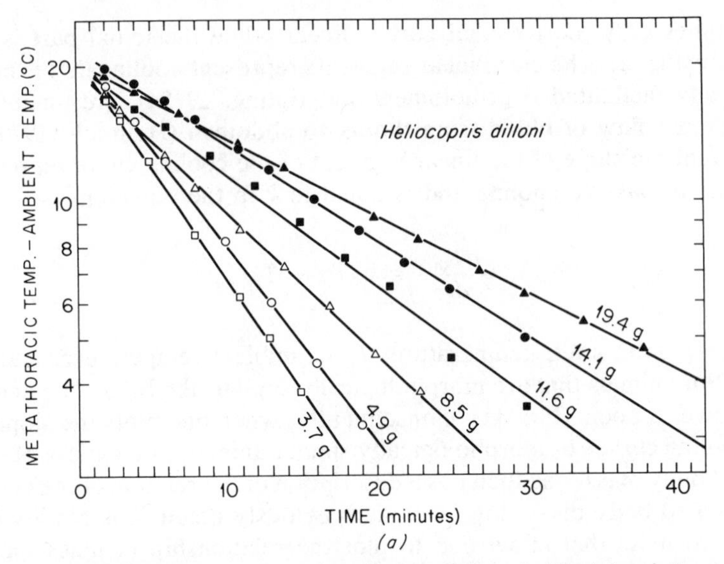

Fig. 3a Cooling curves of an African dung beetle, **Helicopris dilloni**, in relation to body mass. (Data from Bartholomew and Heinrich, 1978.) See also Fig. 3b.

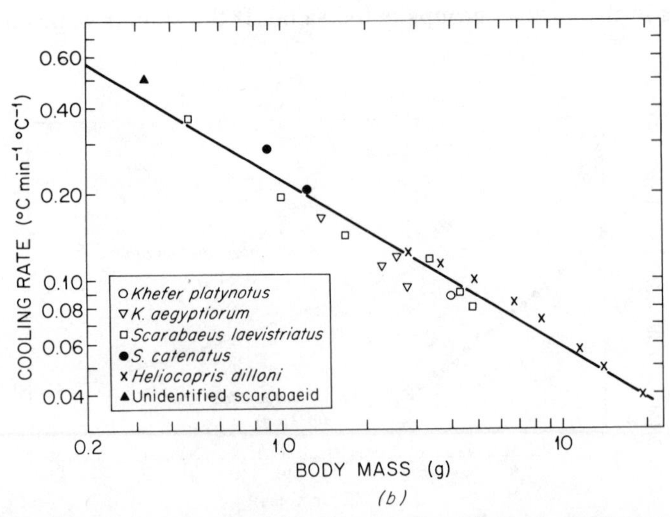

Fig. 3b The least squares regression of the log-transformed data for cooling rate on body mass for five species of scarabaeid beetles. (Data from Bartholomew and Heinrich, 1978.)

same intercepts when extrapolated to a given mass, then the relative rates of heat loss in the various groups can be said to be equivalent.

5.2 Comparison of Heat Loss in Heterothermic Insects, Birds, and Mammals

Data are available on the mass-specific thermal conductance C of birds and small mammals as determined by cooling rates and by heat production (Herreid and Kessel, 1967; Lasiewski et al., 1967). Over the range of sizes measured the thermal conductances of birds and mammals do not differ significantly. In both groups it scales with mass approximately to the -0.5 power (Fig. 4).

To facilitate comparison of the effectiveness of homeothermic vertebrates and heterothermic insects in retaining the heat they produce we can compare birds and mammals with sphinx moths (family Sphingidae) which have evolved a body shape strikingly similar to that of hummingbirds. To do this we convert the cooling constants for sphingids to mass-specific thermal conductance which is the measurement commonly employed in the literature on birds and mammals. In the size range where the moths and the vertebrates overlap (2–5 g) the sphingids have some-

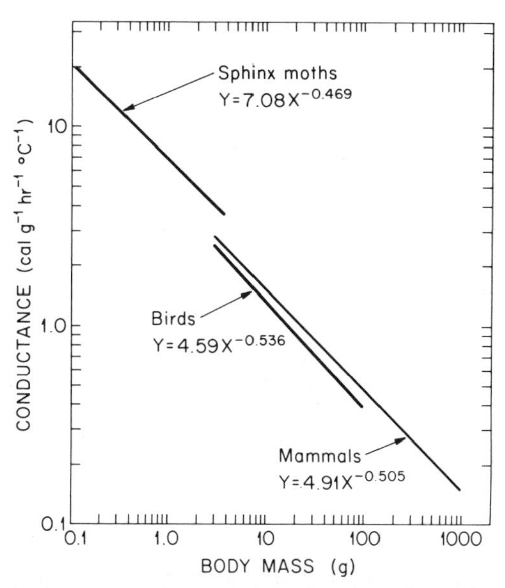

Fig. 4 Mass-specific thermal conductance as a function of total body mass. (The regressions for birds and mammals are from Herreid and Kessel, 1967, and the regression for sphinx moths is from Bartholomew and Epting, 1975a.)

what higher thermal conductances than birds and mammals, but the slopes of the two curves are strikingly similar (Fig. 4). Comparison of insects and vertebrates on the basis of total body mass, however, biases the picture. This is because sphinx moths, like other types of endothermic insects so far studied, maintain high temperatures only in the thorax and keep the abdomen at or near ambient temperature. The abdomen is warmed only when they use it to facilitate heat loss by flushing it with hemolymph warmed by thoracic muscles (Heinrich, 1970). Consequently it is more realistic to calculate the thermal conductance of sphingids on the basis of the mass of the thorax rather than the mass of the body as a whole (Bartholomew and Epting, 1975b; May, 1976b). When computed on the basis of the mass of the thorax, the mass-specific thermal conductance of sphingids is virtually indistinguishable from that of birds and mammals (Fig. 5).

The almost exact equivalence of the curves for mass-specific thermal conductance in sphingids, birds, and mammals is remarkable, particularly when one considers the structural differences among feathers, fur, and moth scales and the profound differences in circulation and integument

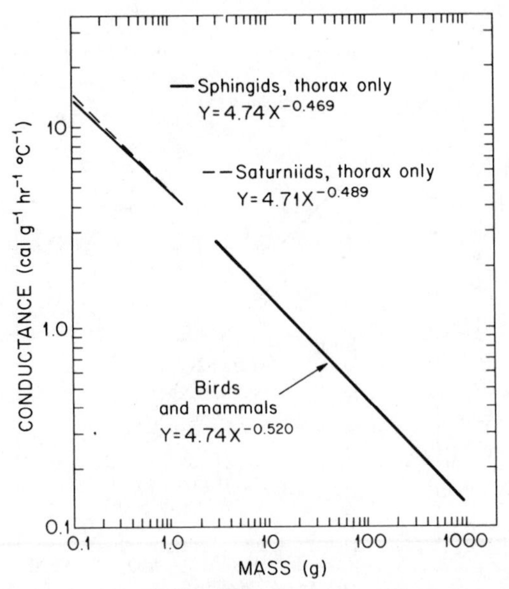

Fig. 5 Mass-specific thermal conductance as a function of body mass. The regression for birds and mammals is based on total mass and is the average of the slopes in Fig. 4. The regression for moths is based on thoracic mass. (Data from Bartholomew and Epting, 1975b.)

between insects and amniotes. However, sphinx moths resemble hummingbirds in habits and shape, and they overlap them in size—the smallest sphingids weigh about 100 mg, but the largest ones weigh 5 g or more. Thus there are gross similarities in body shape on the one hand and differences in structure and organization on the other. Nevertheless, the scaling arguments presented above and summarized in Fig. 5 clearly show that the existence of endothermy in sphinx moths, even those weighing only $\frac{1}{10}$–$\frac{1}{20}$ as much as the smallest birds and mammals, cannot be explained by their having relatively more effective insulation, hence a greater capacity for heat retention.

The next obvious question to ask is, How do the cooling constants of sphinx moths compare with those of other insects? Data on cooling constants in still air are available for live and dead insects of several taxa. In all the insect taxa that have been examined, the cooling constant is inversely related to body mass as exemplified by the dung beetles (Scarabaeidae) shown in Fig. 3*b*. In the absence of forced convection, moths, beetles, bees, flies, and dragonflies all have similar cooling constants (May, 1976b). Endothermy in unstirred air (but not endothermy during flight when forced convection is an important factor) does not appear to be dependent on the type of insulation an insect employs—a dense, deep pile of scales in sphinx and saturn moths, short sparse hairs in tabanids, internal air spaces in dragonflies, or the air spaces under the elytra in beetles. Based on presently available data, it appears that for a given body size one taxon of insects is about as well equipped as another for heat retention during preflight warm-up.

I conclude that the ability of endothermic insects, despite their small size, to achieve a substantial difference in temperature between body and environment, is not a function of capabilities for heat retention that are superior to those of vertebrates. It should be noted, however, that the independence of insects from the constraints imposed by pulmonary and cardiovascular pumping systems simplifies some aspects of heat sequestering and maintains regional differences in body temperature (see Section 10).

6 HEAT PRODUCTION AND BODY SIZE

If the remarkable capacity of insects to depend on endothermy despite their small size cannot be attributed to especially noteworthy capacities for heat retention, it necessarily depends on their having rates of heat production much greater than those that can be achieved by birds and mammals.

6.1 Heat Production and Energy Metabolism

The measurement of heat production is physiologically straightforward. All one needs to do is measure the rates of carbon dioxide production and oxygen consumption. From the ratio between these rates, which is called the respiratory quotient (RQ), one can estimate the mix of carbohydrates, proteins, and fats being combusted so one can determine the heat equivalence of the oxygen being consumed. For purposes of the present discussion we assume that the animals involved have an RQ of about 0.8 and that, for each cubic centimeter of oxygen consumed, 4.8 cal or 20.1 J are released. These values are reasonable approximations for aerobic metabolism in both insects and vertebrates, and they allow convenient comparisons of the metabolic power input of different taxa. By simple conversion factors it is possible to express energy metabolism in terms of oxygen consumed per unit time, as calories or joules released per unit time or as watts.

6.2 The Scaling of Energy Metabolism with Body Size

When measured under similar controlled conditions, the energy metabolism of animals in a given taxon scales with some fractional power (typically 0.75) of body mass. Although its theoretical base is obscure, this familiar empirical generalization is one of the best documented in all of physiology and is one of the central and integrating themes in comparative animal energetics. It is particularly useful when one wishes to compare organisms, such as insects and vertebrates, that differ greatly in size. The relation of energy metabolism (and also many other physiological variables) to total body mass in the members of a given taxon in the same physiological state usually can be fitted to the allometric equation $Y = aX^b$ (see Section 4.3).

To compare the rate of energy metabolism of endothermic insects to that of birds and mammals, one plots the long-transformed data for $\dot{V}_{O_2}$ and mass, using the same units of measurement for each group, and then compares the slopes and intercepts for the different groups. The mechanics of the comparison are simple. The difficulty lies in obtaining data from these very dissimilar groups of animals under physiological conditions that are sufficiently similar so that biologically valid comparisons can be made. For example, insects are characteristically ectothermic and poikilothermic when at rest, while birds and mammals are characteristically endothermic and homeothermic; the response of insects to fasting is very different from that of birds and mammals; insects have no zone of thermal neutrality, hence no basal metabolic rate.

These difficulties, however, can be at least partially circumvented by judicious specification of the types of data to be used and the choice of appropriate taxa for comparison. Toward this end I shall examine selected aspects of the energetics of heterothermic insects and terrestrial vertebrates: (1) the scaling of energy metabolism during flight in birds, bats, and insects, (2) the scaling of energy metabolism in resting insects and in resting, fasting reptiles, (3) comparison of energy metabolism during flight in birds and insects of the same size, (4) comparison of the metabolic scope for activity of heterothermic insects with that of heterothermic birds and mammals of similar size.

6.2.1 Birds and Mammals

Historically most studies of the energy metabolism of birds and mammals have been directed at measurements of basal metabolic rate (BMR) which is defined as the rate of energy metabolism of a fasting, resting homeotherm in its zone of thermal neutrality. (The zone of thermal neutrality is the range of ambient temperatures over which the oxygen consumption of a homeotherm is minimal and above and below which it increases.) In the present context, data on BMR have limited comparative utility because endothermic insects are not homeotherms and so have neither a zone of thermal neutrality nor a BMR. In the past decade it has become technically feasible to measure $\dot{V}_{O_2}$ in birds and bats during flight. $\dot{V}_{O_2}$ during flight scales with mass in a manner remarkably similar to BMR. Energy metabolism during flight and basal metabolism both increase with mass to approximately the 0.75 power, but for an individual bird or bat the former is about an order of magnitude higher than the latter (Fig. 6, Table 1).

The relation of energy metabolism to size among insects is less clear. Historically all insects have been treated as ectotherms. Direct measurements on the energetics of free flight of insects as a function of size are limited, and there has been no consensus about the scaling of their energy metabolism when at rest. Compared with the situation in vertebrates, information on the dependence of $\dot{V}_{O_2}$ on body mass in resting adult insects is suprisingly fragmentary. The data for insects as a whole cannot be characterized by any single value for b in the equation $Y = aX^b$. Values varying from 0.67 to 1.0 have been reported for different orders and families (see May, 1976a; Bartholomew and Casey, 1977b, 1978, for references). For purposes of the present discussion of endothermic insects we use the estimates of Bartholomew and Casey (1977a) for beetles and of Bartholomew and Casey (1978) for moths.

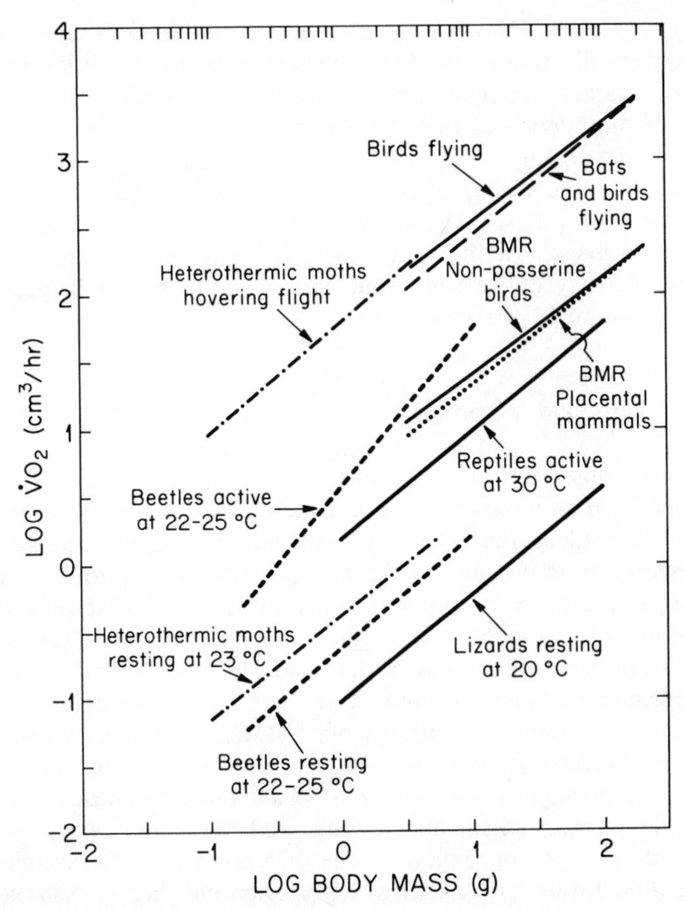

Fig. 6 The regressions of energy metabolism on body mass of heterothermic insects and selected terrestrial vertebrates at rest, during terrestrial activity, and during flight. See Table 1 for the equations for the lines and the sources of the data.

6.3 Comparison of Heterothermic Insects with Terrestrial Vertebrates

When measured at 22–24°C, the rate of oxygen consumption at rest of heterothermic moths belonging to several families and ranging in mass from 60 mg to 5 g scales with body mass to approximately the 0.78 power. During hovering flight the rate of oxygen consumption of moths from the same sample increases almost 150-fold, but the slope of the regression changes very little (Fig. 7). Both these exponents closely approximate

Table 1 Units and coefficients for the allometric equation $Y = aX^b$, relating oxygen consumption (Y) to body mass (X) during rest and activity in selected groups of terrestrial animals[a]

	Y	*a*	*X*	*b*	Source
Heterothermic moths at rest at 23°C	cm³hr	0.403	g	0.76	Bartholomew and Casey, 1978
Heterothermic moths during flight	cm³/hr	59.43	g	0.82	Bartholomew and Casey, 1978
Sphingid moths at rest at 23°C	cm³/hr	0.42	g	0.814	Bartholomew and Casey, 1978
Sphingid moths during flight	cm³/hr	72.28	g	0.768	Bartholomew and Casey, 1978
Beetles at rest at 22–25°C	cm³/hr	0.23	g	0.86	Bartholomew and Casey, 1977b
Beetles during terrestrial activity at 22–25°C	cm³/hr	3.76	g	1.17	Bartholomew and Casey, 1977b
Lizards at rest at 20°C	cm³/hr	0.096	g	0.80	Bennett and Dawson, 1976
Lizards at rest at 37°C	cm³/hr	0.424	g	0.82	Bennett and Dawson, 1976
Reptiles active at 30°C	cm³/hr	1.40	g	0.82	Bennett and Dawson, 1976
BMR passerine birds	kcal/day	129.0	kg	0.72	Lasiewski and Dawson, 1967
BMR nonpasserine birds	kcal/day	78.3	kg	0.72	Lasiewski and Dawson, 1967
Birds during flight	cm³/min	1.02	g	0.73	Hart and Berger, 1972
Birds and bats during flight	W	52.54	kg	0.78	Thomas, 1975
BMR placental mammals	cm³/min	11.6	kg	0.76	Stahl, 1967

[a] See Fig. 6 for curves.

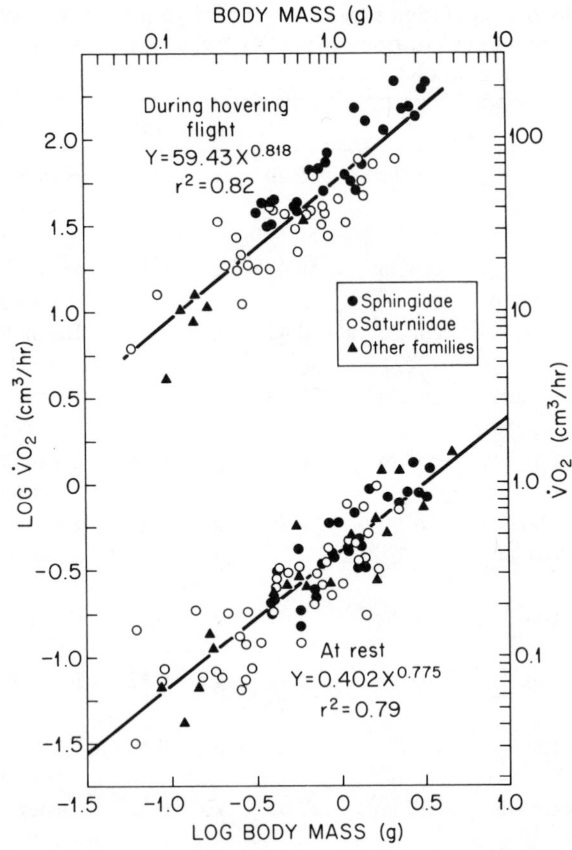

Fig. 7 The oxygen consumption of heterothermic moths at rest and during hovering flight. (Data from Bartholomew and Casey, 1978.)

those reported for resting and active terrestrial vertebrates, but the difference between the active and resting rates of the moths is more than 10 times greater than that for any one group of vertebrates. Indeed, it is about the same as the difference in $\dot{V}_{O_2}$ between resting reptiles and flying birds. This situation of course is related to the fact that the moths being considered are heterothermic (Table 1); at rest they are poikilothermic like reptiles, but during flight they are endothermic and sustain thoracic temperatures equal to or higher than those of birds.

From Fig. 6 it is obvious that $\dot{V}_{O_2}$ during flight scales with mass in a very similar manner in heterothermic moths and in flying birds and bats. These data can be used to examine the mass-specific rates of oxygen consumption in moths and birds if one divides both sides of the equation relating $\dot{V}_{O_2}$ to mass m by mass:

$$\dot{V}_{O_2} m^{-1} = am^{(b-1)}$$

When rate of mass-specific oxygen consumption is plotted as a function of body mass, the relation of energy consumption during flight in the two groups is readily visualized (Fig. 8). The slopes of the regressions for the two groups differ somewhat, −0.18 for moths versus −0.27 for birds, but the difference in rates of energy metabolism in the size range where their masses overlap is negligible. For example, in flight a 2-g bird uses 50.8 $cm^3/g \cdot hr$, while a 2-g moth uses 52.4 $cm^3/g \cdot hr$—a difference of only 3%. The salient feature of this comparison, however, is that these values mark the *upper* limit for metabolic rates of flying birds and approach the *lower*

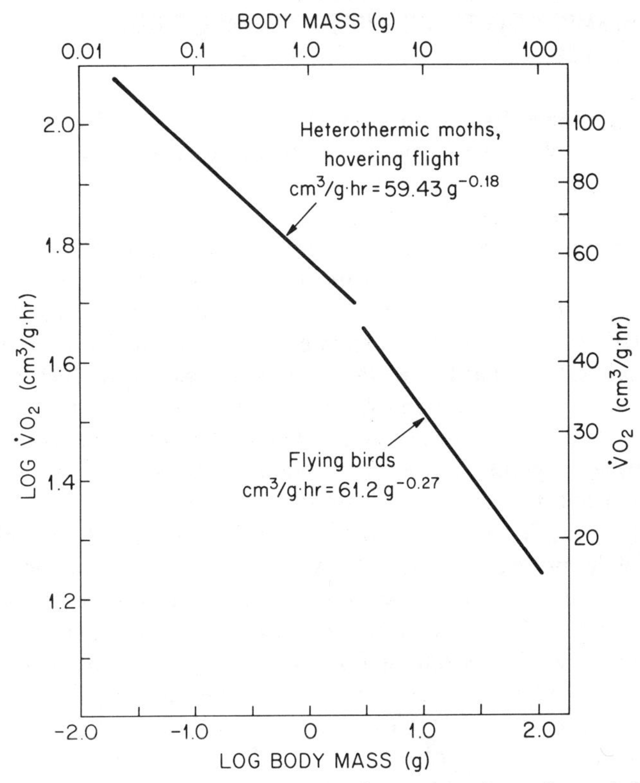

Fig. 8 Mass-specific oxygen consumption in heterothermic moths and flying birds. (Equations from Bartholomew and Casey, 1978, and Hart and Berger, 1972.)

limit of metabolic rates for flying endothermic moths. A typical 60-mg endothermic moth has a metabolic rate of 99 $cm^3/g \cdot hr$—twice the highest predicted rate for any vertebrate homeotherm.

The scaling relationships summarized in Fig. 6 show that during flight endothermic moths ranging in size from 60 mg to several grams maintain the same general relationship between body mass and energy metabolism as birds and mammals ranging in weight from 2–3 g to several hundred grams. I have previously shown by similar procedures of scaling that the quality of insulation in birds, mammals, and endothermic moths is essentially equivalent. Consequently, I confidently infer that the ability of these moths to maintain body temperatures as high or higher than those of birds and mammals is attributable to their capacity for sustaining rates of energy metabolism much higher than those of birds and mammals.

7 ENERGY METABOLISM OF SIMILARLY ADAPTED HETEROTHERMIC INSECTS AND HETEROTHERMIC BIRDS OF SIMILAR SIZE

Scaling arguments based on pooled data from broadly inclusive taxonomic categories are instructive particularly when dealing with major physiological trends. However, scaling arguments necessarily ignore specific cases, and specific cases are of special interest in the study of convergent evolution because it is on them that natural selection operates. In examining size and endothermy, direct comparison of endothermic moths with birds of the same size is essential. Such a comparison is particularly attractive in this instance because critical examples can be drawn from sphinx moths on the one hand and hummingbirds on the other. Consequently one cannot only compare birds and insects of the same size, but heterothermic birds and heterothermic insects of the same size occurring together in the same place at the same time and sharing similar food habits and similar modes of feeding.

In many localities in the American tropics one can find both sphinx moths and hummingbirds that weigh 2–4 g and feed on nectar while hovering in midair. Flight is the most energy-demanding form of locomotion. The aerodynamic constraints imposed by the combination of rapid forward flight while searching for food and by sustained hovering at flowers while feeding on nectar have contributed to the evolution of body size, body shape, and wing shape in the sphingids and hummers so similar that it is easy for a field observer to confuse the two.

Direct measurements of oxygen consumption during sustained hovering

Table 2 Factorial aerobic metabolic scope in sphinx moths and hummingbirds weighing approximately 3 g

		$\dot{V}_{O_2}$ (cm³/hr)			Scope	
Taxon	Mass (g)	A: BMR	B: Inactive at T_b = 21–24°C	C: Hovering flight	C/A:	C/B
Sphinx moths[a]						
Manduca rustica	2.92	—	0.89	133.79	—	150
Pachylia ficus	3.23	—	0.84	193.44	—	230
Oryba archeminides	3.39	—	1.23	210.48	—	171
Hummingbirds						
Calypte costae	3.0	8.4[b]	1.17[b]	127.2[b]	15	109
Archilochus alexandri	2.8–3.6	11.84[b]	1.04[b]	188.9[c]	16	181

[a]From Bartholomew and Casey, 1978.
[b]From Lasiewski, 1963.
[c]From Epting, 1975.

flight are available for both hummingbirds and sphinx moths with a body mass of approximately 3 g (Table 2). The energy cost of hovering flight of the insects is indistinguishable from that of the birds; oxygen consumption in both sphinx moths and hummingbirds approximates 50 cm³/g·hr, a value which is not significantly different from that predicted by scaling (Table 2, Fig. 8). Although values higher than 50 cm³/g·hr for the energy metabolism of hovering hummingbirds are in the literature (Pearson, 1950; Lasiewski, 1963), none of the measured values for birds approach the values for sphinx moths with a mass of 1 g or less. In hovering sphinx moths $\dot{V}_{O_2}$ in cm³/g·hr = 72.3 g$^{-0.23}$ (Bartholomew and Casey, 1978). This equation predicts that the $\dot{V}_{O_2}$ of 1.0-g, 0.5-g, and 0.2-g sphinx moths is 72, 85, 104 cm³/g·hr, respectively. In specific cases values are of course often higher than these average values. For example, Bartholomew and Casey (1978) report 117 cm³/g·hr for a 0.36-g *Enyo ocypete* and 105 cm³/g·hr for a 0.42-g *Perigonia lusca*.

On the basis of the direct measurements and scaling relationships summarized above, I conclude that the capacity of moths less than $\frac{1}{10}$ the size of the smallest birds to sustain elevated body temperatures depends primarily on their remarkable capacity for high rates of energy metabolism and only secondarily on their capacity for reducing rates of heat loss. I hypothesize that the same is true for other types of insects.

8 INSECT HETEROTHERMY VERSUS AVIAN AND MAMMALIAN HETEROTHERMY

The selective advantages of heterothermy are easy to identify (Bartholomew, 1972). When aroused and active, heterothermic animals share the obvious advantages inherent in endothermic homeothermy, such as the capacity for prolonged maintenance of high rates of aerobic metabolism when functionally useful and a substantial physiological independence of environmental variability, including the ability to perform complex neurophysiologically mediated activity at low environmental temperatures. When unaroused or dormant, heterothermic animals are poikilotherms and share the frugal and economical patterns of energy expenditure characteristic of ectotherms and thus readily adjust to periods of inclement weather or food shortage.

All endothermic insects are heterotherms. It seems both conservative and reasonable to hypothesize that the capacity for intermittent endothermy has repeatedly and independently evolved in species belonging to various orders and families of the class Insecta. Similarly there seems to be no reasonable alternative to assuming that the capacity for heterothermy has evolved several times in both birds and mammals. Daily torpor (see Bartholomew, 1972, for references) occurs commonly in insect-eating bats (suborder Microchiroptera) and the smaller flying foxes (suborder Megachiroptera); it is well documented in several families of rodents, and among birds it occurs in swifts, hummingbirds, colies, and goatsuckers.

8.1 Metabolic Scope for Activity

The occurrence of heterothermy in groups of animals that are both taxonomically and ecologically diverse makes it particularly attractive to the student of comparative and ecological physiology. For example, consider *metabolic scope for activity,* an idea first introduced in relation to the aerobic metabolism of fish (Fry, 1947). Metabolic scope was originally defined as the difference between minimum and maximum rates of energy metabolism under a standard set of conditions. However, when comparing animals of very different size or different rates of standard metabolism, it is useful to use factorial scope, the ratio of standard to active metabolism, rather than the difference between the two.

All animals of course have the capacity for greatly increasing their energy metabolism above the resting level. Among homeothermic mammals, the range of factorial scope is approximately 5–20. Among lizards it is about 6–10. In birds and bats energy metabolism is 10–15 times greater during flight than during rest. In heterothermic insects, however, factorial scope is routinely above 100. Among sphinx moths it may exceed 200 (Table 2), and for heterothermic moths as a group measured at ambient temperatures of 22–24°C it averages about 150 (Fig. 7). This great difference of course is a function of the fact that, when at rest these insects are ectothermic and have body temperatures near ambient, but while in flight they are endothermic and often have body temperatures in excess of 40°C. The oxygen consumption of moths at rest scales with that of resting reptiles; the oxygen consumption of flying moths in hovering flight scales with that of flying birds (Fig. 6).

The fact that homeothermic vertebrates have a BMR (the rate of oxygen consumption of a fasting animal at rest in its thermal neutral zone) below which their energy metabolism does not fall means that their resting rate is always high—typically about 6–10 times that of a reptile of about the same size at the same temperature. This high resting level limits the amount by which they can increase their aerobic metabolism. Heterothermic birds and mammals, when they abandon homeothermy, no longer have a BMR (Fig. 9). Their rate of energy metabolism when dormant depends primarily on body temperature which in turn passively follows ambient temperature at least down to near 0°C. Thus a dormant

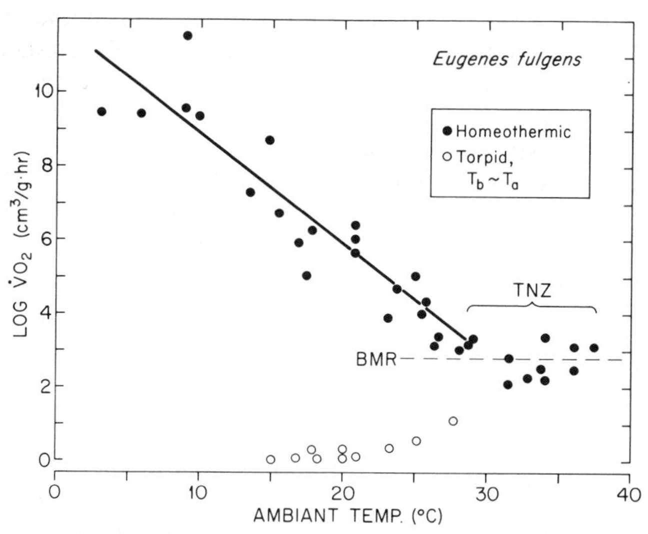

Fig. 9 Mass-specific oxygen consumption as a function of ambient temperature in Rivoli's hummingbird (mass ~ 8 g) when homeothermic and when dormant with body temperature approximately equal to ambient temperature. TNZ, Thermal neutral zone. (Data from Lasiewski and Lasiewski, 1967.)

vertebrate heterotherm resembles a resting ectothermic vertebrate or a heterothermic insect at rest. There is, however, a major and critical difference. At body temperatures of 20°C or less, heterothermic birds and mammals are torpid and nearly helpless, while heterothermic insects and most ectothermic vertebrates are responsive and capable of integrated and effective activity.

Aerobic scope usually increases with size in homeothermic vertebrates. The relation of aerobic scope to size is not clear in heterothermic vertebrates. We can, however, compare heterothermic birds with heterothermic moths of about the same size—3 g. Data are available on 3-g hummingbirds and on several 3-g sphinx moths measured at ambient temperatures near 20°C. Aerobic scope in the 3-g hummingbird calculated on the basis of $\dot{V}_{O_2}$ during torpor averages about 145. This is about 10 times greater than the value for factorial scope when the resting value for $\dot{V}_{O_2}$ is set equal to the observed BMR in the same species (Table 2), about 90% that of a 3-g heterothermic moth calculated from the equations for $\dot{V}_{O_2}$ during rest and flight (Table 1), and about 80% of that measured in 3-g sphinx moths (Table 2). In both the hummingbird and the sphinx moths, the values for $\dot{V}_{O_2}$ at rest in the ectothermic state near 20°C are 0.3–0.4 ml O_2/g·hr. However, it is important to note that at this level of mass-specific oxygen consumption the birds are torpid while the moths are not.

Similar but less well-documented comparisons can be made between beetles and mammals. The factorial scope of small homeothermic mammals is typically 5–7 (Wunder, 1970) and, in the case of the masked shrew, *Sorex cinereus,* the smallest species for which data are available, it is only 3 (Table 3). This limited scope is presumably a function of the extremely high BMR of the smallest mammals.

Aerobic scope during terrestrial activity in beetles increases with body mass (Table 1, Fig. 6), ranging from an average of 16 in 1-g beetles to an average of 33 in 10-g beetles. This increase correlates with the fact that large beetles are often endothermic during terrestrial activity and that in general the bigger the beetle the greater the difference between active and resting body temperature (Bartholomew and Casey, 1977b). One can assume that aerobic scope would be 3 or 4 times the values mentioned above if measurements of oxygen consumption during flight were available for beetles, which they are not. Data are available for $\dot{V}_{O_2}$ of a 6-g scarab beetle, *Strategus aloeus,* that had warmed itself to flight temperature (40°C) but did not fly. Its scope was over 100, which is more than 20 times that of homeothermic mammals of similar size but resembles that of a small heterothermic mammal (*Perognathus longmembris*) over the same range of body temperature (Table 3). Again, it should be emphasized that at the lower body temperature (20°C) the mammal is virtually immobilized while the beetle is relatively responsive and well-coordinated.

Table 3 Factorial aerobic metabolic scope in large beetles and small mammals

	Inactive			Active		
Taxon	Mass (g)	$\dot{V}_{O_2}$ ($cm^3/g \cdot h$)	T_b (°C)	$\dot{V}_{O_2}$ ($cm^3/g \cdot h$)	T_b (°C)	Scope
Sorex cinereus[a] (masked shrew)	3.6	9.0	36–40	30.0	37–39	3.3
Perognathus longimembris (pocket mouse)	8–8.5	0.17[b]	20–22[b]	11.3[c]	34–36[c]	66
Stenodontes molarium[d] (cerambycid beetle)	4.1	0.3	23	10.3	33.7	34
Strategus aloeus[d] (scarab beetle)	6.3	0.17	23	19.27	40.2	113

[a]From Morrison et al., 1959.
[b]From A. R. French, unpublished.
[c]From Chew et al., 1965.
[d]From Bartholomew and Casey, 1977a.

9 THE PROCESS OF AROUSAL—SWITCHING FROM ECTOTHERMY TO ENDOTHERMY

The preflight warm-up of heterothermic insects is strikingly similar to the arousal of mammals and birds from the dormancy associated with daily torpor or the arousal of mammals from the profound hypothermia associated with hibernation. During warm-up in both insects and vertebrates, heat production is maximized, heat loss is minimized, and body temperature rises from near ambient to the high operating levels characteristic of full activity. The wing whirring and isometric contractions of the flight muscles of insects during warm-up are analogous to the shivering of vertebrates during arousal from torpor, but nonshivering thermogenesis, which is important in the warming of mammalian hibernators, has not yet been demonstrated in insects.

9.1 Size and Rate of Warm-up

Small heterothermic birds (Fig. 10) and mammals warm up more rapidly than large ones (Bartholomew et al., 1957; Morrison, 1960; Lasiewski, 1967). Inasmuch as insects are generally smaller than vertebrates and warm up faster than vertebrates, the generalization that rate of warm-up in heterothermic animals is inversely related to body mass can be ex-

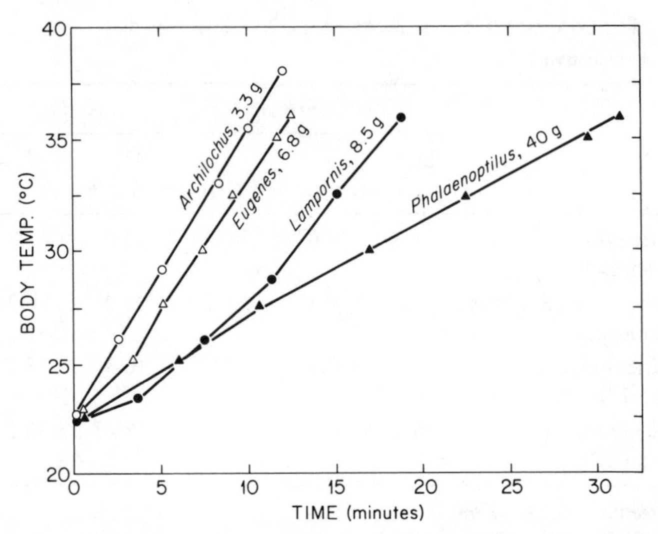

Fig. 10 Body temperature during arousal from dormancy in three hummingbirds and a poor-will at an ambient temperature of 20°C. Note that the rate of increase in T_b is inversely related to body mass. (Data from Lasiewski and Lasiewski, 1967.) See also Fig. 11.

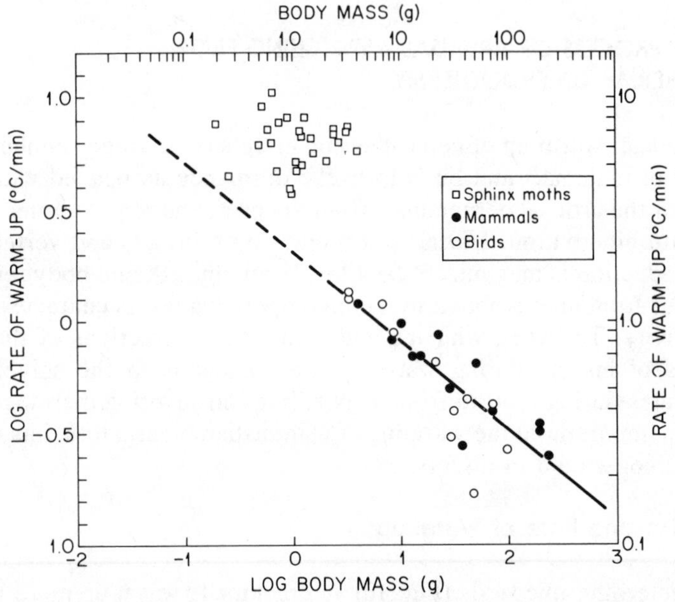

Fig. 11 The relation between rates of increase in body temperature and body mass during warm-up in heterothermic insects and in heterothermic birds and mammals. Ambient temperatures, 20–25°C. (Data for birds and mammals from Heinrich and Bartholomew, 1971, and data for sphinx moths from Bartholomew and Epting, unpublished.)

tended to include heterothermic insects (Fig. 11) as well as heterothermic vertebrates (Heinrich and Bartholomew, 1971; Bartholomew, 1972). However, the generalization does not apply within the class Insecta. The data presently available indicate that the rate of endothermic warm-up in insects is either unrelated to body mass (Heinrich and Bartholomew, 1971; Heinrich and Casey, 1973; Bartholomew and Epting, 1975) or positively (but weakly) correlated with body mass (May, 1976b).

On the basis of allometric considerations May (1976b) has suggested that among insects mass-specific heat loss increases more rapidly with decreasing size than mass-specific heat production, and that as a consequence the rate of warm-up should be faster in large insects than in small ones. This issue cannot be resolved until data on rates of energy metabolism as a function of body temperature are obtained from insects of various sizes during warm-up. In any event, it is clear that heterothermic insects warm up much more rapidly than heterothermic vertebrates (Fig. 12).

10 INTERPRETATION OF THE CHARACTERISTICS OF HETEROTHERMY IN INSECTS AND IN VERTEBRATES

Heterothermy has probably arisen independently and repeatedly in the class Insecta, several times in the class Mammalia, and also several times in the class Aves. The multiple convergence on a common physiological pattern by diverse and morphologically very different types of animals is a priori evidence that heterothermy is strongly favored by natural selection. The selective advantages of heterothermy are not difficult to identify. A heterothermic animal has the best of both worlds. When appropriate, it has the familiar advantages associated with homeothermy, such as sustained high rates of aerobic metabolism and substantial physiological and behavioral independence of variations in thermal conditions in the environment. However, when circumstances dictate, it can exploit the familiar advantages of poikilothermy and ectothermy, such as frugality of energy expenditure and the capacity for waiting out periods of prolonged food shortage or inclement environmental conditions.

10.1 Size and Heterothermy

Because of their great mass, large vertebrates experience only slow changes in mean body temperature. Animals weighing hundreds of kilograms are necessarily very stable thermally, and whether ectotherm or endotherm they can reasonably be thought of as *inertial homeotherms* (McNab, 1978).

Heterothermic animals, however, are characterized by rapid, drastic, and repeated changes in body temperature; such liability of body temperature is apt to occur only in animals weighing less than 100 g. Even the relatively slow and infrequent shift from low to high body temperatures found in mammalian hibernators is strongly constrained by body mass. No large mammals experience the profound adaptive hypothermia characteristic of physiological hibernation in small mammals. One clear correlate of this situation is the inverse relation of rate of warm-up to mass in birds and mammals (Fig. 11). The largest mammals that experience physiological hibernation are marmots, sciurid rodents weighing 3–8 kg. As Morrison (1960) pointed out almost 20 years ago, it is not physiologically feasible nor energetically necessary for large mammals, like bears, to undergo profound hypothermia, even on a seasonal basis, partly because of the long duration of warm-up and partly because their large mass makes it possible for them to fast for several months without exhausting their fat reserves even though their metabolism remains at or near the basal level. The capacity for short-term, cyclic hypothermia of the sort associated with daily torpor ordinarily is usually restricted to birds and mammals weighing 50 g or less. However, all insects are small enough for heterothermy to be temporally feasible and energetically advantageous.

Many heterothermic insects warm up repeatedly during a single day. For example, when a worker of the bumblebee, *Bombus terricola* (mean mass, 200 mg), feeds on a panicle of goldenrod (*Solidago* sp.) which has hundreds of closely spaced florets, it climbs, rather than flies, from one floret to the next, and its thoracic temperature falls below flight temperature and varies directly with ambient temperature. Before it can fly from one goldenrod plant to another, it must warm itself to at least 30°C, and this cycle of body temperature change may be repeated several times in a single hour (Heinrich, 1972). Similar short, repeated alternations between ectothermy and endothermy appear to be commonplace in the smaller heterothermic moths. Whenever they alight, they cool down and then must warm up again before they take off and fly. Such short-term heterothermic cycles have not been observed in vertebrates, although they may occur in some bats.

10.2 Facultative Endothermy versus Adaptive Hypothermia

Insects as a group are ectothermic and poikilothermic. Most insects are continuously ectothermic, and even heterothermic species are ectothermic when at rest. The endothermy of insects is intermittent and associated with special activities usually related to flight and always dependent on the flight motor. Consequently, it is useful to think of heterothermic insects as animals that are basically ectotherms but intermittently shift to endothermy. Thus their pattern of body temperature can be described as *facultative endothermy*. The utility of a basically ectothermic temperature regime with endothermy limited to periods of intense activity is obvious when one considers the small size of most insects. For an animal weighing less than 1 g, the energy expenditure necessary to produce enough heat to maintain body temperature at 20°C or more above ambient is too great to be practical for more than brief periods.

Birds and mammals are basically endothermic homeotherms. At rest they usually keep body temperatures between 35 and 40°C, and they function normally only when they can maintain body temperature within this range. Small birds and mammals that are heterothermic are fundamentally endotherms that have evolved a capacity for *adaptive hypothermia*. They can temporarily abandon homeothermy, allow energy metabolism to decrease below the basal rate, maximize heat loss, and allow body temperature to fall to, or near to, ambient. At body temperatures of 20°C or less, the animal enters a state of torpor or dormancy much more profound than deep sleep. This of course contrasts with the situation in heterothermic insects which are coordinated and responsive at body temperatures of 20°C. Moreover, the duration of hypothermia in heterothermic birds and mammals is relatively brief. It is usually less than a day in birds, and in mammals even the prolonged seasonal dormancy associated with hibernation is periodically interrupted by arousals during which body temperature and metabolism return to typical mammalian levels and the normal functions for physiological maintenance are resumed. During these periods of arousal, the mammal is normally active but usually does not leave its hibernaculum. In contrast, heterothermic insects apparently can remain poikilothermic indefinitely and then shift almost instantaneously from ectothermy to endothermy.

Thus low ectothermically controlled body temperatures and low rates of energy metabolism that vary directly with ambient temperature represent the usual resting condition for heterothermic insects. Heterothermic birds and mammals at rest maintain high, endothermically controlled body temperatures and high rates of energy metabolism. Moreover, birds and mammals have a basal metabolic rate below which their energy metabolism cannot fall without a loss of normal capacities for reaction and decline in coordination. Consequently, when birds and mammals cease being endothermic, they enter a state of dormancy. When this dormancy is prolonged more than a few days, as during hibernation in mammals, the animal must arouse periodically for routine physiological maintenance.

From the preceding, it can be seen that heterothermic birds and mammals are imperfect ectotherms and that their escape from the high energy cost of endothermy, and the associated reduction in metabolism far below the basal rate, is a complex and demanding process.

The relative complexity of the shift from endothermy to ectothermy in heterothermic birds and mammals compared with its relative simplicity in heterothermic insects is correlated with the profound differences in the organization of the systems for respiratory exchange in insects and terrestrial vertebrates. The supply of oxygen to and the removal of carbon dioxide from tissues in vertebrates involves coordinated interactions between the pulmonary and the cardiovascular system, both of which are extremely complex and both of which serve functions (or involve structures) not related to respiration.

In birds and mammals, even at rest, the pulmonary and cardiovascular pumps have to maintain a high "idling speed" to support the oxygen demands associated with endothermy. The high idling rates of the pulmonary pump and the circulatory pump are functionally linked with the BMR and also appear to be necessary for making rapid shifts from a resting condition to aerobically supported intense activity. The high idling rates of the cardiovascular and respiratory support systems account, at least in part, for the existence of a BMR, which in turn contributes directly to the modest factorial scope of birds and mammals as compared with that of endothermic insects. When the cardiovascular and respiratory pumps are slowed, allowing escape from the constraint of maintaining a BMR, the organism becomes dormant.

Among insects, respiratory exchange depends on large numbers of tracheae which allow movement by diffusion of oxygen and carbon dioxide directly to and from the sites of metabolic activity. Little is required in the way of support systems; the greater the rate of oxygen utilization, the steeper the diffusion gradient and the more rapid the movement of oxygen molecules to the sites of aerobic metabolism. Although the tracheal systems of some large heterothermic insects require pumps during maximum activity, at rest such auxiliary support is not essential. Thus insects bypass the necessity for operating energetically costly respiratory support systems when at rest or modestly active. Moreover, by virtue of possessing a system where oxygen supply is directly and almost automatically controlled by demand, they are able to support the extremely high rates of aerobic metabolism required for sustained, intense muscular activity and the endothermy associated with it.

11 SUMMARY

The smallest endothermic vertebrates weigh 2–3 g, but many endothermic insects weigh less than 0.5 g and some weigh as little as 0.04 g. How is it possible for these insects to achieve endothermic control of body temperature despite their small size? The rates of heat production and rates of heat loss of endothermic insects, birds, and mammals were examined by scaling their respective mass-specific thermal conductances and rates of oxygen consumption against body mass and by comparing the rates of energy metabolism of species of the same size in the three groups. The slopes and intercepts of the regressions of mass-specific thermal conductance of sphinx moths (calculated on the basis of thoracic mass) are almost identical to those of birds and mammals. Thus the existence of endothermy in sphinx moths (and presumably other groups of insects, because all endothermic insects so far have similar cooling constants) is not explained by their having a greater capacity for heat retention than vertebrate endotherms. In heterothermic moths and in birds and bats the rate of energy metabolism during flight scales with mass to approximately the 0.75 power. In the range of sizes where moths, birds, and bats overlap, their rates of oxygen consumption during flight are virtually the same—at a body mass of 2 g $\dot{V}_{O_2}$ is about 50 $cm^3/g \cdot hr$. However, this value approaches the upper limit of energy metabolism for flying birds and is near the lower limit of energy metabolism for flying moths. It appears that the capacity of insects to sustain endothermically elevated body temperatures depends on their very high rate of energy metabolism rather than on a special ability for heat retention.

Factorial aerobic scope in the smallest homeothermic mammals is between 4 and 10. In small birds it is about 15. In endothermic moths and beetles it is 100 or more. This difference is a function of the fact that the insects are ectothermic when inactive and have the low metabolic rates characteristic of ectotherms. The only vertebrates that have comparably large metabolic scopes are heterothermic forms such as small bats, hummingbirds, and some small rodents, which under appropriate conditions become essentially ectothermic when at rest. However, it should be noted that heterothermic mammals and birds with body temperatures of 20–25°C or less are torpid and virtually helpless, while at the same body temperatures heterothermic insects, although unable to fly, are capable of integrated and effective terrestrial activity.

Among heterothermic vertebrates rate of warm-up decreases with size. Among heterothermic insects it is either independent of size or increases slightly with size. Insects warm up much more rapidly than heterothermic birds or mammals of the same size.

Heterothermic insects are ectotherms with a faculty for intermittent endothermy. Heterothermic birds and mammals are endotherms with a capacity for intermittent adaptive hypothermia. This contrast correlates with the differences between the respiratory systems of vertebrates and insects.

REFERENCES

Bakken, G. S. (1976a). An improved method for determining thermal conductance and equilibrium body temperature with cooling curve experiments. *J. Therm. Biol.* **1,** 169–175.

Bakken, G. S. (1976b). A heat transfer analysis of animals: Unifying concepts and the application of metabolism chamber data to field ecology. *J. Theor. Biol.* **60,** 337–384.

Bartholomew, G. A. (1972). Aspects of timing and periodicity of heterothermy. In *Hibernation and Hypothermia: Perspectives and Challenges*, F. E. South, et al., Eds., pp. 663–680. Elsevier: New York.

Bartholomew, G. A. (1977). Body temperature and energy metabolism. In *Animal Physiology: Principles and Adaptations*, 3rd ed., M. S. Gordon, Ed., pp. 364–449. MacMillan: New York.

Bartholomew, G. A. and Casey, T. M. (1977a). Endothermy during terrestrial activity in large beetles. *Science* **195,** 882–883.

Bartholomew, G. A. and Casey, T. M. (1977b). Body temperature and oxygen consumption during rest and activity in relation to body size in some tropical beetles. *J. Therm. Biol.* **2,** 173–176.

Bartholomew, G. A. and Casey, T. M. (1978). Oxygen consumption of moths during rest, pre-flight warm-up, and flight in relation to body size and wing morphology. *J. Exp. Biol.* **76,** 11–25.

Bartholomew, G. A. and Epting, R. J. (1975a). Rates of post-flight cooling in sphinx moths. In *Perspectives of Biophysical Ecology*, D. M. Gates and R. B. Schmerl, Eds., pp. 405–415. Springer-Verlag: New York.

Bartholomew, G. A. and Epting, R. J. (1975b). Allometry of post-flight cooling rates in moths: A comparison with vertebrate homeotherms. *J. Exp. Biol.* **63,** 603–613.

Bartholomew, G. A. and Heinrich, B. (1978). Endothermy in African dung beetles during flight, ball making, and ball rolling. *J. Exp. Biol.* **73,** 65–83.

Bartholomew, G. A., Howell, T. R., and Cade, T. J. (1957). Torpidity in the white-throated swift, anna hummingbird, and poor-will. *Condor* **59,** 145–155.

Carey, F. G. (1973). Fishes with warm bodies. *Sci. Am.* **228,** 36–44.

Carey, F. G. and Teal, J. M. (1966). Heat conservation in tuna fish muscle. *Proc. Nat. Acad. Sci. U.S.* **56,** 1464–1469.

Chew, R. M., Lindberg, R. G., and Page, H. (1967). Temperature regulation in the little pocket mouse, *Perognathus longimembris*. *Comp. Biochem. Physiol.* **21,** 487–505.

Epting, R. J. (1975). Power input for hovering flight in hummingbirds and its effect on the time and energy budgets of foraging. Ph.D. Dissertation, University of California, Los Angeles.

Fry, F. E. J. (1947). Effects of the environment on animal activity. *Publ. Ont. Fish. Res. Lab.* no. 68, pp. 1–62.

Graham, J. B. (1975). Heat exchange in the yellowfin tuna, *Thunnus albacares*, and skipjack tuna, *Katsuawanus pelamis*, and the adaptive significance of elevated body temperatures in scombrid fishes. *Fish. Bull.* **78,** 219–229.

Hainesworth, F. R. and Wolf, L. L. (1979). The economics of temperature and torpor in non-mammalian organisms. In *Strategies in Cold*, L. Wang and J. W. Hudson, Eds., pp. 147–184. Academic: New York.

Hart, J. S. and Berger, M. (1972). Energetics, water economy and temperature regulation during flight. *Proc. XV Ornithol. Congr.*, pp. 189–199. Brill: Leiden.

Heath, J. E. and Josephson, R. K. (1970). Body temperature and singing in the katydid, *Neoconocephalus robustus* (Orthoptera, Tettigoniidae). *Biol. Bull.* **138,** 272–285.

Heinrich, B. (1970). Nervous control of the heart during thoracic temperature regulation in a sphinx moth. *Science* **169,** 606–607.

Heinrich, B. (1972). Energetics of temperature regulation and foraging in a bumblebee, *Bombus terricola* Kirby. *J. Comp. Physiol.* **77,** 49–64.

Heinrich, B. and Bartholomew, G. A. (1971). An analysis of pre-flight warm-up in the sphinx moth, *Manduca sexta*. *J. Exp. Biol.* **55,** 223–239.

Heinrich, B. and Casey, T. M. (1973). Metabolic rate and endothermy in sphinx moths. *J. Comp. Physiol.* **82,** 195–206.

Herreid, C. F. and Kessel, B. (1967). Thermal conductance in birds and mammals. *Comp. Biochem. Physiol.* **21,** 405–414.

Hutchison, V. H., Dowling, H. G., and Vinegar, A. (1966). Thermoregulation in a brooding female Indian python, *Python molurus bivittatus*. *Science* **151,** 694–695.

Lasiewski, R. C. (1963). Oxygen consumption of torpid, resting, active, and flying hummingbirds. *Physiol. Zool.* **36,** 122–140.

Lasiewski, R. C. and Dawson, W. R. (1967). A re-examination of the relation between standard metabolic rate and body weight in birds. *Condor* **69,** 13–23.

Lasiewski, R. C. and Lasiewski, R. J. (1967). Physiological responses of the blue-throated and Rivoli's hummingbirds. *Auk* **84,** 34–48.

Lasiewski, R. C., Weathers, W. W., and Bernstein, M. H. (1967). Physiological responses of the giant hummingbird, *Patagona gigas*. *Comp. Biochem. Physiol.* **23,** 797–813.

McNab, B. K. (1978). The evolution of endothermy in the phylogeny of mammals. *Am. Nat.* **112,** 1–21.

May, M. L. (1976a). Thermoregulation and adaptation to temperature in dragonflies (Odonata: Anisoptera). *Ecol. Monogr.* **46,** 1–32.

May, M. L. (1976b). Warming rates as a function of body size in periodic endotherms. *J. Comp. Physiol.* **111,** 55–70.

Morrison, P. (1960). Some interrelations between weight and hibernation function. *Bull. Mus. Comp. Zool.* **124,** 75–91.

Morrison, P., Ryser, F. A., and Dawe, A. R. (1959). Studies on the physiology of the masked shrew *Sorex cinereus*. *Physiol. Zool.* **32,** 256–271.

Pearson, O. P. (1950). The metabolism of hummingbirds. *Condor* **52,** 145–152.

Stahl, W. R. (1967). Scaling of respiratory variables in mammals. *J. Appl. Physiol.* **22,** 453–460.

Thomas, S. P. (1975). Metabolism during flight in two species of bats, *Phyllostomus hastatus* and *Pteropus gouldii. J. Exp. Biol.* **63,** 273–293.

Wunder, B. (1970). Energetics of running activity in Merriam's chipmunk, *Eutamias merriami. Comp. Biochem. Physiol.* **33,** 821–836.

4

Behavioral Mechanisms of Thermoregulation

TIMOTHY M. CASEY

1 INTRODUCTION

The capacity of insects to control body temperature by behavioral means has long been recognized. Anecdotal references to behavioral thermoregulation of insects date back to biblical times (see May, 1979) and, in 1929, Gottfried Fraenkel documented the dramatic behavioral thermoregulation of the desert locust. Since that time, many detailed laboratory and field studies on both immature and adult insects have indicated that a wide variety of insect species from several taxa routinely control body temperature by alterations of their behavior.

Behavioral thermoregulation of insects has been previously discussed in the broader contexts of response to temperature (Precht et al., 1963; Cloudsley-Thompson, 1970; Uvarov, 1977), adaptation to hot, dry conditions (Cloudsley-Thompson, 1968; Edney, 1974), and water balance (Edney, 1978), as well as in reviews of insect thermoregulation (Heinrich, 1974; May, 1979). The purpose of this chapter is specifically to review behavior patterns that result in maintenance of a relatively constant body temperature over a range of air temperatures. The small size of insects allows them to exploit a wide range of microenvironments unavailable to larger animals, and thermal heterogeneity is necessary for insects to regulate their body temperatures behaviorally. Since such conditions are difficult to duplicate in the laboratory, where possible I will discuss behavior of insects in the field.

Thermoregulation is here distinguished from temperature selection behavior in which an animal chooses a particular ambient temperature and body temperature follows passively. While temperature selection is clearly adaptive and virtually all animals exhibit some capacity to seek out appropriate thermal regimes, this behavior is characteristic of poikilothermic rather than homeothermic animals. Thermal preferenda for many insect species have been determined in thermal gradient apparatus (see Uvarov, 1977; Herter, 1953; Thiele, 1977, for details) and in general the temperatures selected by the insects conform to those obtained for the same species in the field. By design, a thermal gradient represents a continuous change from a region of high temperature to one of low temperature with all other stimuli (hopefully) removed. Such a homogeneous environment may closely match the environment of aquatic insects, or soil dwellers, but is atypical of the environment faced by most terrestrial insects. Consequently, although specific mechanisms of movement of insects in thermal gradients may be characterized, these mechanisms may or may not have relevance to the behavior of insects in the field. Behavior of insects in laboratory thermal gradients is discussed elsewhere (Laudien, 1963; Uvarov, 1977) and will not be reviewed here.

The distinction between physiological and behavioral thermoregulation in insects is often difficult because many insects can be ectothermic or endothermic, depending upon the circumstances. In reptiles, which are essentially complete ectotherms, physiological thermoregulation involves the capacity to control rates of heat exchange via blood circulation or evaporative cooling. In insects, the capacity to control rates of metabolic heat production could have additional important consequences for their thermal balance. Large changes in rates of heat production often result in profound changes in behavior (and vice versa).

2 REGULATION OF BODY TEMPERATURE

A variety of insects from several taxa control body temperature either exclusively by behavioral means or in combination with endogenous heat production. The level at which body temperature is controlled may differ dramatically among different species. Factors determining body temperature include local environmental variables and morphological constraints associated with size and shape, as well as physiological factors such as the temperature for optimum flight motor function or maximum feeding rates. In this section, I will present a series of examples of insects that thermoregulate at least partially by their behavior and, where possible, indicate precision of thermoregulatory control. Subsequent sections will analyze the mechanisms of thermoregulation.

The degree to which insects regulate their body temperatures in the field is often described by relating body temperature to air temperature. Such plots indicate the level at which T_b is regulated, the precision with which regulation occurs based on the variability of T_b at a given T_a, and the degree of independence of the body temperature from the air temperature (a slope of zero indicates perfect regulation, while a slope of one indicates no regulation). However, air temperature is often a poor index to the thermal conditions in the environment, particularly in strong sunlight (see below). Although T_b of insects has been measured for many years, the data available for insects in the field over a range of environmental temperatures are surprisingly fragmentary and confined mostly to large insects from a few taxa (Fig. 1). A similar plot is shown (Kammer, this volume) for insects that control T_b by physiological mechanisms.

Because of their economic importance, locusts have been most extensively studied (see Uvarov, 1977, for review). Regressions a, b, and c, in Fig. 1 indicate T_b of three locust species from different environments. In general, locusts maintain similar degrees of regulation in warm and cool habitats. Their range of T_{th} is high, and the level is similar to that of other

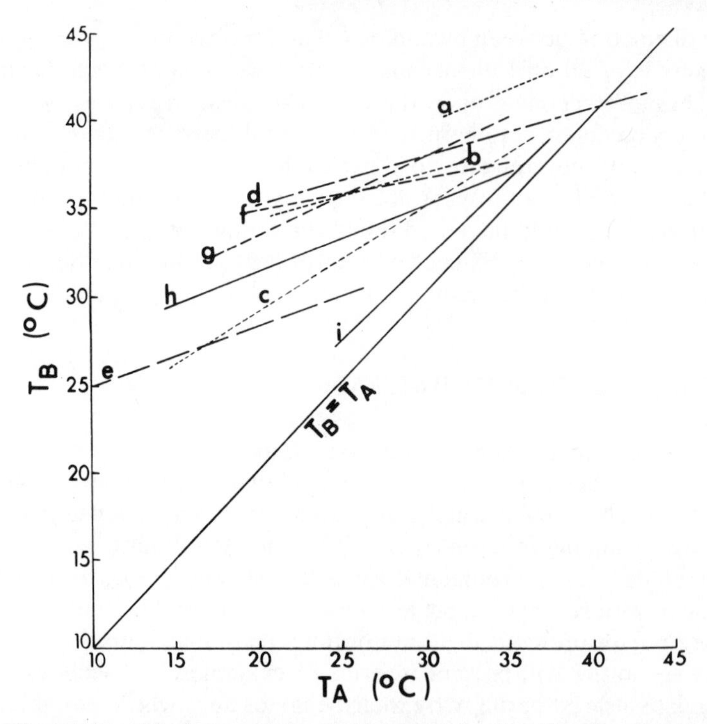

Fig. 1 The relation of body temperature to air temperature in various insect species while relying on behavior patterns to control body temperatures. (a–c) Locusts, **Schistocerca gregaria** (Waloff, 1963), **Locustana pardalina** (Smit, 1960, cited by Uvarov, 1977), and **Psoloessa delicatula** (Anderson et al., 1978); (d) the desert cicada **Diceroprocta apache** (Heath and Wilkins, 1970); (e) the syrphid fly, **Syrphus** spp. (Heinrich and Pantle, 1975); (f and g) the dragonflies **Libellula** spp. and **Pachydiplax longipennis** (May, 1976b); (h and i) the sphingid caterpillars, **Hyles lineata** and **Manduca sexta** (Casey, 1976a).

large flying insects that rely on behavior to control body temperature. Although locusts produce fairly large quantities of heat as a by-product of flight activity, they have little capacity to regulate thoracic temperature physiologically during flight (Weis-Fogh, 1956; Church, 1960a,b) and rely exclusively on behavioral thermoregulation (Uvarov, 1977).

Several early studies on dragonflies (Odonata: Anisoptera) suggested that they probably regulated body temperature, and Corbet (1963) divided them into two categories, "perchers" and "fliers," based on presumed mechanisms of thermoregulation. A recent extensive study by May (1976b) confirms that perchers rely on their behavior to control body temperature, while fliers utilize physiological mechanisms. May's study presents T_b versus T_a plots for a variety of perching and flying dragonflies, and two species, *P. longipennis* and *Libellula* sp., for which extensive data are available, are shown in Fig. 1. The effectiveness of behavioral thermoregulation of different percher dragonflies is directly related to body size (May, 1976b).

Similarly, field observations on butterflies have suggested that they probably thermoregulate behaviorally (Clench, 1966). Thoracic temperatures of butterflies during flight range from a few degrees above the air temperature to as much as 15°C above T_a, and regulated T_{th} of different species is related to wing loading (Heinrich, 1972; Douglas, 1977). In most cases, T_{th} is elevated behaviorally (Clench, 1966; Watt, 1968; Kevan and Shorthouse, 1970; Heinrich, 1972; Douglas, 1977), but in several species behavioral thermoregulation occurs simultaneously with shivering (Vielmetter, 1954; Kammer, 1970).

Syrphid flies (Diptera: Syrphidae) show a rather remarkable degree of regulation given their small size and low thermal inertia (Heinrich and Pantle, 1975) by a combination of behavioral movements, related to the location and frequency of perching in sun or shade, and by shivering. The range of body temperatures maintained is much lower than that of other flying insects, undoubtedly as a result of physical factors associated with their small size.

Different species of cicadas (Homoptera: Cicadidae) also show different capacities for control of body temperature. The desert cicada, *Diceroprocta apache* (Fig. 1), and the cactus dodger, *Cacama valvata,* show a high degree of regulation (Heath and Wilkin, 1970; Heath et al., 1972), while the T_b of the periodic "17-year" cicada, *Magicicada cassini,* in Oklahoma generally parallels the air temperature (Heath, 1967).

Larvae of holometabolous insects do not produce heat in sufficient quantity to elevate their body temperature. Regulation by caterpillars is dependent entirely on external heat sources. The desert caterpillar, *Hyles lineata* (Lepidoptera: Sphingidae), regulates its body temperature, while another species, *Manduca sexta,* from the same family, in the same habitat, does not (regressions h and i, Fig. 1). Larvae of *Colias* spp., while having distinct preferred body temperatures, have only limited capacity to regulate T_b behaviorally (Sherman and Watt, 1973).

Tenebrionid beetles inhabiting the Namib Desert regulate body temperatures entirely via behavior (Edney, 1971; Hamilton, 1973; Henwood, 1975a). In particular, Hamilton (1975) has shown that several species of diurnal desert tenebrionid beetles confined to the substrate show mean body temperatures above 40°C for much of the day.

3 FACTORS AFFECTING BODY TEMPERATURE

3.1 Avenues of Heat Exchange

As a result of a large number of studies in both the laboratory and the field (Parry, 1951; Digby, 1955; Church, 1960a,b; Stower and Griffiths, 1966; Hadley, 1970; and others), a series of generalizations can be made concerning the relative importance of various avenues of heat exchange in determining the body temperature of insects. For most insects, except species that produce substantial quantities of heat via flight muscles, radiative heat gain and convective heat loss are the major avenues of heat exchange. Although evaporative heat loss has occasionally been implicated as a major factor in controlling body temperature (Edney, 1953; Edney and Barass, 1962; Seymour, 1974), under most circumstances the quantities of heat lost by evaporation are an order of magnitude or more lower than those lost by convection. Heat exchange by conduction is usually considered negligible because of the small surface contact between an insect and solid substrates, but this parameter can be manipulated behaviorally by insects (see next section). In sunshine, the temperature excess ($T_b - T_a$) varies directly with radiation intensity (Parry, 1951; Digby, 1955; Shepard, 1958; Stower and Griffiths, 1966). Since the spectral composition of sunlight includes a variety of wavelengths, both visible and infrared, the absorptivity of the cuticle (see, for example, Watt, 1969; Henwood, 1975), the presence or absence of pubescence (Church, 1960a; Heinrich, 1971) and surface coloration may all be important in determining the magnitude of radiation absorbed. Convective heat exchange varies directly with some fractional power of body size and varies inversely with a fractional power of wind velocity (Digby, 1955; Church, 1960a). At relatively high wind velocities, the convection coefficient of insects and cylinders is proportional to the 0.5 power of the velocity. At lower wind speeds, 0–30 cm/sec, however, the convection coefficient and the equilibrium temperature $T_b - T_a$ are essentially independent of speed (Digby, 1955).

3.2 Body Size and Shape

Body size, of course, is extremely important in determining the equilibrium body temperature and the rate at which it is achieved. As body mass decreases, the ratio of body surface to body volume increases and, since heat exchange occurs at the body surface, a higher relative surface area will facilitate a more rapid rate of heat exchange. Other things being equal, a small insect will reach equilibrium body temperature more

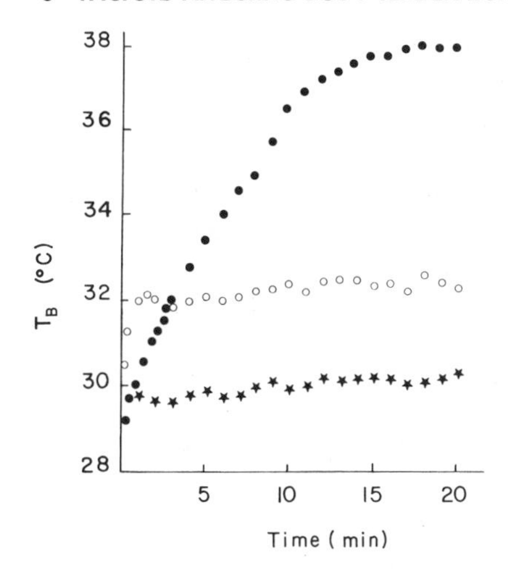

Fig. 2 Effect of solar radiation on the internal temperature of green solitarious hoppers of *S. gregaria*. Open circles represent first instar; solid circles represent fifth instar; stars indicate air temperature. (After Stower and Griffiths, 1966.)

rapidly than a large one under similar conditions (Fig. 2). The large insect will have a higher equilibrium body temperature than the small one because of higher heat capacity and will be less subject to transient changes in body temperature as a result of higher thermal inertia.

Body shape also appears to be important in determining equilibrium body temperature of animals in sunshine, presumably because of alterations in surface/volume ratios within a given size range. Elongate insects, such as locusts, tend to have a lower temperature excess for a given mass than compact, spherically shaped insects (bees, flies). The evolution of a dipteran, hymenopteran, or coleopteran body shape may be associated with maximizing the temperature excess for a given mass by reducing the surface-volume ratio (Digby, 1955).

3.3 Energy Balance

The body temperature of an animal is a complex function of various factors which determine rates of heat gain and heat loss. Under steady-state conditions, heat gain exactly balances heat loss, and the body temperature can be calculated by quantifying heat exchanges via radiation, metabolism, conduction, convection, and evaporation (Gates, 1962; Birkebak, 1966; Bartlett and Gates, 1967; Porter and Gates, 1969). This approach allows determination of the relative importance of various avenues of heat exchange in determining body temperature (Edney, 1953; Parry, 1951; Digby, 1955; Hadley, 1970; Edney, 1971; Stower and Griffiths, 1966) and provides a quantitative estimate of the importance of various postures for controlling body temperature (Digby, 1955; Stower and Griffiths, 1966; Tracy et al., 1979).

Parry (1951) suggested that factors were so numerous and complex that body temperatures of insects in nature probably could not be predicted with any degree of accuracy. However, in several recent studies, measured body temperatures of arthropods fall within the range of T_b predicted, based on behavioral observations of the animals in the field, accurate microclimate data, and solution of the energy balance equation (Stower and Griffiths, 1966; Hadley, 1970; Edney, 1971; Smith and Miller, 1973; Henwood, 1975a; Anderson et al., 1979). Coefficients of convection and radiation are either determined experimentally in the laboratory (see, for example, Henwood, 1975a) or approximated by known physical characteristics of cylinders or spheres (Parry, 1951).

The utility of such an analysis is demonstrated in Fig. 3 (Anderson et al., 1979) in which the air temperature, predicted body temperature range, and measured body temperature of two different grasshopper species from different habitats are shown. The range of predicted body temperatures for *Eritittix simplex* is small. The habitat is well watered, and thermal heterogeneity of the habitat is reduced because evaporation causes the substrate temperature to remain near T_a. In *Psoloessa delicatula* occupying a dry habitat, a much wider range of body temperatures is possible because of increasing substrate temperatures. In both species, the measured body temperatures fall within the range of temperatures predicted by energy balance analysis, but thermoregulation occurs only in *P. delicatula*.

The solution of the energy balance equation has led to the construction of ecological models of considerable predictive power (Porter et al., 1973, 1975; Tracy, 1975). However, several factors limit the utility of such models for insects. As a result of their small size and the large range of microclimates available to them, a very wide range of body temperatures is possible at any given time of the day (Heath, 1967; Casey, 1976a; May, 1976b, 1977). Convective heat exchange is particularly difficult to quantify for insects in the field, because wind conditions are difficult to measure with sufficient precision, are highly variable in both velocity and direction, and are often turbulent. Consequently, laboratory-derived convective coefficients of animals or models of animals having similar size, shape, and thermal characteristics often vary significantly from those obtained for animals in the field (Bakken and Gates, 1975). As a result of their low thermal inertia, small gusts of wind may change body temperature by several degrees Celsius in a matter of seconds (see, for example, Strelnikov, 1936). Finally, and perhaps most significantly, insects are usually associated with some substrate, whether it is the ground, vegeta-

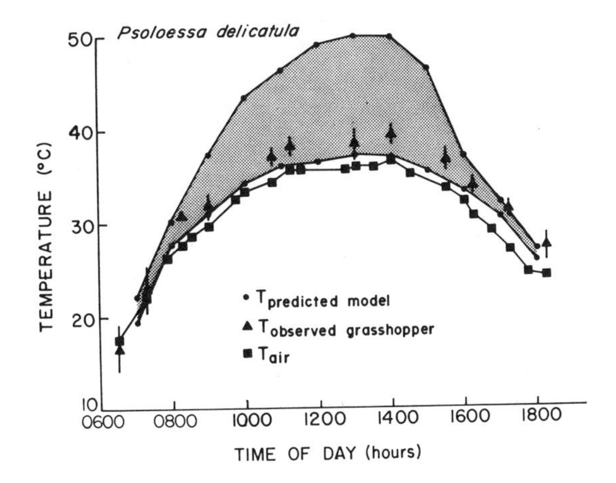

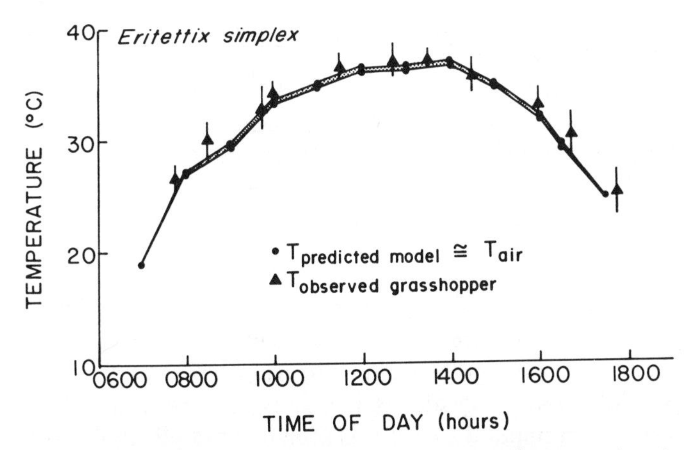

Fig. 3 Temperature relations of **P. delicatula** and **E. simplex** in their natural environments in relation to the time of day. Upper curve connecting solid circles represents predicted maximum T_b achievable by the grasshoppers; lower curve connecting solid circles represents the predicted minimum body temperature achievable by the grasshoppers. (From Anderson et al., 1978.)

tion, or other insects. Wind velocity profiles and temperature profiles (Fig. 4) indicate that, as a result of boundary-layer phenomena, different parts of the body are exposed to different wind velocities and air temperatures. Depending upon the thickness of the boundary layer, wind velocity could be sufficiently low that free, rather than forced, convection effects apply (Digby, 1955). It is clear that such effects are sufficiently

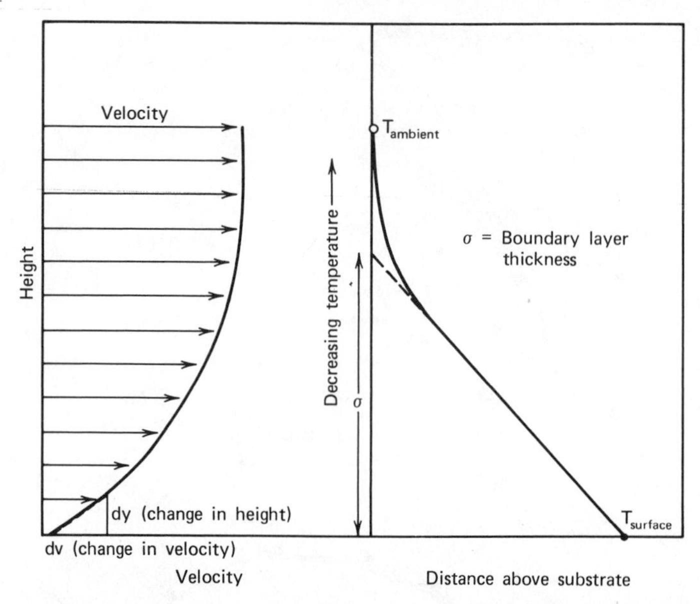

Fig. 4 A schematic diagram of the changes in wind velocity and the associated changes in temperature with increasing height above the surface. (From Douglas, 1977, modified from Gates, 1962.)

complex that detailed analysis should be performed for a variety of insect species in several habitats before any but the grossest generalization about the effects of substrates on the body temperature of insects can be made.

In view of the foregoing discussion, it is obvious that relating body temperature to air temperature (Fig. 1) is an oversimplification which does not truly indicate the degree of thermal stress on the animal. Recently, George Bakken presented a biophysical framework for using models of animals in the field (Bakken and Gates, 1975; Bakken, 1976). This approach was suggested first by Parry (1951) but not used for studies of insect T_b despite the fact that the temperatures of dead insects or other inanimate objects (models) are often compared with the responses of living insects (see for example, Heinrich, 1971; Sherman and Watt, 1973; Heinrich and Casey, 1978). This formulation is useful because solution of the energy balance equation to predict T_b is tedious and requires a great deal of microclimate data as well as derived coefficients which involve several assumptions. The model can be placed in the microclimate in any posture desired, and its temperature measured rather than calculated. A measure of the "true" environmental temperature is gained by measuring the operative environmental temperature T_e. "T_e may be identified as the temperature of an inanimate object of zero heat capacity with the same size, shape, and radiative properties as the animals and exposed to the same microclimate. It is also equivalent to the temperature of a blackbody cavity producing the same thermal load on the animal as the actual non-blackbody microclimate, and therefore, may be regarded as the true environmental temperature seen by the animal" (Bakken and Gates, 1975). Temperature measured using such a model provides more information than a "black bulb" temperature because the model has the same physical characteristics as the animal. For most small nonflying insects, the body temperature is approximately equal to T_e (i.e., the model temperature equals the insect body temperature). Studies on behavioral thermoregulation could be aided if a series of models of the animal were placed in a variety of locations and orientations throughout the habitat to determine the range of T_e available to the animals. In addition, differences in temperature of the model and the animal in the same microclimate indicate the degree of regulation by physiological means. Bakken and Gates (1975) define this difference as the *physiological offset temperature* $T\Delta$ and $T_b = T_e + T\Delta$.

To date, studies on insects utilizing a biophysical framework to analyze body temperatures have utilized a lumped parameter analysis in which conductance values, environmental temperatures, and body temperature are averaged over the entire body (Bakken and Gates, 1975, for further discussion). While this may be adequate for a caterpillar, it may represent a significant oversimplification for adults of many insect species. It is well known that the thorax and abdomen of adult insects are thermally independent (Church, 1960b; Heinrich, 1971a; Casey, 1976b). As a result of differences in surface area, shape, size, and insulation, particularly in butterflies and dragonflies, body temperatures in different body parts could differ significantly. For example, a dragonfly oriented with the long axis perpendicular to solar radiation exhibited equilibrium abdominal temperatures 3°C lower than the equilibrium thoracic temperature (Heinrich and Casey, 1978). Greater differences might occur if the thorax is irradiated while the abdomen is shaded, which commonly occurs at high T_a when dragonflies face the sun. If a significant temperature difference is established between thorax and abdomen, physiological heat transfer via blood circulation could occur (Heinrich, 1970). A distributed parameter analysis, characterizing heat exchange of both the thorax and abdomen, while more complex, would increase the understanding of regional body temperature differences.

4 ORIENTATION AND POSTURAL ADJUSTMENTS

Insects routinely vary their body orientation in response to changes in the thermal environment. During periods when body temperatures or air temperatures are below the range normally associated with activity, insects orient the long axis of the body perpendicular to solar radiation. This behavior, often called positive orientation, or basking, maximizes radiative heat gain by increasing the surface area of the body exposed to sunlight. This pattern is common in cicadas (Heath, 1967, 1970, 1972), dragonflies (Corbet, 1963; May, 1976b, 1977; Heinrich and Casey, 1978), some butterflies (Vielmetter, 1958; Clench, 1966; Kevan and Shorthouse, 1970; Heinrich, 1972; Rawlins and Lederhouse, 1978), caterpillars (Sherman and Watt, 1973; Casey, 1976), tenebrionid beetles (Hamilton, 1971, 1973, 1975; Henwood, 1975a), and undoubtedly many other species.

In most cases, the dorsal surface of the body is exposed. Locusts, however, often exhibit an elaborate basking posture (see Fraenkel, 1929) in which the metathoracic leg and wing on the side of the body facing the sun are moved out of the way, thus facilitating heating. In addition, the lateral side of the body rather than the dorsal surface is exposed to solar radiation because the lateral portion has a much larger surface area.

Dragonflies are particularly adept at orientation with respect to solar radiation. Figure 5 illustrates several types of positive and negative postures of dragonflies (May, 1976b). Corbet (1963) earlier described these postures and attributed thermoregulatory function to them, but reported no data on body temperature. May's study demonstrates how these postures vary at different times of the day in relation to environmental heat load. Dragonflies tethered in the sun, unable to alter their posture, were unable to regulate body temperature. The obelisk posture (Fig. 5) of *P. longipennis* is particularly interesting and reasonably common among various perching dragonflies. This posture only occurs at high ambient temperatures. When exposed to a radiant heat source in the laboratory, *P. longipennis* assumed the obelisk posture, causing body temperature to fall 2.5°C (May, 1976b).

Size and shape are important in determining the efficacy of orientation for body temperature control. Experiments by Edney (1971) indicate that a change in orientation of 90° for the tenebrionid beetle *Onymachris rugatipennis* results in a change of about 4°C (Fig. 6). Elongate insects expose a smaller portion of the total body surface for heating, while the long axis of the body is oriented parallel to sunlight. For example, a shift from positive to negative orientation in a final-instar caterpillar, *H. lineata,* results in about a 10-fold reduction in the surface area exposed to solar radiation (Casey, 1976a). Moreover, negative orientation occurs most

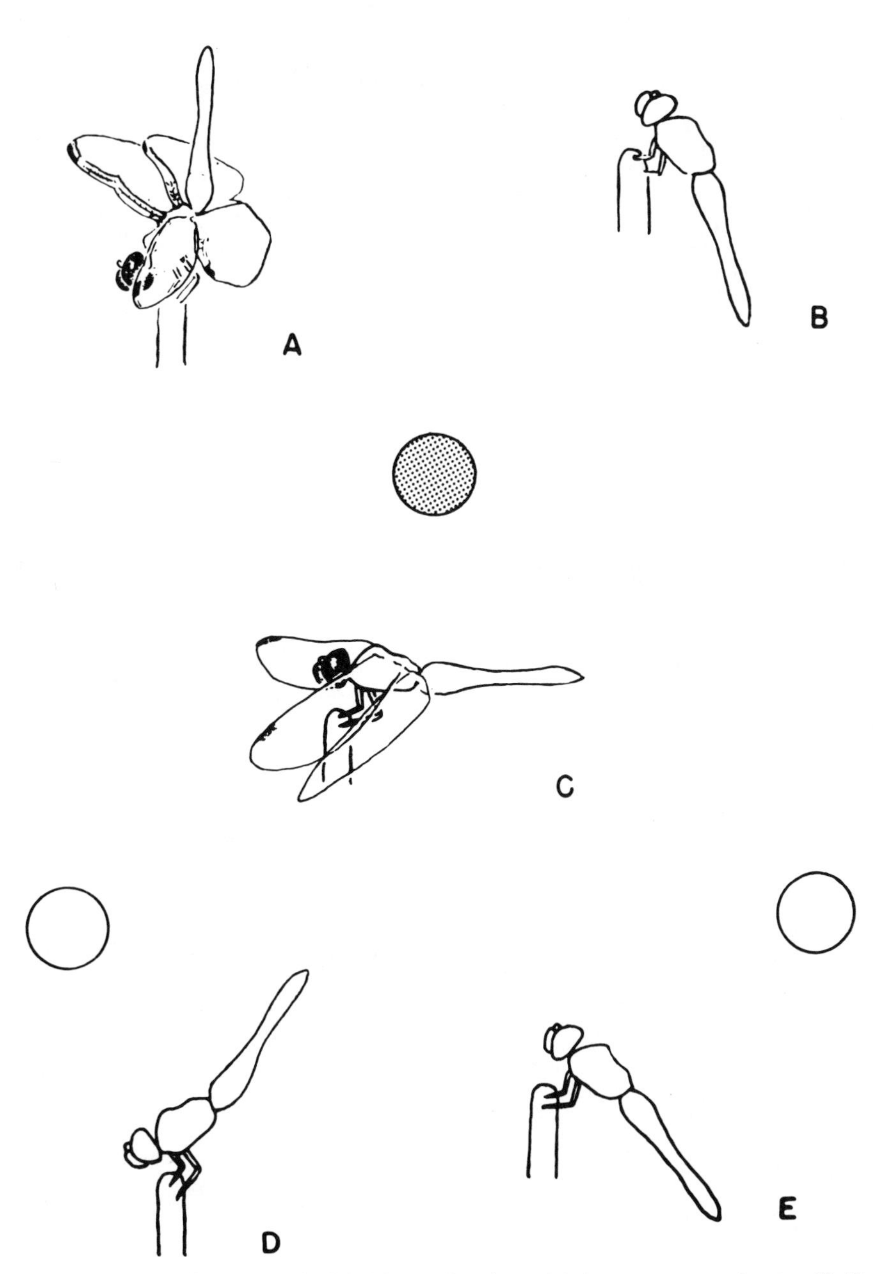

Fig. 5 (A and B) Postures in libellulid dragonflies that minimize exposure to the sun. (C–E) Postures that maximize exposure to the sun. Stippling indicates sun above or behind plane of paper. Typical wing positions shown in (A) and (C). (From May, 1976.)

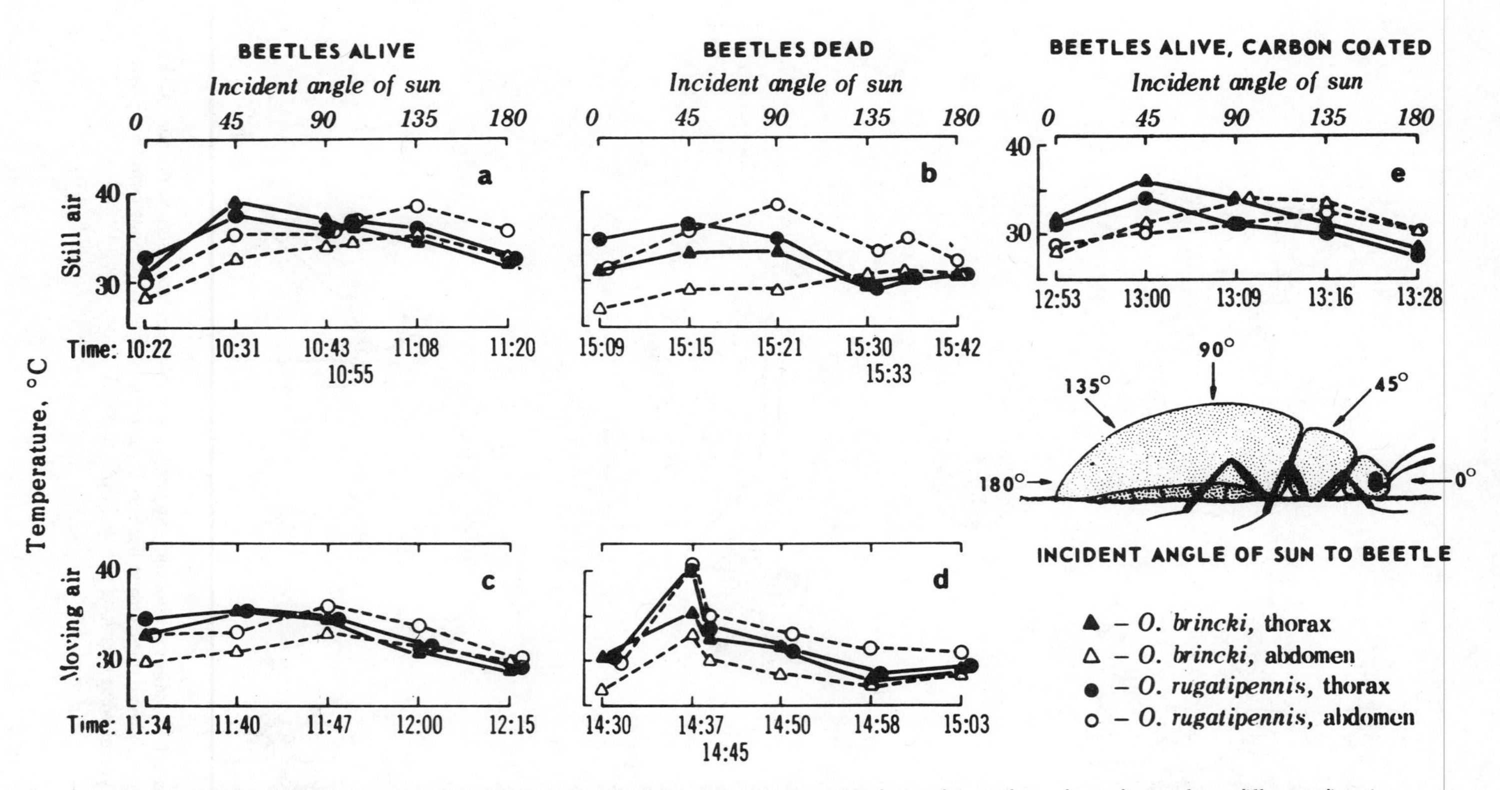

Fig. 6 Thoracic and abdominal temperatures of **Onymacris brincki** and **O. rugatipennis** when subjected to solar radiation from different directions. When the incident angle was varied from 0 to 180°, the abdomen of **O. brincki** (white elytra) remained cooler than the thorax until the incident angle was 135–180°, and even then was scarcely warmer. In **O. rugatipennis**, the abdomen became warmer than the thorax at 45–90° and was considerably warmer. The effects are clearest in (b) (beetles dead in still air). In (e), carbon painted on the white elytra of **O. brincki** eliminated most of the differences between its temperature and that of **O. rugatipennis.** (From Edney, 1971.)

often around midday, when the sun is overhead and the long axis of the body is vertical. Under these circumstances, not only is radiative heat gain minimized, but convective heat transfer is also maximized because of increased surface area exposed to the wind. Digby (1955) measured up to 50% greater convective heat loss in elongate insects.

Behavior of insects in the field suggests that orientation in the environment is varied with respect to wind velocity as well as solar radiation to maintain body temperature. For example, in the sphinx caterpillar, *H. lineata* (Fig. 7), at midday, when T_a is 36°C, virtually the entire population is in vertical orientation on vegetation, resulting in a mean T_b only 2°C above T_a. However, at midday on a cool day (T_a = 28°C), most of the population remains horizontal with the long axis of the body oriented into the wind. Under these circumstances, the mean body temperature is 7°C above T_a (Casey, 1976a). Locusts also orient with respect to the wind, as well as with respect to solar radiation (Waloff, 1963).

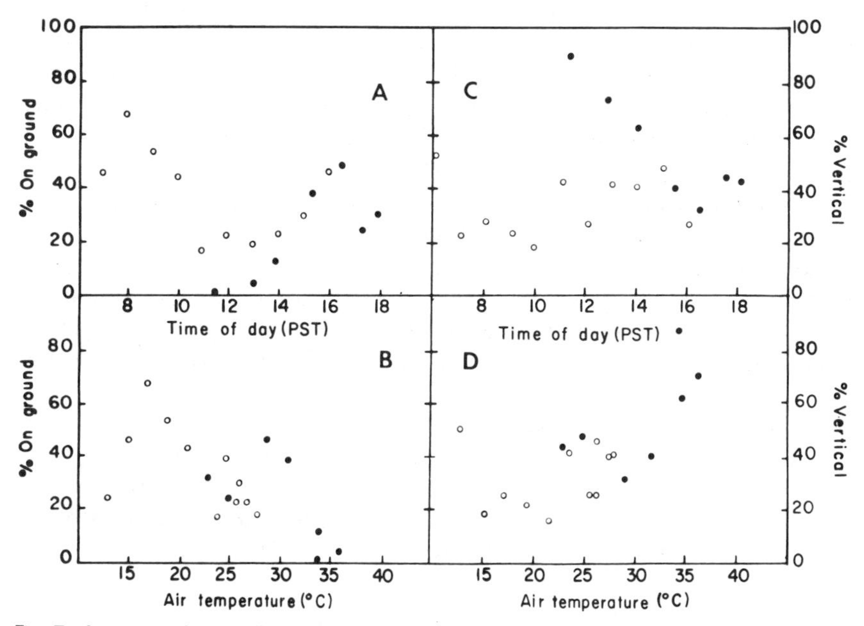

Fig. 7 Location of caterpillars of H. lineata in the habitat as a function of time of day and air temperature. (A and B) Percent of population on the ground. (C and D) Percent of the population in a vertical posture (perpendicular to the ground) on vegetation. Open circles represent a cool day (T_a max. = 28°C); solid circles represent a hot day (T_a max. = 36.2°C). Minimum of 28 animals counted for each point. (From Casey, 1976.)

5 INSECTS OF THE GROUND

The importance of substrate conditions in determining the body temperature of insects has been demonstrated in many studies. During midmorning and late afternoon, ground temperatures exceed air temperatures but are below the temperatures that limit activities of insects. Locusts (Fraenkel, 1929; Chapman, 1959; Waloff, 1963), dragonflies (May, 1976b), butterflies (Vielmetter, 1958), caterpillars (Casey, 1976), and cicadas (Heath, 1964) routinely occupy the ground at certain times of the day.

Insects, such as flightless tenebrionid beetles, which are tied to the substrate (particularly those of small body size) are at a distinct disadvantage in regulating their body temperature because they lie within the boundary layer of the surface where air temperatures are near substrate temperatures and wind velocity is low relative to wind velocity a few centimeters above the ground (Fig. 4). In addition, infrared radiation and reflected solar radiation from the ground are more important to the heat budget of ground dwellers. Although they warm up more quickly than insects in vegetation, by midmorning temperatures at the ground surface become too warm and the insects must seek cooler microhabitats underground (Holm and Edney, 1973; Hamilton, 1975). Nevertheless, insects on the ground often exhibit a surprising degree of control over their body temperature. For example, Fig. 8 illustrates the "stilting" posture of locusts described by Waloff (1963). The effectiveness of such behavior is related to the distance between the ground and the ventral surface of the body and therefore to the length of the legs. It is significant that a variety of surface-dwelling desert arthropods have evolved long legs. Stilting is characteristic of several Old and New World tenebrionid beetles (Hadley, 1970; Edney, 1971; Hamilton, 1971; Henwood, 1975a), locusts (Waloff, 1963; Anderson et al., 1979), and scorpions (Hadley, 1972). It is highly effective for several reasons. By stilting, the insect (1) reduces contact between the body and the ground, thereby minimizing conduction, (2) moves the body into cooler temperatures as there is a sharp gradient in ambient temperature within a few millimeters of the ground, and (3) enhances convective heat transfer by exposing a greater portion of body surface to higher wind velocities. Edney's (1971) data for tenebri-

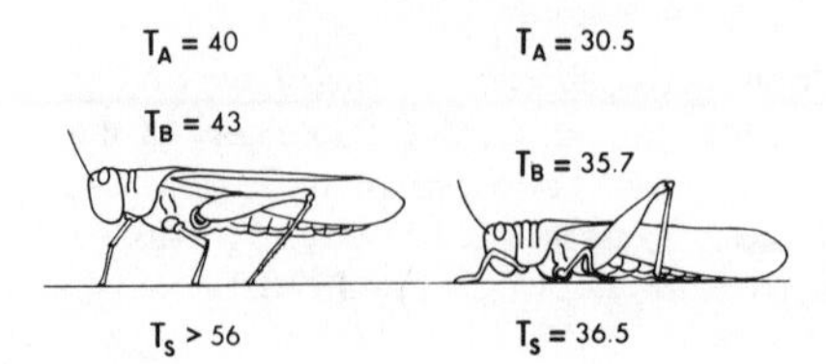

Fig. 8 The stilting posture (left) and crouching posture (right) of the desert locust (*S. gregaria*) with body temperature (T_b), substrate temperature (T_s) and air temperature (T_a) when the postures were exhibited. (After Waloff, 1963.)

onid beetles suggest that body temperature can be shifted by about 4°C by this behavior. Data obtained by Waloff for locusts are more dramatic (Fig. 8*a*). In two experiments, the average T_b of stilting locusts was 43°C (T_a = 40°C) when substrate temperature exceeded 56°C.

Insects usually bask on the ground to increase their body temperature. "Crouching" behavior of locusts in the desert occurs most often during late afternoons when air temperatures are beginning to fall but the ground remains warm and wind velocities are relatively high. The close appression of the ventral surface of the body to the ground undoubtedly facilitates heat uptake by conduction, while convective heat exchange is reduced. Body temperatures of locusts exhibiting this posture were about 5°C above air temperature and within 1°C of substrate temperature (Waloff, 1963). Crouching postures have also been shown to occur in the grasshopper, *P. delicatula* (Anderson et al., 1979), and a semicrouching posture in the dragonfly, *Hagenius brevistylus,* substantially increases body temperature (Tracy et al., 1979).

Shuttling between sun and shade has been reported for several insect species. This behavior has been interpreted as a type of on-off control of body temperature (see Heath, 1967), where movement into the shade is associated with a body temperature near an upper set point and movement into the sun occurs at T_b near a lower set point. Such behavior maintains the body temperature within a certain range and prevents overheating. Experiments on the tenebrionid beetle *O. rugatipennis* by Edney (1971, Fig. 9) suggest that, if shuttling between sun and shade is

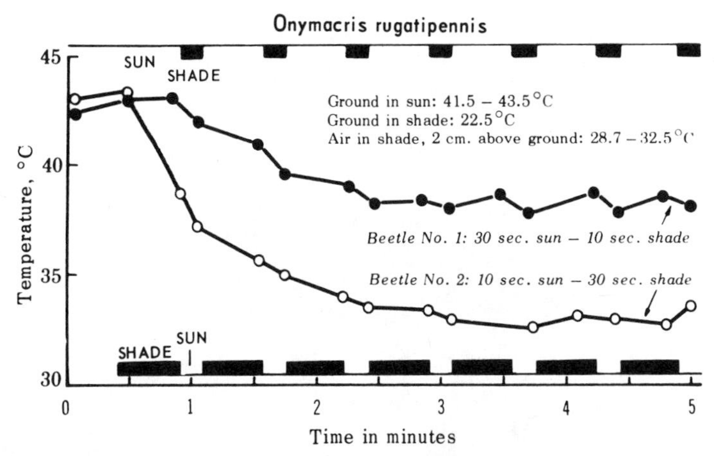

Fig. 9 The effect of rapid alteration between sunshine and shade on the body temperature of the tenebrionid beetle O. rugatipennis. One beetle (●) exposed to 30 sec sunshine alternating with 10 sec of shade, reached a temperature of 38°C, while another (○) under the reverse conditions, reached a temperature about 4–5°C lower. (From Edney, 1971.)

frequent enough, body temperature can be held practically constant. Unfortunately, data are not presently available on the timing of shuttling or on the body temperature of free-roaming insects exhibiting this behavior.

6 ROLE OF THE WINGS IN THERMOREGULATION

The butterflies are among the most conspicuous of the basking insects. It has long been suggested that butterflies utilize their wings to control body temperature. Clench (1966) summarized earlier field observations on the behavior of butterflies, supplied further evidence from his own observations, and classified butterflies according to the position of their wings during basking. Dorsal baskers, as illustrated by *Papilio polyxenes* (Fig. 10*a*), position the dorsal surface of the thorax perpendicular to the rays of the sun, with the fore- and hindwing to the sides of the body. Under high solar loads and high ambient temperatures, the wings are folded over the body. Lateral baskers such as *Colias eurytheme* (Fig. 10*b*) fold the wings tightly over the back and orient the body broadside. Lateral baskers vary the yaw (leaning from side to side) and roll (movement left or right) but have little control of pitch (head up or down) during basking (Watt, 1968). A third type intermediate between the dorsal and lateral baskers is exhibited by skippers (Hesperiidae). Skippers' hindwings are spread flat to

Fig. 10 Basking postures of butterflies. (A) Dorsal basking by the swallowtail butterfly **Papilio**. (From Rawlins and Lederhouse, 1979.) (B) Lateral basking by **C. eurytheme**. (Photograph courtesy of W. Watt.) (C) Body basking by the skipper, **Hylephila phylaeus**. (Photograph courtesy of B. Heinrich.)

the side of the body while the forewings are perpendicular to the hindwings (Fig. 10*c*).

Vielmetter (1954, 1958), working with the dorsal basker *Argynnis paphia* L., showed that, in the laboratory under various radiant heat loads, variation in wing angle resulted in regulation of thoracic temperature between 34 and 37°C. Field observation showed that the butterflies altered their wing position in relation to heat load, although the mechanism for regulation of T_{th} was unclear. Clench (1966) suggested that, as the wings heated up, heat was picked up by hemolymph flowing in the wing veins and returned to the thorax, thereby facilitating thoracic heating. However, although the wings of several species of butterflies contribute in varying degrees to facilitating thoracic heating, Clench's suggestion involving heat transfer from the wings to the body via hemolymph circulation has so far been refuted (Watt, 1968, 1969; Kammer and Bracchi, 1973; Heinrich, 1972).

Several recent studies have further clarified the role of the wings in basking butterflies. The dorsal basking butterfly, *Papilio machaon,* exhibited a 30% reduction in the equilibrium temperature attained by the thorax, and a decreased rate of thoracic heating, when the wings were shaded (Wasserthal, 1975). Experiments in which the wings were partially shaded (Fig. 11) indicate that most of the effect of the wings in facilitating heat uptake by the thorax occurred within the first 15% of the wing near the body. The overlap of the wings at their base provides an increased mass in this region, resulting in greater heat capacity and higher temperature. The major mechanism appears to be an alteration in convective

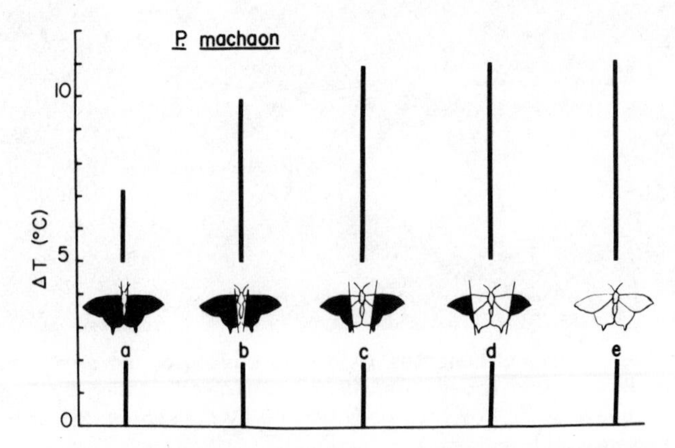

Fig. 11 The thoracic temperature excess achieved by the butterfly P. machaon during basking with varying degrees (a–e) of wing shading. (After Wasserthal, 1975.) The thoracic temperature excess ($\Delta T = T_{th} - T_a$) is indicated by vertical lines.

heat exchange as a result of warm air accumulating beneath the wing. When the wings are cut at the base and placed approximately 1 mm from the thorax, thoracic heating curves are nearly identical to those of size-matched intact individuals. Therefore conductive heat transfer between the wing base and the thorax is negligible (Douglas, 1977).

In lateral basking lepidopterans, the wing base is also important in thoracic heating. Temperature profiles of irradiated wings of the butterfly *Colias philodice* show the greatest temperature increases at the wing base (Douglas, 1977; see also Clark et al., 1973). However, in view of the wing position during basking, and the large area of contact between the wing and the body, it is likely that, unlike the situation in dorsal baskers, a large proportion of the heat absorbed at the wing base is transferred directly to the thorax by conduction.

Wings appear to have a function in thermoregulation in other insects. In the cicada, *Magicicada cassini,* the wings are held out horizontally away from the body as in dorsal basking lepidopterans, although it is unclear whether the wings facilitate thoracic heating rates or equilibrium T_{th} directly or are merely moved to the sides of the body to maximize the surface of the thorax and abdomen exposed to solar radiation. At high T_a when the cicada is heat-stressed, the wings are folded over the body, reducing direct solar radiation impinging on the dorsal surface. Movement of wings between sunlight and the dorsal surface resulted in a 5°C decrease in thoracic temperature (Heath, 1970).

The position of the wings of basking dragonflies is important. Dragonflies of several species bask with the wings bent forward and downward (Fig. 5), and this facilitates heating, particularly over a warm substrate (May, 1976b; Tracy et al., 1979). In the black dragonfly, *Haganius brevistylus,* this posture is associated with rapid warming rates. The wings apparently shield the thorax, thereby reducing convective heat loss (Tracy et al., 1979).

7 THE ROLE OF SURFACE COLOR

The surface coloration of insects may have important consequences for their thermal balance. Although most living tissue is essentially a black-body in the infrared region of the spectrum, approximately 50% of the energy in solar radiation lies within the visible region. Consequently, different surface colors could potentially result in large differences in the quantity of radiant energy absorbed. The results of Digby (1955) have been used to argue that coloration is an important determinant of body temperature (Watt, 1968; Hamilton, 1973) and also to discount the significance of differences in the coloration of insects (Stower and Griffiths, 1966; Edney, 1971, 1974).

Similarly, studies on the effects of coloration in locusts are numerous and contradictory. Digby's (1955) findings that thoracic temperatures of black locusts were significantly greater than lighter forms corroborated the earlier studies of Buxton (1924), Hill and Taylor (1933), and Strelnikov (1936). In these three studies, black locusts had average temperatures 3–6°C higher than their light counterparts. However, Pepper and Hastings (1952) found higher but insignificant differences between buff and black morphs of *Melanoplus differentialis*. In addition, Stower and Griffiths (1966), in the most extensive series of measurements to date, were unable to demonstrate a consistent difference in equilibrium body temperature or rate of increase in body temperature in paired comparisons between red-dark extreme morphs and green extreme morphs of *S. gregaria* at several radiation levels. It is possible, of course, that the differences in color between green and red-dark forms were not great enough to account for different absorption of visible wavelength solar radiation, but the findings are disconcerting in light of the findings of earlier studies.

The occurrence of tenebrionid beetles having either white or jet black elytra (see Edney, 1971, Plate 1) in the hot, arid Namib Desert of southwest Africa has led to several studies on the thermal significance of different-colored elytra. The species having white elytra exhibit significantly lower measured abdominal temperatures than those having black elytra (Fig. 6; see also Hadley, 1970). Direct measurements of reflectance indicate that white elytra absorb significantly less radiation than black ones (Henwood, 1975a), and Edney (1971) calculated body temperatures consistent with measured values assuming reasonable differences in the reflective properties of white and black cuticles. Hamilton (1973) has suggested that these differences in coloration relate to differences in the thermal ecology of the species, with black beetles becoming active earlier in the day because of increased radiant heat absorption, and white beetles remaining active during the hotter parts of the day because of reduced heat absorption. Henwood's (1975b) findings that the cuticle of black beetles selectively absorbs a greater portion of the incident radiation when the sun is at low angles, such as during early morning and late afternoon, are consistent with this hypothesis. Cloudsley-Thompson (1970) discounts the thermal significance of surface color in tenebrionid beetles and suggests that coloration is aposematic. While it is clear that white versus black cuticles of tenebrionid beetles have different absorbing capacities for solar radiation, further ecological data are necessary before the adaptive significance of this coloration is clearly established (for further discussion, see Hamilton, 1973; Edney, 1974).

Perhaps the best demonstration of the adaptive significance of surface coloration in thermoregulation of insects is the work of Watt (1968, 1969) on lateral basking butterflies, *Colias* spp. In certain butterflies, only the wing base on the ventral portion of the fore- and hindwings (i.e., the side of the wing facing the sun when the butterflies are in basking postures) is melanized. Dark forms achieve equilibrium temperatures about 15% greater than size-matched light forms under the same conditions, and these differences in temperature are correlated with greater measured absorbance of radiation by the wing base at wavelengths in the visible region (Watt, 1968). Similar differences in equilibrium T_b occur in dorsal basking *P. machaon* before and after the black scales on the dorsal surface of the wing base are removed. Most dorsal baskers, regardless of wing coloration, have black wing bases (Wasserthal, 1975). *Colias* species inhabiting different altitudes exhibit similar flight temperatures, and species at higher altitudes (lower mean T_a) exhibit significantly more melanin in the wing base (Watt, 1969). Finally, the presence of melanin in the wing base varies seasonally in *Colias* butterflies. In summer, at high mean ambient temperatures, the degree of melanized wing bases in *Colias* spp. from a given habitat is much lower than in the spring and fall (Watt, 1969). A similar response is seen in the butterfly *Natholis iole*. When reared under long-day photoperiods in the laboratory, adults emerge in an immaculate (nonmelanized) form, while shorter photoperiods result in adults with melanized wing bases. As in *Colias* spp., melanized adults of *N. iole* exhibit more rapid heating rates and higher equilibrium temperatures than immaculate forms (Douglas and Grula, 1978; Fig. 12).

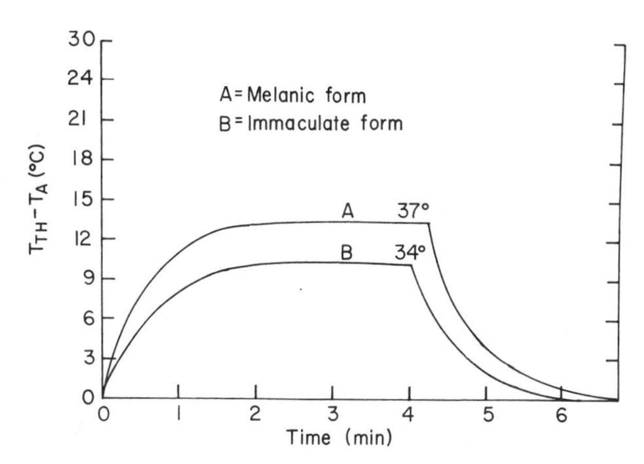

Fig. 12 Heating and cooling curves for sized-matched melanic and immaculate forms of N. iole. (After Douglas and Grula, 1978.)

The coloration patterns discussed above are structural and therefore cannot be manipulated physiologically in response to environmental changes. However, their effects on body temperature are varied with appropriate behavior. A butterfly, for example, exposes its wing bases to direct solar radiation only during basking. Under conditions of heat stress, the wing bases are not exposed to solar radiation and therefore make little or no contribution to thoracic heating. At high ambient temperatures, the tenebrionid beetle *Stenocara phalanguim* orients the long axis of its body parallel to solar radiation with its white abdomen facing the sun (Henwood, 1975a). Basking cicadas expose a black dorsum, but at high T_a expose a white venter (Heath and Wilkin, 1970). Consequently, structural coloration can be manipulated for regulating temperature under different environmental circumstances.

There are several reported cases of physiological color changes in relation to temperature. Australian grasshoppers (*Kosciuscola tristis*) change color in response to different temperatures. Below 15°C while in a basking posture the hoppers are black, while at higher temperatures the surface color becomes blue (Key and Day, 1954a). The response is related to the migration of pigment granules in cuticular cells in response to temperature (Key and Day, 1954b). Similar color changes occur in several species of damselflies (Odonata: Zygoptera), although their significance in thermoregulation is equivocal. Different studies on the phenomenon indicate that darker color facilitates thoracic heating (O'Farrell, 1963) or has no effect (Vernon, 1973; May, 1976a). An alternative hypothesis suggests that the small temperature effect may provide information about the direction of solar radiation so that orientation behavior is more precise (Vernon, 1974).

8 MICROHABITAT SELECTION

The ability of insects to find and occupy areas of thermal heterogeneity within their environment is among the most common and most effective ways of behaviorally controlling body temperature. Movement within their habitat allows insects to attain body temperatures anywhere from a few degrees below to as much as 15°C above air temperature.

Perhaps the most extensively studied case of habitat selection for thermoregulation is that of migratory locusts (see Uvarov, 1977, for detailed discussion). However, the behavior described below could with modification be applied to butterflies, caterpillars, cicadas, and dragonflies.

In the early morning before dawn, locusts are perched on vegetation with their backs facing east. Air temperatures at this time are usually sufficiently low that the locusts are completely inactive. Their position above the ground allows them to intercept the rays of the sun several minutes before it strikes the ground. Air temperatures rise more rapidly than ground temperatures in the early morning, and higher body temperatures are achieved by basking on the vegetation. After the ground has warmed to higher temperatures about midmorning, the hoppers move from the vegetation to the ground. The stimulus for movement to the ground is unclear. Chapman (1955) suggests that a general increase in nonspecific movement occurs as a consequence of elevated T_b through basking. Occupying bare ground during this part of the day allows higher body temperature to be achieved than could occur in another portion of the habitat. When body temperature is elevated, flight or other activity commences. During midday, if T_a is high, activity is reduced and animals take up positions high in the vegetation where, as discussed previously, increased convection and reduced solar radiation prevent overheating. Locusts also seek shade at very high ambient temperatures and remain on the ground throughout the day on cool or cloudy days.

In a few cases, the presence of insects themselves alters the microhabitat sufficiently to affect their body temperature. For example, locusts (both gregarious and nongregarious species) often form quiescent aggregations comprising from a few to several hundred individuals. The aggregations are so tightly packed that individuals within a group are often in direct contact with several other locusts. This behavior has obvious thermoregulatory significance, because these groups are invariably located in an open habitat on bare ground which is significantly warmer than the air temperature. After a locust becomes settled in the group, it usually assumes a basking posture (Chapman, 1955). If the ground group is partially shaded, locusts located in the shade will move to a new, sunny location, while the animals in the sunlit position remain equiescent (Ellis and Ashall, 1957). Group basking should result in higher equilibrium body temperatures than those obtained by individuals, because the surface/volume ratio of the group is much smaller than the sum of the surface/volume ratios of its individuals and convective heat exchange is reduced.

Seymour (1974) verified the thermal significance of aggregation in the gregarious larvae of the sawfly *Perga dorsalis*. During the day, this species forms stationary aggregations including from 15 to 250 individuals. The groups disperse during the evening to feed on *Eucalyptus* leaves. Since the groups remain stationary throughout the day, they may be found either in sunlight or in shade. Aggregating results in higher T_b of insects in sunlight than that achieved by solitary individuals through a reduction in convective heat loss (Seymour, 1974). Above a body temperature of 30°C, the clumped larvae raise their abdomens to facilitate heat loss via convection. Dead larvae placed in this position had abdominal temperatures as much as 5°C higher than thoracic temperatures. At still higher T_b, larvae spread a semiliquid excretia over the surface of the body, facilitating evaporative heat loss from the body surface.

Some arctic insects bask in the flowers of *Dryas integrifolia* M. Vahl and *Papaver radicatum* Rottb. These flowers track the sun throughout the 24 hr of daylight at Lake Hazen, Ellesmere Island (81°49′ N latitude), acting as parabolic reflectors in focusing heat from the sun onto their sporophylls (Hocking and Sharplin, 1965; Kevan, 1975). Mosquitos and flies perched on the flowers attained body temperatures of 6–15°C above ambient temperature and spent up to 13 min in each flower while foraging for nectar. In flowers without petals, the temperature excess of the insects is lower than in normal or desporophyllate flowers (Kevan, 1975).

In most cases of microhabitat selection, insects maintain a relatively constant body temperature by positioning themselves in thermal conditions where they are above the air temperature. If T_a exceeds the normal activity range, insects of many species seek shade and activity ceases until the air temperature declines. Once in the shade of course body temperature will follow air temperature unless evaporative cooling occurs. Consequently, particularly in very hot environments, insects having access to shade may well be poikilothermic during midday. At high T_a, several desert arthropods burrow in the sand (Cloudsley-Thompson, 1968; Hamilton, 1973; Henwood, 1975a), although it is unclear to what extent activity and thermoregulation occur in surface dwellers that burrow at midday to avoid heat stress.

The physiological mechanisms governing behavioral responses associated with microhabitat selection and basking are difficult to evaluate, because in most cases the nature and location of the receptors are poorly understood (Uvarov, 1977). Moreover, it is apparent that a variety of sensory stimuli may be utilized to yield a thermoregulatory behavior pattern. A few examples from the large literature on locusts should illustrate the point. The percentage of second-instar *Locusta migratoria* assuming a prependicular basking posture under fluorescent light increased from 55% at 1 lux to 100% at 2000 lux. When the compound eyes were blackened, the locusts lost the ability to orient to the lights (Cassier, 1965). However, locusts are also able to orient perpendicular to a radiant heat source in total darkness (Volkonsky, 1939), hence both visual and thermal stimuli are capable of eliciting a thermoregulatory response. Negative orientation of locusts at midday may be modified by wind which affects body surface temperature (Waloff, 1963). In the formation of ground groups, locusts find and occupy the warmest spot on a laboratory cage floor, implying tarsal sensitivity to heat (Chapman, 1955). Both phototactic and photokinetic responses are involved in grouping responses (Cassier, 1965). Fraenkel and Gunn (1961) suggest that the thermoregulatory behavior of locusts in the field contains elements of both taxis and kinesis. Therefore the uses of taxes and kineses (Fraenkel and Gunn, 1961), whether thermal or photo, are of limited utility in characterizing thermoregulatory behavior as a result of the redundancy of sensory input. The relative importance of various stimuli, the nature and location of receptors, and their neurophysiological properties in most cases are poorly understood.

9 ALTERATION OF FLIGHT BEHAVIOR

During flight, all but the smallest insects produce heat sufficient to elevate thoracic temperature. Diurnal flying insects must also contend with added heat input from solar radiation (see, for example, Cena and Clark, 1972). If insects cannot regulate their heat loss during continuous flight, they are forced to fly only at certain times of day, when the combination of exogenous and endogenous heat results in a thoracic temperature within the range of temperatures appropriate for flight. Flight duration and the proportion of time on the perch are directly related to ambient temperature in the dragonflies, *Micrathyria* (May, 1977), and in *Libellula saturata* (Heinrich and Casey, 1978). At intermediate air temperatures, these species are capable of essentially continuous flight, while at lower and higher temperatures flight duration declines and perching duration increases. Basking occurs at low air temperatures in dragonflies that have temporarily ceased flight, while negative orientation to solar radiation occurs in perched individuals in the high-T_a range (May, 1976b, 1977, 1978; Heinrich and Casey, 1978). It is therefore unlikely that thermoregulation occurs during flight; behavioral regulation occurs only while animals are perched. At high T_a, *L. saturata* is reluctant to fly and, when forced off its perch, will only fly for a few seconds. Similarly, the desert cicada, *Diceroprocta apache,* reduces its flight activity during midday and, if forced to fly at T_a typical of Phoenix, Arizona, will do so only for a few seconds. Estimates based on calculated heat production of flying cicadas suggest that T_{th} will rise to the level where motor control is lost within $7\frac{1}{2}$ sec (Heath and Wilkins, 1970).

Behavioral thermoregulation during continuous flight could be achieved by alternating between flapping and gliding flight. Insects having low wing loading, such as butterflies, dragonflies, and locusts, exhibit the capacity to glide for considerable periods of time. In dragonflies, most species glide for longer durations at high T_a (Hankin, 1921; Corbet, 1963). For example, the percentage of time spent gliding by the dragonfly *Tramea carolina* increased linearly from about 20 to 45% as T_a increased between 20 and 35°C (May, 1978). Similarly, gliding is common in locust swarms (Rainey and Waloff, 1951). At air temperatures above 40°C up to 90% of the locusts in immature swarms glided for periods of up to 45 sec (Roffey, 1963). Under windy conditions, adult migrating locusts can glide almost continuously. Increased T_{th}, due to heat production associated with the flight effort, has been suggested as the mechanism for the shift from flapping flight to soaring in migrating locusts (Uvarov, 1977), implying a thermoregulatory mechanism. Obviously, a shift from powered flight to gliding flight will result in a reduction in endogenous heat production, and increased durations of gliding at high T_a are consistent with the hypothesis of thermoregulation. However, atmospheric instability, formation of thermals, and updrafts are most likely to occur during the hot period of the day, making gliding flight more effective about midday. It is therefore unclear whether gliding is an active thermoregulatory mechanism utilized to control body temperature or merely a flight pattern that can only be sustained at certain times of day. From an energetic standpoint, high flight costs should represent a selective advantage favoring gliding whenever it is possible, particularly in dragonflies and locusts which spend long periods of time in continuous flight.

In addition to gliding, migrating locusts may fly at a variety of altitudes. The decrease in temperature with altitude should allow flying locusts to "select" the appropriate T_a that allows T_{th} to fall within the range optimal for muscle function. Weis-Fogh (1967) suggests that the ability to fly at different altitudes has important consequences for maintaining water balance during migratory flights. A similar rationale could be applied for thermoregulation. However, little evidence is available to suggest that locusts actively control their altitude during flight (see Uvarov, 1977).

10 AREAS OF FUTURE STUDY

In the last decade, study of the behavioral thermoregulation of insects has proceeded from the descriptive to the quantitative. Experimental studies have corroborated most and negated a few of the predictions about thermoregulation based on earlier field observations (Fraenkel, 1929; Corbet, 1963; Clench, 1966). However, considering the diversity of the insects, the wide range of environments they inhabit, and the narrow range of taxa for which good data are available, it is likely that new mechanisms of behavioral thermoregulation will be reported for some time to come. Comparative data are needed from a wide variety of taxa. In addition, since closely related insects show differing capacities for thermoregulation while occurring simultaneously in the same habitat (May, 1977), in different habitats (Anderson et al., 1979), and in the same habitat during different seasons (Casey, 1976), further work is also needed on the groups of insects for which data are currently available. Field observations of behavior, coupled with measured body temperatures, quantitative microenvironment data, and biophysical analysis should continue to be a fruitful approach (Hadley, 1975).

Virtually all data available on body temperature of insects under natural conditions are single measurements (grab-and-stab method) for individuals which, under the best circumstances, are in a specific posture and/or positioned in a particular microenvironment. It would be useful to have continuous measurements of body temperature for diurnal insects under the constantly changing thermal conditions that characterize their environment. Such data would give a better idea of the precision of temperature control that can be achieved by behavioral thermoregulation and help evaluate the hypothesis of temperature control based on an upper and lower set point. Continuous measurements would also determine to what degree insects control their body temperature during specific activities such as feeding and locomotion.

The size and shape of insects result in an intimate association between them and various aspects of their thermal environment. The effects of morphology on the biophysics of heat exchange need to be examined more carefully in many cases, particularly with respect to postural changes. While the general effects of size and shape are understood (Digby, 1955), the role of surface structures such as spines, bristles, and pile should be examined, because these factors can significantly alter the characteristics of the unstirred layer surrounding the body of the insect, resulting in significant effects on convective heat exchange. In addition, behavior of insects in relation to environmental boundary layers must be evaluated quantitatively, both from the standpoint of providing better understanding of the flow of heat by various avenues and also with regard to identifying the type(s) of stimuli governing such behavior.

Future studies on behavioral thermoregulation of insects must include identifying biological, as well as physical, constraints placed on a species by a given environment. The body temperature of an insect cannot always be predicted from the knowledge of its size and shape and the range of operative environmental temperatures within its habitat. It has become increasingly apparent that regulation of body temperature by insects is often related to ecological factors, such as food abundance and quality, predation pressure, competition, and reproductive behavior, as well as to the conditions of the physical environment (see Heinrich, 1977; and this volume, for further discussion). As a result of high cost or low benefits, there are situations in nature where it is more advantageous for an insect to be poikilothermic rather than regulate its body temperature.

11 SUMMARY

Insects from several taxa regulate their body temperatures by behavioral means; radiation and convection being the major avenues of heat exchange. Equilibrium body temperatures of ectothermic (as well as endothermic) insects are related to their size and shape. As size decreases, body temperatures achieved under a given set of circumstances decrease. Within a given size range, spherically shaped insects exhibit higher equilibrium body temperatures than elongate insects, while the latter achieve a greater range of temperatures by varying their orientation to a radiant heat source.

Orientation to sun and wind direction is a means of varying rates of radiative and convective heat exchange. At low T_a, basking insects increase their body temperatures by orienting the long axis of the body perpendicular to solar radiation and parallel to the prevailing wind. When heat-stressed, heat gain and equilibrium body temperature can be minimized by orienting the long axis parallel to sunlight and at right angles to the wind.

The position of the wings during basking affects both the rate of heat gain and the equilibrium body temperature. In butterflies, radiant heating of the wing bases affects the temperature of the air beneath them, thus altering convective heat exchange. Patterns of coloration, resulting in differences in the absorbance of solar radiation, also affect equilibrium body temperature.

Insects exploit the thermal heterogeneity of their environment in part by postural changes as well as by shuttling between sunshine and shade and by moving to and from the ground at different times of the day. Insects on the ground can usually maintain higher T_b than those on vegetation, and they can exert some control of T_b by stilting (heat avoidance) or crouching (heat-seeking) postures. By basking in large aggregations, insects themselves can affect the microenvironment.

Insects that do not control T_{th} during flight must often restrict flight to periods of the day when the combination of exogenous and endogenous heat gain yield appropriate thoracic temperatures. At intermediate T_a flight can be nearly continuous, but at low T_a flight duration generally decreases and becomes interspersed with basking behavior. At high T_a

flight may cease entirely. Some dragonflies glide for progressively longer periods as T_a increases during the day, suggesting that alteration in flight behavior between gliding and powered flight is a means of preventing thoracic overheating while on the wing.

The study of behavioral thermoregulation has proceeded from the descriptive to the quantitative. Biophysical analyses, used to quantify all avenues of heat exchange between the insect and its surroundings, coupled with a detailed characterization of its microclimate, have allowed calculation of body temperature by the solution of energy balance equations. The thermoregulatory significance of substrate boundary layers and morphological characteristics can be evaluated quantitatively using a biophysical approach. The biophysical approach has also been useful in predicting the range of body temperatures that are possible in a given habitat.

ACKNOWLEDGMENTS

It is a pleasure to thank Drs. Matthew Douglas, Eric Edney, Bernd Heinrich, Michael May, and C. Richard Tracy for stimulating discussions. I thank Drs. Bernd Heinrich, Bob Lederhouse, John Rawlins, and Ward Watt for providing photographs, C. R. Tracy for access to previously unpublished material, and Matt Douglas for allowing me to cite from his Ph.D dissertation. Thanks also to Terrie Williams for the drawings. Partial support to the author was provided by NSF Grant PCM77-16450.

REFERENCES

Anderson, R. V., Tracy, C. R., and Abramsky, Z. (1979). Habitat selection in two species of grasshoppers: The role of thermal and hydric stress. *Oikos*. (in press).

Bakken, G. S. (1976). A heat transfer analysis of animals: Unifying concepts and the application of metabolic chamber data to field ecology. *J. Theor. Biol.* **60,** 337–384.

Bakken, G. S. and Gates, D. M. (1975). Heat-transfer analysis of animals: Some implications for field ecology, physiology, and evolution. In *Perspectives of Biophysical Ecology,* D. M. Gates and R. B. Schmerl, Eds., pp. 255–290. Springer-Verlag: Berlin.

Bartlett, P. N. and Gates, D. M. (1967). The energy budget of a lizard on a tree trunk. *Ecology* **48,** 315–322.

Birkebak, R. C. (1966). Heat transfer in biological systems. *Int. Rev. Gen. Exp. Biol.* **2,** 269–344.

Buxton, P. A. (1924). Heat, moisture and animal life in deserts. *Proc. Roy. Soc.* **B96,** 123–131.

Casey, T. M. (1976a). Activity patterns, body temperature, and thermal ecology in two desert caterpillars (Lepidoptera: Sphingidae). *Ecology* **57,** 485–497.

Casey, T. M. (1976b). Flight energetics in sphinx moths: Heat production and heat loss in *Hyles lineata* during free flight. *J. Exp. Biol.* **64,** 545–560.

Cassier, P. (1965). Contribution a l'etude des reactions photomenotaxique du criquet migrateur: *Locusta migratoria migratorioides* (R. + F.), leur determinisme sensoriel. *Insectes Soc.* **12,** 363–382.

Cena, K. and Clark, J. A. (1972). Effect of solar radiation on temperatures of working honey bees. *Nat. New Biol.* **236,** 222–223.

Chapman, R. F. (1955). Some temperature responses of nymphs of *Locusta migratoria migratorioides* (R. + F.), with special reference to aggregation. *J. Exp. Biol.* **32,** 126–139.

Chapman, R. F. (1959). Field observations of the behavior of hoppers of the red locust (*Nomadacris semifasciata* Serville). *Anti-Locust Bull.* **33,** 1–51.

Church, N. S. (1960a). Heat loss and the body temperature of flying insects. I. Heat loss by evaporation of water from the body. *J. Exp. Biol.* **37,** 171–185.

Church, N. S. (1960b). Heat loss and the body temperature of flying insects. II. Heat conduction within the body and its loss by radiation and convection. *J. Exp. Biol.* **37,** 186–213.

Clark, J. A., Cena, K., and Mills, N. J. (1973). Radiative temperatures of butterfly wings. *Z. Angew. Entomol.* **73,** 327–332.

Clench, H. K. (1966). Behavioral thermoregulation in butterflies. *Ecology* **47,** 1021–1034.

Cloudsley-Thompson, J. L. (1964). Terrestrial animals in dry heat: Arthropods. In *Handbook of Physiology*, J. Field, Ed., American Physiological Society, Sec. 4, pp. 451–465. Williams and Wilkins: Baltimore.

Cloudsley-Thompson, J. L. (1970). Terrestrial invertebrates. In *Comparative Physiology of Thermoregulation*, G. C. Whittow, Ed., 1:15–77. Academic: New York.

Corbet, P. S. (1963). *A Biology of Dragonflies*. Quadrangle: Chicago: 247 pp.

Digby, P. S. B. (1955). Factors affecting the temperature excess of insects in sunshine. *J. Exp. Biol.* **32,** 279–298.

Douglas, M. M. (1977). The behavioral and biophysical strategies of thermoregulation in temperate butterflies. Ph.D. Dissertation, University of Kansas.

Douglas, M. M. and Grula, J. W. (1978). Thermoregulatory adaptations allowing ecological range expression by the pierid butterfly, *Nathalis iole Boisduval. Evolution* **32**(4), 776–783.

Edney, E. B. (1953). The temperature of woodlice in the sun. *J. Exp. Biol.* **30,** 331–349.

Edney, E. B. (1971). The body temperature of tenebrionid beetles in the Namib Desert of southern Africa. *J. Exp. Biol.* **55,** 253–272.

Edney, E. B. (1974). Desert arthropods. In *Desert Biology*, vol. II, G. W. Brown, Ed., pp. 311–384. Academic: New York.

Edney, E. B. (1977). *Water Balance in Land Arthropods*. Springer-Verlag: New York.

Edney, E. B. and Barass, R. (1962). The body temperature of the tsetse fly, *Glossina moristans* Westwood (Diptera, Muscidae). *J. Insect Physiol.* **8,** 469–481.

Ellis, P. E. and Ashall, C. (1957). Field studies on diurnal behavior, movement and aggregation in the desert locust (*Schistocerca gregaria* Forskal). *Anti Locust Bull.* **25,** 1–94.

Fraenkel, G. (1929). Untersuchungen über Lebensgewohnheiten, Sinnes Physiologie und Socialpsychologie der wandernden Larven der Afrikanischer Wanderheuschrecke *Schistocerca gregaria* (Forsk). *Biol. Zentralbl.* **49,** 657–680.

Fraenkel, G. (1930). Die Orienterrung von *Schistocerca gregaria* zu strahlender Wärme. *Z. Vergl. Physiol.* **13,** 300–313.

Fraenkel, G. S. and Gunn, D. L. (1961). *The Orientation of Animals*. Dover: New York.

Gates, D. M. (1962). *Energy Exchange in the Biosphere*. Harper and Row: New York.

Hadley, N. F. (1970). Micrometeorology and energy exchange in two desert anthropods. *Ecology* **51,** 434–444.

Hadley, N. F. (1972). Desert species and adaptation. *Am. Sci.* **60,** 338–347.

Hadley, N. F. (1975). Environmental physiology of desert organisms: Synthesis and comments on future research. In *Environmental Physiology of Desert Organisms*, N. F. Hadley, Ed., pp. 269–276. Dowden, Hutchinson & Ross: Stroudsburg, Pa.

Hamilton, W. J. (1971). Competition and thermoregulatory behavior of the Namib Desert tenebrionid beetle genus *Cardiosis*. *Ecology* **52,** 810–822.

Hamilton, W. J. (1973). *Life's Color Code*. McGraw-Hill: New York.

Hamilton, W. J. (1975). Coloration and its thermal consequences for diurnal desert insects. In *Environmental Physiology of Desert Organisms*, N. F. Hadley, Ed., pp. 67–89. Dowden, Hutchinson & Ross: Stroudsburg, Pa.

Hankin, E. H. (1921). The soaring flight of dragonflies. *Proc. Camb. Phil. Soc. Biol. Sci.* **20,** 460–465.

Heath, J. E. (1964). Reptilian thermoregulation: Evaluation of field studies. *Science* **146,** 784–785.

Heath, J. E. (1967). Temperature responses of the periodical "17-year" cicada, *Magicicada cassinii* (Homoptera, Cicadidae). *Am. Midl. Nat.* **77,** 64–76.

Heath, J. E., Hanegan, J. L., Wilkin, P. J., and Heath, M. S. (1971). Adaptation of the thermal responses of insects. *Am. Zool.* **11,** 147–158.

Heath, J. E. and Wilkin, P. J. (1970). Temperature responses of the desert cicada, *Dicero-procta apache* (Homoptera, Cicadidae). *Physiol. Zool.* **43,** 145–154.

Heath, J. E., Wilkin, P. J., and Heath, M. S. (1972). Temperature responses of the cactus dodger, *Cacama valvata* (Homoptera, Cicadidae). *Physiol. Zool.* **45,** 238–246.

Heinrich, B. (1970). Thoracic temperature stabilization by blood circulation in a free flying moth. *Science* **168,** 580–582.

Heinrich, B. (1971a). Temperature regulation of the sphinx moth, *Manduca sexta*. I. Flight energetics and body temperature during free and tethered flight. *J. Exp. Biol.* **54,** 141–152.

Heinrich, B. (1971b). Temperature regulation of the sphinx moth, *Manduca sexta*. II. Regulation of heat loss by control of blood circulation. *J. Exp. Biol.* **54,** 153–166.

Heinrich, B. (1972). Thoracic temperatures of butterflies in the field near the equator. *Comp. Biochem. Physiol.* **43A,** 459–467.

Heinrich, B. (1974). Thermoregulation in endothermic insects. *Science* **185,** 747–756.

Heinrich, B. (1977). Why have some animals evolved to regulate a high body temperature? *Am. Nat.* **111,** 623–640.

Heinrich, B. and Casey, T. M. (1978). Heat transfer in dragonflies: "fliers" and "perchers." *J. Exp. Biol.* **74,** 17–36.

Heinrich, B. and Pantle, C. (1975). Thermoregulation in small flies (*Syrphus* sp.): Basking and shivering. *J. Exp. Biol.* **62,** 599–610.

Henwood, K. (1975a). A field-tested thermoregulation model for two diurnal Namid Desert tenebrionid beetles. *Ecology* **56,** 1329–1342.

Henwood, K. (1975b). Infrared transmittance as an alternative adaptive strategy in the desert beetle *Onymacris plana*. *Science* **189,** 993–994.

Herter, K. (1953). *Der Temperaturisinn der Insekten*. Duncker and Numblot: Berlin.

Hill, L. and Taylor, H. J. (1933). Locusts in sunlight. *Nature* **132,** 276.

Hocking, B. and Sharplin, C. D. (1965). Flower basking by arctic insects. *Nature* **206,** 215.

Holm, E. and Edney, E. B. (1973). Daily activity of Namib Desert arthropods in relation to climate. *Ecology* **54,** 45–56.

Kammer, A. E. (1970). Thoracic temperature, shivering and flight in the monarch butterfly, *Danaus plexippus* L. *Z. Vergl. Physiol.* **68,** 334–344.

Kammer, A. E. and Bracchi, J. (1973). Role of the wings in the absorption of radiant energy by a butterfly. *Comp. Biochem. Physiol.* **45A,** 1057–1063.

Kevan, P. G. (1975). Sun-tracking solar furnaces in High Arctic flowers: Significance for pollination and insects. *Science* **189,** 723–726.

Kevan, P. G. and Shorthouse, J. D. (1970). Behavioral thermoregulation by High Arctic butterflies. *Arctic* **23,** 268–279.

Key, K. H. L. and Day, M. F. (1954a). A temperature controlled physiological colour response in the grasshopper *Kosciuscola tristis* Sjost. (Orthoptera: Acrididae). *Aust. J. Zool.* **2,** 309–339.

Key, K. H. L. and Day, M. A. (1954b). The physiological mechanism of colour change in the grasshopper *Kosciuscola tristis* Sjost. (Orthoptera: Acrididae). *Aust. J. Zool.* **2,** 340–363.

Laudien, H. (1973). Activity, behavior, etc. In *Temperature and Life,* H. Precht, J. Christophersen, H. Hensel, and W. Lacher, Eds., pp. 441–447. New York: Springer-Verlag.

May, M. L. (1976a). Physiological color change in New World damselflies (Zygoptera). *Odonatologica* **5,** 165–171.

May, M. L. (1976b). Thermoregulation and adaptation to temperature in dragonflies (Odonata: Anisoptera). *Ecol. Monogr.* **46,** 1–32.

May, M. L. (1977). Thermoregulation and reproductive activity in tropical dragonflies of the genus *Micrathyria*. *Ecology* **58,** 787–798.

May, M. L. (1978). Thermal adaptations of dragonflies. *Odonatologica* **7,** 27–47.

May, M. L. (1979). Insect thermoregulation. *Ann. Rev. Entomol.* **24,** 313–349.

O'Farrell, A. F. (1963). Temperature-controlled physiological colour change in some Australian damselflies (Odonata: Zygoptera). *Aust. J. Sci.* **25,** 437–438.

Parry, D. A. (1951). Factors determining the temperature of terrestrial arthropods in sunlight. *J. Exp. Biol.* **28,** 445–462.

Pepper, J. H. and Hastings, E. (1952). The effects of solar radiation on grasshopper temperature and activities. *Ecology* **33,** 96–103.

Porter, W. P. and Gates, D. M. (1969). Thermodynamic equilibria of animals. *Ecol. Monogr.* **39,** 227–244.

Porter, W. P., Mitchell, J. W., Bechman, W. A., and DeWitt, C. B. (1973). Behavioral implications of mechanistic ecology (thermal and behavioral modeling of desert ectotherms and their microenvironment). *Oecologia* **13,** 1–54.

Porter, W. P., Mitchell, J. W., Beckman, W. A., and Tracy, C. R. (1975). Environmental constraints on some predator-prey interactions. In *Perspectives of Biophysical Ecology*, D. M. Gates and R. B. Schmerl, Eds., Ecological Studies 12, pp. 347–364, Springer-Verlag. New York.

Precht, H., Christophersen, J., Hensel, H., and Larcher, W. (1973). Temperature and Life. Springer-Verlag, New York, 779, pp.

Rainey, R. C. and Waloff, Z. (1951). Flying locusts and convection currents. *Anti-Locust Bull.* **9,** 51–70.

Roffey, J. (1963). Observations on gliding flight in the desert locust. *Anim. Behav.* **15,** 359–366.

Seymour, R. S. (1974). Convective and evaporative cooling in sawfly larvae. *J. Insect Physiol.* **20,** 2447–2457.

Shepard, R. F. (1958). Factors controlling the internal temperatures of spruce budworm larvae, *Choristoneura fumiferana*. *Can. J. Zool.* **36,** 779–786.

Sherman, P. W. and Watt, W. B. (1973). The thermal ecology of some *Colias* butterfly larvae. *J. Comp. Physiol.* **83,** 25–40.

Smit, C. J. B. (1960). The behavior of the brown locust in the solitary phase. *Tech. Commun. Dep. Agric. Tech. Serv. Pretoria* **1,** 1–132.

Smith, W. K. and Miller, P. C. (1973). The thermal ecology of two South Florida fiddler crabs: *Uca rapax* Smith and *Uca pugilator* Bosc. *Physiol. Zool.* **46,** 186–207.

Stower, W. J. and Griffiths, J. F. (1966). The body temperature of the desert locust. *Entomol. Exp. Appl.* **9,** 127–178.

Strelnikov, H. F. (1936). Effect of solar radiation and the microclimate upon the body temperature and behavior of the larvae of *Locusta migratoria* L. (in Russian with an English summary). *Trudy. Zool. Inst. Leningr.* **2,** 637–733.

Thiele, H. N. (1977). *Carabid Beetles in Their Environments*. Springer-Verlag: Berlin.

Tracy, C. R. (1975). Water and energy relations of terrestrial amphibians: Insights from mechanistic modeling. In *Perspectives of Biophysical Ecology*, D. M. Gates and R. B. Schmerl, Eds., Ecological Studies 12, pp. 325–346. Springer-Verlag: New York.

Tracy, C. R., B. J. Tracy, and D. S. Dobkin. (1979). The role of pasturing in the behavioral thermoregulation by black dragons (*Hagenius brevistylus* Selys: Odonata). *Physiol. Zool.* **52,** 565–571.

Uvarov, B. (1977). *Grasshoppers and Locusts, A Handbook of General Acridology*, vol. 2. London: Centre for Overseas Pest Research.

Vernon, J. E. N. (1973). Physiological control of chromatophores of *Austrolestes annulosus* (Odonata). *J. Insect Physiol.* **19,** 1689–1703.

Vernon, J. E. N. (1974). The role of physiological colour change in the thermoregulation of *Austrolestes annulosus* (Selys) (Odonata). *Aust. J. Zool.* **22,** 457–459.

Vielmetter, W. (1954). Die Temperaturregulation des Kaisermantels in der Sonnenstrahlung. *Naturwissenschaften* **41,** 535–536.

Vielmetter, W. (1958). Physiologie des Verhaltens zur Sonneneinstrahlung bei dem Tagfalter *Argynnis paphia* L. I. Untersuchungen in Freiland. *J. Insect. Physiol.* **2,** 13–37.

Volkonsky, M. A. (1939). Sur la photo-akinese des acridiens. *Arch. Inst. Pasteur Alger.* **17,** 194–220.

Waloff, Z. (1963). Field studies on solitary and *transiens* desert locusts in the Red Sea area. *Anti-Locust Bull.* **40,** 1–93.

Wasserthal, L. T. (1975). The role of butterfly wings in regulation of body temperature. *J. Insect Physiol.* **21,** 1921–1930.

Watt, W. B. (1968). Adaptive significance of pigment polymorphism in *Colias* butterflies. I. Variation of melanin pigment in relation to thermoregulation. *Evolution* **22,** 437–458.

Watt, W. B. (1969). Adaptive significance of pigment polymorphisms in *Colias* butterflies. II. Thermoregulation of photoperiodically controlled melanin variation in *Colias eurytheme*. *Proc. Nat. Acad. Sci. U.S.* **63,** 767–774.

Weis-Fogh, T. (1956). Biology and physics of locust flight. II. Flight performance of the desert locust (*Schistocerca gregaria*). *Phil. Trans. R. Soc. Lond.* **B239,** 459–510.

Weis-Fogh, T. (1967). Respiration and tracheal ventilation in locusts and other flying insects. *J. Exp. Biol.* **47,** 561–587.

5
Physiological Mechanisms of Thermoregulation

ANN E. KAMMER

1 INTRODUCTION

An insect regulating its thoracic temperature maintains a dynamic equilibrium between the rate of heat production due to its metabolism and the rate of heat exchange with its environment. The thermal environment is complex and difficult to measure under field conditions, but for physiological studies it is usually sufficient to consider only the ambient temperature. Theoretically, the rate of heat production or the rate of heat loss or both can be altered to maintain thoracic temperature within a limited range. In this chapter I review the physiological mechanisms by which these rates can be adjusted and also the underlying neural control mechanisms.

1.1 Context

Before considering physiological control mechanisms in detail, it is worth remembering the context in which they operate. Physiological mechanisms function in conjunction with behavior that influences heat exchange with the environment. For example, the monarch butterfly *Danaus plexippus* (L.) raises its thoracic temperature by shivering (Kammer, 1970a), as well as by basking in the sunlight. Its readiness to shiver is influenced by acclimation temperature (Kammer, 1971). Dragonflies that routinely perch may both bask and wing-whirr (shiver) (May, 1976), and syrphid flies also both bask and shiver (Heinrich and Pantle, 1975). Physiological mechanisms also function in conjunction with behavior that influences the storage of heat produced during activity. For example, at high ambient temperatures flight may be intermittent, or at low ambient temperatures flight may alternate with bouts of warm-up (e.g., bumblebees, Heinrich, 1975). Although behavior could be considered within the province of physiology (Wigglesworth, 1972), it is discussed separately in this book (Casey, this volume).

Physiological and behavioral mechanisms controlling heat production and heat exchange also operate within limits determined by anatomy and by environmental conditions. Important anatomical constraints include size (Bartholomew, this volume), relative muscle mass, and insulation (Church, 1960b). An insulating coat of scales or pile is especially well developed in large moths and bumblebees, respectively; in dragonflies air sacs outside the flight muscles serve as insulation. These determinants of thermal conductance in insects are not subject to neural control as is the fur of mammals.

Environmental conditions forming the limits within which an insect can thermoregulate can be viewed as a multidimensional climate space (Porter and Gates, 1969). Ambient temperatures and insolation are two important dimensions of this space and, although insolation cannot be varied, the dimensions of this space are nevertheless to some extent flexible. For example, a widely distributed sphinx moth, *Hyles lineata*, is active at dusk and at night when air temperatures are moderate, but at higher latitudes or elevations it flies during the daytime (Douglas, 1978). In the tropics (Costa Rica) the day-flying sphingid moths are all small. The largest ones fly late at night, when it is coolest (P. Opler, personal communication to B. Heinrich). These interactions among morphology, environmental constraints, and activity provide the context for physiological mechanisms of temperature regulation.

1.2 Insects That Regulate Physiologically

Given the appropriate environmental conditions, many large insects when active maintain their thoracic temperature relatively constant and higher than ambient. Such regulation during flight was first suggested in an important study by Adams and Heath (1964a) on the hawkmoth *Hyles* (=*Celerio*) *lineata*. However, their results were confounded by the effects of the experimental conditions on the behavior of the moths which, restrained by the thermocouple leads, usually flew only briefly. A subsequent study (Heath and Adams, 1965) convincingly demonstrated thoracic temperature regulation during continuous flight at different ambient temperatures.

One of the first thoroughly studied examples was the hawkmoth *Manduca sexta*, which regulates its thoracic temperature within 39–42°C over ambient temperatures from 17 to 30°C (Heinrich, 1971a). Bumblebees have also been thoroughly investigated, primarily by Heinrich (1972a–c, 1974b, 1975, 1976), and they are capable and versatile thermoregulators. For example, in free flight *Bombus edwardsii* queens maintained a thoracic temperature of 38°C at an ambient temperature as low as 2°C, and over a temperature range of 2–36°C the slope of a regression line relating thoracic temperature to ambient temperature was 0.27 for *B. vosnesenskii* queens (Heinrich, 1975). Other insects are less competent regulators; for example, for some species of dragonflies the slopes of regression lines relating thoracic temperatures to ambient temperatures in the field are 0.4–0.6 (May, 1976). These and other large insects such as beetles regulate during flight.

During terrestrial activity, regulation, or even simply elevation of the thoracic temperature above ambient, is uncommon. However, bumblebees maintain a constant, high thoracic temperature during brief periods of foraging interspersed between frequent flights (Heinrich, 1972b) and during incubation of their brood (Heinrich, 1974b). Endothermy (but not regulation) has been observed in some dung beetles walking or rolling dung balls (Bartholomew and Heinrich, 1978) and in other large beetles (Bartholomew and Casey, 1977a,b).

In all cases so far examined, endothermy depends on activity of the flight muscles, which are massive structures with high metabolic rates (Kammer and Heinrich, 1978). Regulation during flight depends not on the control of heat production by these muscles but on the control of heat dissipation, whereas during terrestrial activity rates of heat production can also be varied and controlled.

2 CONTROL OF HEAT LOSS

During the last decade a new picture of insect temperature regulation has emerged. Previously attention was focused on control of heat production, in part because, in an animal with such a large surface/volume ratio, the metabolic rate sufficient to produce a body temperature greater than ambient is impressively high. During flight, however, the rate of heat production is necessarily high because of the intense and continuous muscular work demanded by this mode of locomotion. As much as 80–90% of the energy produced by muscle contraction appears as heat in the thorax (Kammer and Heinrich, 1978). It is now clear that the rate of heat production depends on the work performed in flight, and there is no independent control of energy production for temperature regulation. How then is body temperature regulated? In 1970 Heinrich showed that heat loss was adjusted to stabilize thoracic temperature in *M. sexta*. Subsequent work showed that several kinds of insects varied heat loss to regulate thoracic temperature during flight, and attention was profitably turned to the mechanisms by which this control was accomplished.

2.1 Sphinx Moth, *Manduca sexta*

The mechanisms involved in regulating thoracic temperature during flight were first thoroughly analyzed by Heinrich (1970a,b, 1971a,b) in *M. sexta*. Measurements of thoracic and abdominal temperatures have shown that during the preflight warm-up thoracic temperature rises rapidly (5°C/min at $T_A = 25$°C) but abdominal temperature remains near ambient (Heinrich and Bartholomew, 1971). During flight at low ambient temperatures, the abdomen may be 20°C cooler than the thorax. The temperature difference between thorax and ambient is substantial, but between abdomen and ambient the difference is much less (Fig. 1). Thus heat loss from the

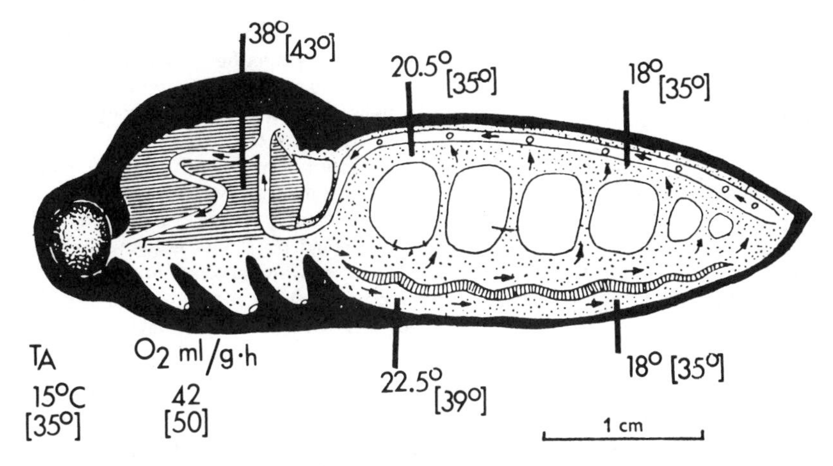

Fig. 1 Manduca sexta, sagittal section illustrating structures important in temperature regulation. During flight heat is produced by the massive thoracic flight muscles (the dorsal longitudinal is shown by horizontal lines). The thorax is insulated by a thick coat of scales (indicated in black) and partially isolated from the abdomen by an air space. Hemolymph flows ventrally from thorax to abdomen and is pumped anteriorad by the dorsally situated heart (flow is indicated by small arrows). In the thorax the aorta loops between the right and left dorsal longitudinal muscles. The heart rate increases as thoracic temperature rises, and the rate of heat transfer from thorax to poorly insulated abdomen increases. Typical temperatures during flight at two different ambient temperatures are given. (For T_A = 15°C, body temperatures without brackets; for T_A = 35°C, body temperatures indicated in brackets.) The oxygen consumption during flight increases only slightly over this temperature range (cf. Fig. 10). Regulation of thoracic temperature (cf. Fig. 2) depends not on increasing rates of heat production at lower T_A but on increasing rates of heat loss at higher T_A. (Modified from B. Heinrich, J. Exp. Biol. 54, 1971.)

abdomen is minimized. At high ambient temperatures the abdomen becomes almost as warm as the thorax, and it radiates heat effectively.

Heat transfer from thorax to abdomen depends on blood flow. Hemolymph is propelled posteriad by the ventral diaphragm and returned from abdomen to thorax by the dorsally situated heart. The abdomen is poorly insulated in comparison with the thorax, which is thickly covered with scales. Hence the abdomen serves as an effective radiator, and at high T_A the rate of heat loss is maximized by a high rate of hemolymph circulation. That the rate of circulation varies during flight at different T_A is shown by the fact that at high ambient temperatures the ventral region of the first abdominal segment, where the blood first enters, is 4°C warmer than the dorsal region of that segment and only 2°C cooler than the thorax, whereas at low ambient temperatures (15°C) the difference between ventral and dorsal regions is smaller, 2°C.

Heartbeat, although not measured in flight, behaves appropriately in

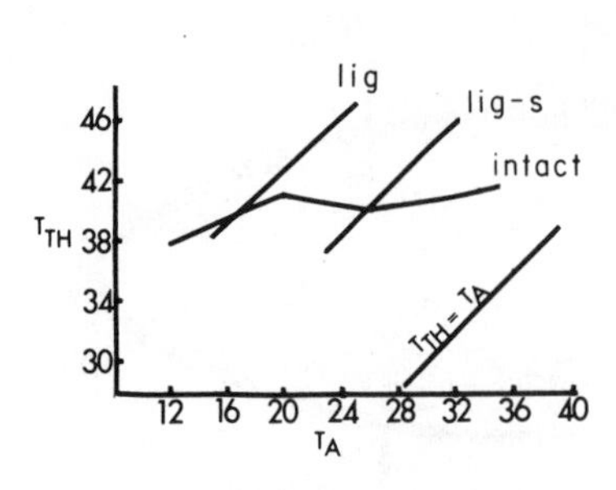

Fig. 2 *Manduca sexta*, thoracic temperatures (T_{Th}) immediately after free flight at different ambient temperatures (T_A). Intact moths maintained a relatively constant T_{Th}; over a range of T_A = 17–35°C, the slope of a line relating T_{Th} to T_A is 0.17 (calculated from Heinrich, 1971a). Moths in which the dorsal vessel was ligated (lig) had T_{Th} 23°C greater than T_A and flew only briefly at $T_A >$ 23°C; they could no longer regulate. When the insulating scales were removed from the thorax of these moths (lig-s), the temperature excess (T_{Th} − T_A) was less. (Modified from Heinrich, 1970a.)

restrained moths artifically heated on the thorax. At thoracic temperatures above 40°C the frequency and amplitude of heart pulsations are markedly greater than at lower T_{Th}. Ligating the heart in the first abdominal segment abolishes the moth's thermoregulating ability in free flight. A moth so treated flies over a limited range of T_A and its thoracic temperature parallels the ambient temperature (Fig. 2). That flight is limited by overheating was shown by removing the scales insulating the thorax. Depilated moths flew at ambient temperatures 7°C higher than similarly ligated but normally insulated moths, and T_{Th} again paralleled T_A. This experiment shows clearly that a functioning heart and hemolymph circulation are essential for temperature regulation during flight in *M. sexta*. It also provides further evidence that the rate of heat production is not regulated during flight.

2.2 Other Moths

Sphingid moths in addition to *M. sexta* regulate thoracic temperature during flight (Heinrich and Casey, 1973; Casey, 1976b), but the regulatory mechanisms have not been analyzed in detail. *Hyles lineata* maintains a thoracic temperature of 40°C over T_A = 16–25°; thoracic temperature increases slightly to 42.5°C at T_A = 32°C (Casey, 1976b). Regulation requires active cooling, since the metabolic rate remains unchanged. Casey (1976b) has calculated that a 1-g moth flying at an ambient temperature of 23°C and 0% relative humidity loses 18% of its heat production by evaporation, 34% by radiation and convection from the thorax, and 18% from the abdomen. The calculations leave 30% of the heat produced unaccounted for. If this surprisingly large figure is correct, then other body surfaces participate in active cooling. Head, legs, and wings are possible sites of additional, variable heat dissipation (Casey, 1976b).

2.3 Dragonflies

Like moths, cruising dragonflies use the abdomen for dissipating excess heat. Some dragonflies, such as *Anax junius*, fly continuously when active. They are active over a wide range of ambient temperatures, and they maintain thoracic temperature moderately independent of ambient temperature (Fig.3) (May, 1976). Although abdominal temperatures were not measured in free flight, transfer of heat from the thorax to the abdomen was observed in stationary dragonflies heated on the thorax (May, 1976; Heinrich and Casey, 1978). When the circulation was interrupted by ligating the heart, heat transfer was abolished (Fig. 4A). In contrast, libellulid dragonflies that regulate temperature primarily by behavioral mechanisms, such as intermittent flight and postural adjustments, have a relatively small abdomen (16% of body mass in comparison with 31–35% in aeshnids) and a smaller heart and show little heat transfer from thorax to abdomen (Heinrich and Casey, 1978).

2.4 Large Beetles

Large beetles have long been known to require a high thoracic temperature for flight (Krogh and Zeuthen, 1941), but temperature regulation during flight has not been studied. Regulation during flight by control of heat loss, particularly from the dorsal surface of the abdomen, can be inferred from measurements of two physical parameters, wing loading (milligrams body mass per square centimeter wing area) and mass (Bartholomew and Heinrich, 1978). Thoracic temperature during flight is independent of wing loading for beetles that are larger than about 2 g and that fly with a T_{Th} greater than 40°C. Since power input and therefore heat production should increase directly with wing loading, it is likely that the more heavily loaded beetles are actively losing more heat as T_{Th} approaches or exceeds 40°C. A similar argument can be made from the observation that thoracic temperature excess during flight is a direct function of mass in small beetles, but in larger beetles (2.5–20 g) T_{Th} approaches 45°C and thoracic temperature excess is independent of mass. Given the observed increase in oxygen consumption with body mass (Bartholomew and Casey, 1977b), a 20-g beetle would produce six times more heat than a 2.5-g beetle per unit time. Since the cooling constants of dead beetles of these sizes differ by a factor of 3.4 while T_{Th} in flight remains the same, the difference predicted from calculated heat production and passive cooling is probably due to active cooling by the larger beetles.

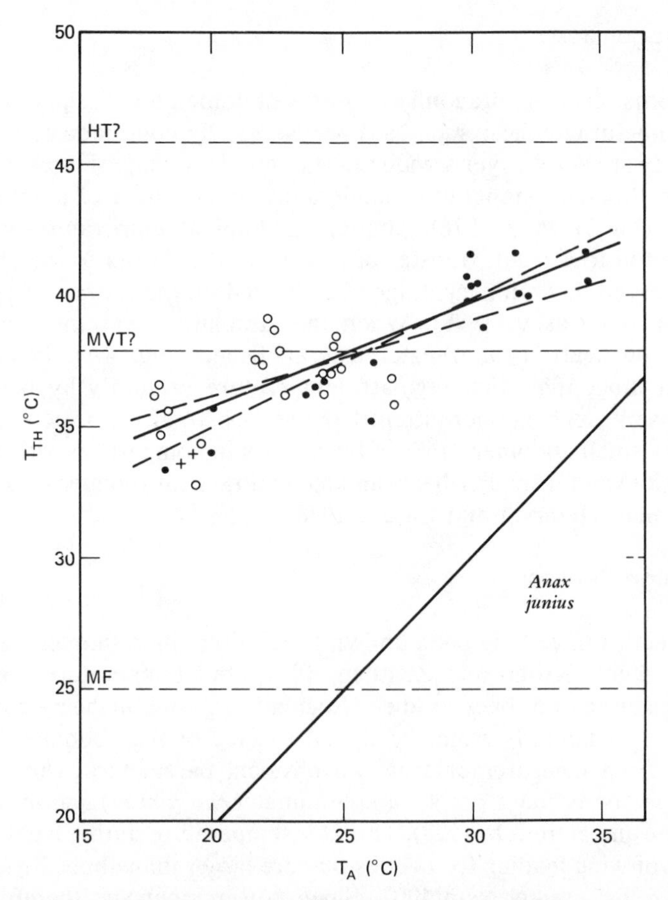

Fig. 3 **Anax junius,** thoracic temperatures (T_{Th}) measured in the field (●, males in flight; ○, females in flight; +, perched; solid line, regression of T_{Th} on T_a from all data; dashed lines indicate regression for males and females separately). Individuals were seen (but not measured) flying after sunset at $T_a = 12°C$. Since the minimum temperature for free, level flight (MF) is 25°C, these large, fast dragonflies are capable endotherms. The slope of the regression line is 0.40, indicating some regulation of T_{Th}. The maximum voluntary tolerance (MVT), determined by heating insects with a lamp and noting the T_{Th} at which they moved away, was 38°C, less than some thoracic temperatures during flight. Heated animals became torpid at a temperature (HT) substantially above T_{Th} in flight. (From May, 1976.) Reprinted from **Ecological Monographs** by permission of Duke University Press. Copyright 1976 by the Ecological Society of America.

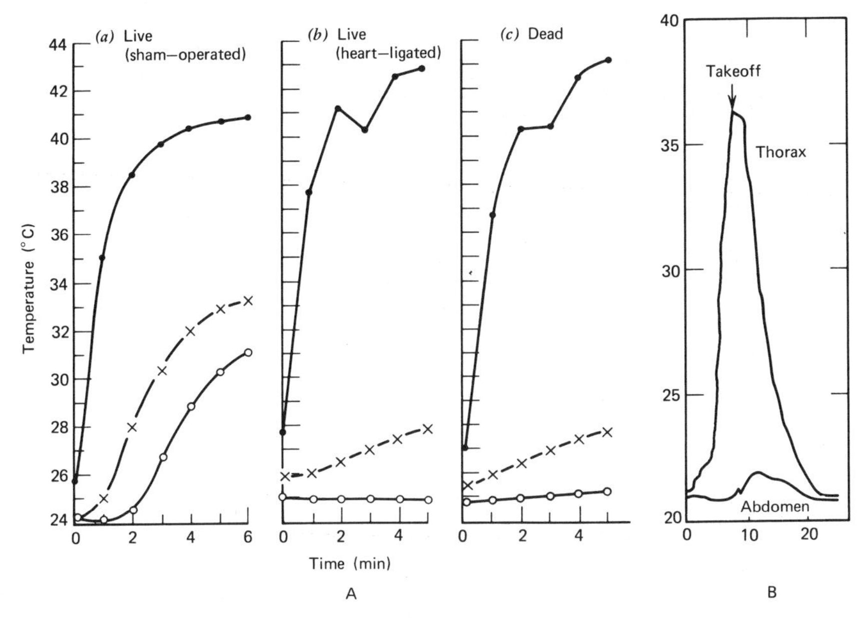

Fig. 4 Mechanisms of temperature regulation. (A) The thorax of a restrained dragonfly, **A. junius**, was heated with a lamp, and body temperatures were measured in the thorax (●), anterior and ventral abdomen (X), and posterior dorsal abdomen (O). The importance of the circulation in transferring heat from thorax to abdomen is indicated by a comparison of temperatures in a sham-operated animal (a) with those in the same animal with its heart ligated with a fine hair (b) or immediately after death (c). (The dip in thoracic temperature in (b) reflects movement of the lamp away from the insect.) (Heinrich and Casey, 1978.) (B) Thoracic and abdominal temperatures during a preflight warm-up (wing whirring) and subsequent cooling in **Anax** sp. at an ambient temperature of 20°C. (The dragonfly was not allowed to fly.) (From May, 1976.) Reprinted from **Ecological Monographs** by permission of Duke University Press. Copyright 1976 by the Ecological Society of America.

2.5 Bumblebees

Bumblebees, like sphinx moths, regulate their thoracic temperatures during flight over a range of ambient temperatures, and they have a well-developed mechanism for circulating hemolymph between thorax and abdomen. At low ambient temperatures, the abdomen may be as much as 20°C cooler than the thorax, but at higher T_A the thoracic and abdominal temperatures become approximately equal (Fig. 5; Heinrich, 1975). As in *Manduca*, hemolymph circulation is altered so that the heat loss from the abdomen to the environment is decreased at lower ambient temperatures.

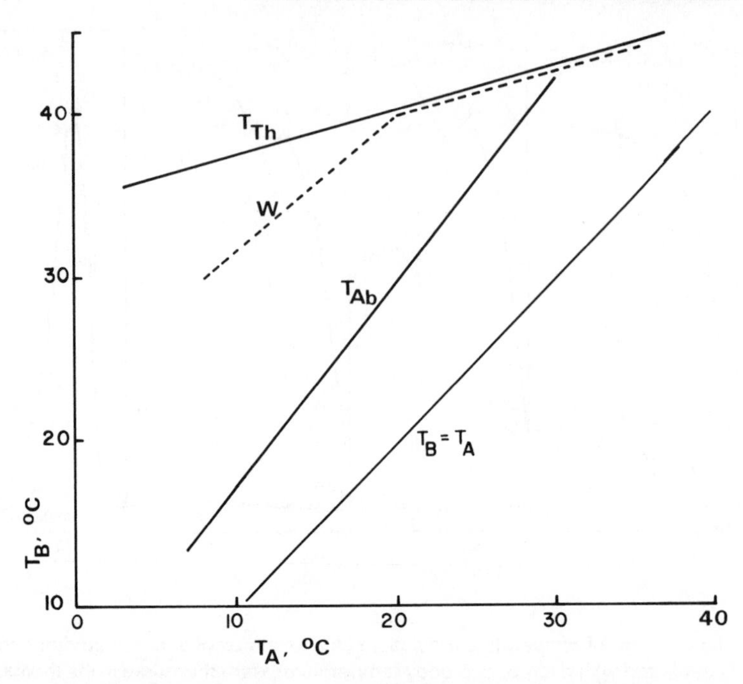

Fig. 5 **Bombus body temperatures, measured after several minutes of free flight, as a function of ambient temperature. T_{Th}, Thoracic temperature of B. vosnesenskii queens; T_{Th} is regulated over a wide range of ambient temperatures (slope = 0.27). T_{Ab}, Abdominal temperatures of the same individuals; more heat is transferred from thorax to abdomen at high T_A than at low T_A. W, Thoracic temperatures of B. edwardsii workers weighing 0.11–0.13 g; the queens of this species weighed 0.25–0.60 g and had thoracic temperatures similar to those represented by line T_{Th}. At ambient temperatures from 18 to 35°C the thoracic temperatures of the smaller workers were indistinguishable from those of the queens. (From B. Heinrich, J. Comp. Physiol. B96, 1975.)**

Unlike the insects discussed above, bumblebees also regulate thoracic temperature when not in flight. For example, while walking on and feeding from nectar-rich, dispersed flowers at low ambient temperatures, *Bombus vagans* maintain their thoracic temperature at levels sufficient for flight ($T_{Th} \sim 32$°C over $T_A = 9$–24°C, Heinrich, 1972a,b). At low ambient temperatures, thoracic and abdominal temperatures differ by 8–12°C, but at high ambient temperatures they differ by only 3°C, suggesting that, as in flight, heat loss from the abdomen is reduced at low T_A.

Additional control of circulation is evinced by queen bees incubating their brood (Heinrich, 1974b). During this behavior queens maintain an abdominal temperature within 5°C of thoracic temperature, even at a T_A of 4°C (Fig. 6), and by positioning their body against the brood they elevate

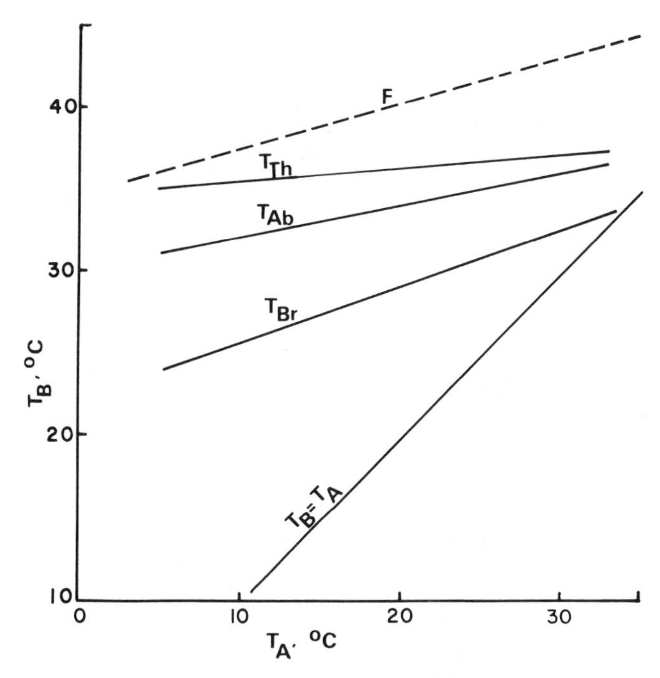

Fig. 6 Bombus vosnesenskii temperatures during incubation of brood. Abdominal temperature (T_{Ab}) is only a few degrees less than thoracic temperature (T_{Th}) over a wide range of ambient temperatures (T_A). Brood temperature (T_{Br}) is higher than ambient while the queen is on the brood, because of heat transfer from thorax to abdomen to brood. These temperatures contrast with those of flight (F, Thoracic temperature in flight; see Fig. 5 for abdominal temperatures in flight) (Heinrich, 1974b).

its temperature above ambient. Although beating rates of the heart and ventral diaphragm have not been measured directly in incubating queens, since insertion of recording wires disrupts the behavior, it is apparent that the rate of heat transfer from thorax to abdomen is always high during brooding.

The temperature measurements discussed above show that the rate of transfer of heat from thorax to abdomen can be low or high, depending on ambient temperature and the behavior of the bumblebees. To explain how bumblebees produce these different rates of heat transfer, Heinrich (1976) examined the circulatory system and blood flow.

In bumblebees, as in other insects, the dorsally situated heart pumps blood in an anterior direction (Fig. 7). Return flow into the abdomen is assisted by rhythmic contractions of the ventral diaphragm. These two flows are separate but contiguous in the narrow petiole that connects the thorax with the abdomen. Using restrained bumblebees, Heinrich (1976)

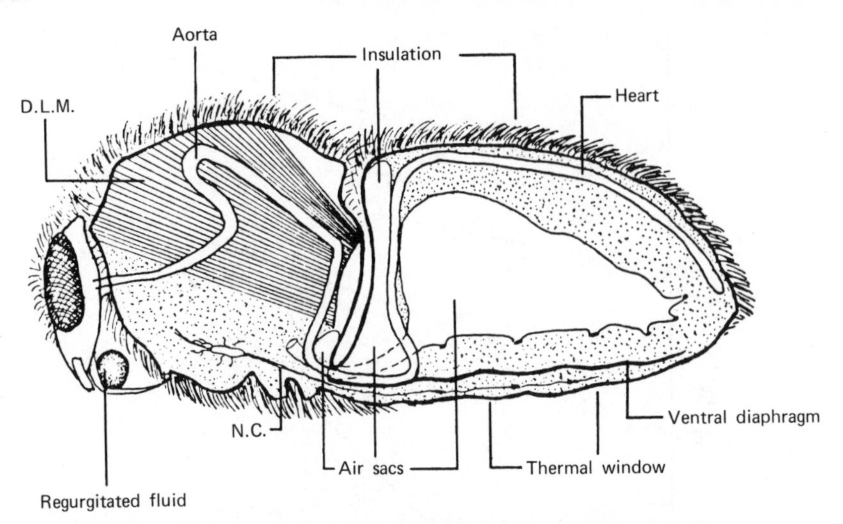

Fig. 7 **Bombus vosnesenskii**, sagittal section illustrating structures important in temperature regulation. As in **Manduca** (Fig. 1), heat is produced during activity by the thoracic flight muscles (the dorsal longitudinal, D.L.M., is shown). The thorax and dorsal surface of the abdomen are insulated by a layer of pile. The heart, which pumps hemolymph primarily in the anterior direction, is situated close to the dorsal surface of the abdomen and loops down posterior to an air sac in the anterior portion of the abdomen. It passes to the thorax through the narrow petiole, where it is dorsal to but in direct contact with the ventral diaphragm. In the thorax the vessel loops between the right and left dorsal longitudinal muscles and then enters the head. The ventral diaphragm is a thin muscular sheet that propels hemolymph posteriorad. The ventral nerve cord (N.C.) surrounded by hemolymph lies below the ventral diaphragm. (From B. Heinrich, **J. Exp. Biol.** 64, 1976.)

showed that, at low thoracic temperatures, the ventral diaphragm beat with a variable rhythm while the heart beat at a higher frequency and with a low amplitude. This pattern of contractions should result in a fairly continuous flow in opposite directions through the two contiguous channels. The resulting countercurrent flow of hemolymph would facilitate exchange of heat between blood leaving the thorax and blood returning to the thorax and it would be expected to result in heat retention within the thorax. On the other hand, when thoracic temperatures were high, the diaphragm pulsated with a faster, regular rhythm. Heartbeat increased greatly in amplitude but decreased in frequency, and at T_{Th} above 42–44°C the heart beat synchronously with the pulsations of the ventral diaphragm. The contractions of both pulsatile organs were also synchronized with abdominal pumping movements. Distinct temperature fluctuations correlated with heartbeat and ventral diaphragm movements were recorded in the second abdominal segment. Heinrich has suggested that under these conditions blood flow is pulsatile and that a pulse of blood impelled into the abdomen alternates with blood returned to the thorax in the heart. Because of this alternation, heat exchange in counterflows should be reduced; warm blood can enter the abdomen, and the rate of heat transfer from thorax to abdomen can thus be increased.

2.6 Evaporative Cooling, Especially by Honeybees

Most insects do not increase the rate of heat loss by controlling the rate of evaporation, presumably because their water supply is limited. Casey (1976b) has calculated that *H. lineata* loses by evaporation about 20% of the heat produced during flight at 0% relative humidity; the loss rate is not regulated. Although evaporative cooling is relatively unimportant as a regulatory mechanism during flight (Church, 1960a), it is employed by a small number of insects primarily as an emergency mechanism to avoid lethal overheating (see May, 1979, and Casey, this volume, for further discussion). The tropical tsetse fly, *Glossina morsitans,* sucks the blood of large mammals and thereby obtains an ample supply of liquid, while exposed to high ambient temperatures in the sunshine. Under these conditions the insect opens its spiracles, and the body temperature may decline as much as 1.6°C (Edney and Barrass, 1962). At high body temperatures a sphingid moth, *Pholus achemon*, repeatedly regurgitated and recovered a drop of fluid in the coils of its proboscis, and this behavior could be a means of cooling by evaporation (Adams and Heath, 1964b). Similarly, bumblebees, *B. vosnesenskii*, forced to fly at high ambient temperatures (35 and 42°C), regurgitated fluid and moved it about on their proboscis (Heinrich, 1976). Evaporation of such a droplet from the proboscis cooled the head by 2°C, but in dead bumblebees thoracic temperature was little affected. This mechanism therefore may not contribute significantly to temperature regulation in bumblebees.

In contrast to bumblebee mechanisms, evaporative cooling enables honeybees to fly continuously at the unusually high ambient temperature of 46°C (Heinrich, 1979b, 1980). Honeybees in flight at this temperature repeatedly regurgitated a droplet of fluid, manipulated it with their tongue, and sucked it back in. Measurements of bees restrained at T_A = 24°C and heated with a lamp showed that the appearance of fluid on the tongue was followed within seconds by a 2–8°C decrease in head temperature and a similar decline in thoracic temperature. Head and thoracic temperatures are strongly coupled, primarily because of passive conductance and also because of blood flow. As a consequence of this cooling mechanism, during flight at T_A = 46°C head temperatures averaged 2°C below ambient and thoracic temperatures only 0.5°C above. Evaporative cooling appears to be the only mechanism available to honeybees for temperature regulation during continuous flight. They do not use the abdomen as a variable radiator of heat. Unlike the arrangement of the heart and ventral diaphragm in the petiole of bumblebees (Fig. 6), the aorta of honeybees is highly coiled in the petiole. This configuration permits efficient countercurrent heat exchange; it would retain heat in the thorax, but it also would be difficult to alter physiologically (Heinrich, 1979b). The different adaptations of honeybees and bumblebees are consistent with the different behaviors of the two genera. In *Bombus*, an individual queen incubates the brood, and when so doing transfers heat to the abdomen, a process that bypasses the countercurrent mechanism. In *Apis*, on the other hand, heat lost from the thorax can warm the hive, and transfer to the abdomen is not necessary. Honeybees do not regulate their thoracic temperature independent of ambient temperature over T_A = 10–26°C during continuous flight (Heinrich, 1979a,b). They may be limited in comparison with bumblebees by their smaller size. Therefore, the more efficient the countercurrent mechanism, the more heat loss from the abdomen is reduced during flight and the better the flight capabilities of these bees at low T_A.

3. CONTROL OF HEAT PRODUCTION

In this section I consider the other side of the dynamic equilibrium of body temperature stabilization, that is, the production of heat. During flight, as during any vigorous exercise, there is an obligatory rapid production of heat, but the metabolic rate is not adjusted to control body temperature during flight. During activity other than flight, however, control of heat production contributes to regulation of body temperature. Muscular activity produces elevated thoracic temperatures during preflight warm-up, during foraging and brood incubation by bumblebees and honeybees, and during ball rolling by dung beetles.

3.1 Metabolic Rate during Flight

Heat production during flight depends on the work done by the flight muscles. Metabolic rate varies during flight as a function of flight speed, lift, and load carried (review, Kammer and Heinrich, 1978). For example, in bumblebees both rate of oxygen consumption and thoracic temperature increase as the load of honey in the honey stomach increases (Figs. 8 and 9) (Heinrich, 1975).

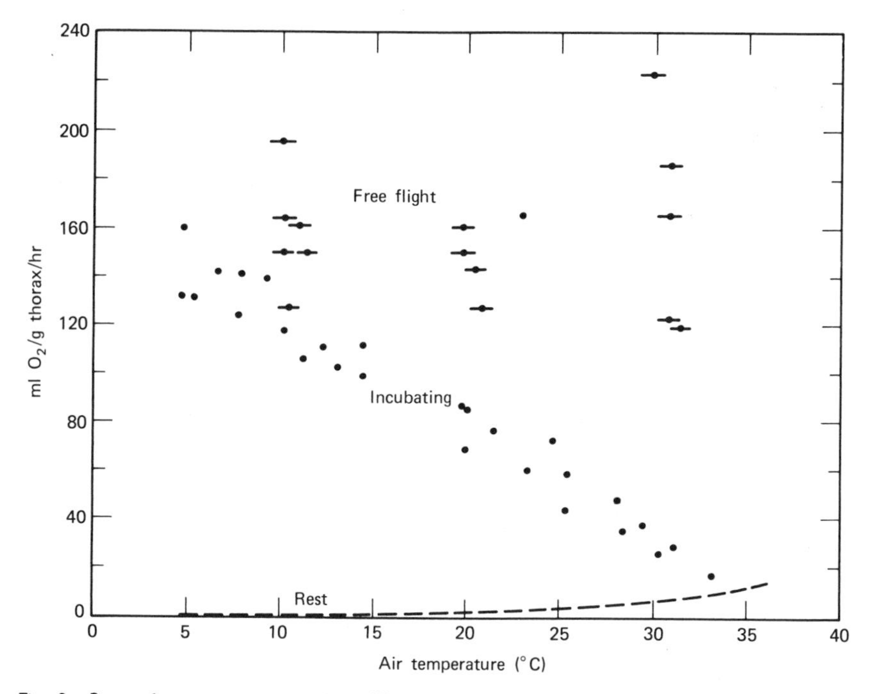

Fig. 8 Rate of oxygen consumption of B. vosnesenskii queens at different air temperatures. During continuous free flight there is no correlation between metabolic rate and air temperature. The range of oxygen consumption rates at any one air temperature is attributed to differences in the load of syrup in the honey stomach (cf. Fig. 9). In contrast, oxygen consumption of queens incubating brood increases at lower air temperatures, indicating an increased rate of heat production. Bumblebees at rest (no flight muscle activity; thoracic temperatures near ambient) had much lower metabolic rates (Heinrich, 1975; Kammer and Heinrich, 1974).

The metabolic rate and resultant thoracic temperature can in some cases be correlated with a morphological feature, wing loading. For example, within a family of moths mean thoracic temperature increases with wing loading (Dorsett, 1962; Bartholomew and Heinrich, 1973; Heinrich and Casey, 1973); smaller-winged moths have a higher wingbeat frequency and, per gram of muscle, produce heat at a greater rate. In these hawkmoths, other factors that influence body temperature, such as motor pattern, insulation, and body shape, are relatively similar. That active cooling mechanisms are not employed to regulate at a lower thoracic temperature suggests that the higher T_{Th} is adaptive.

Relating wing loading to metabolic rate in different families and orders

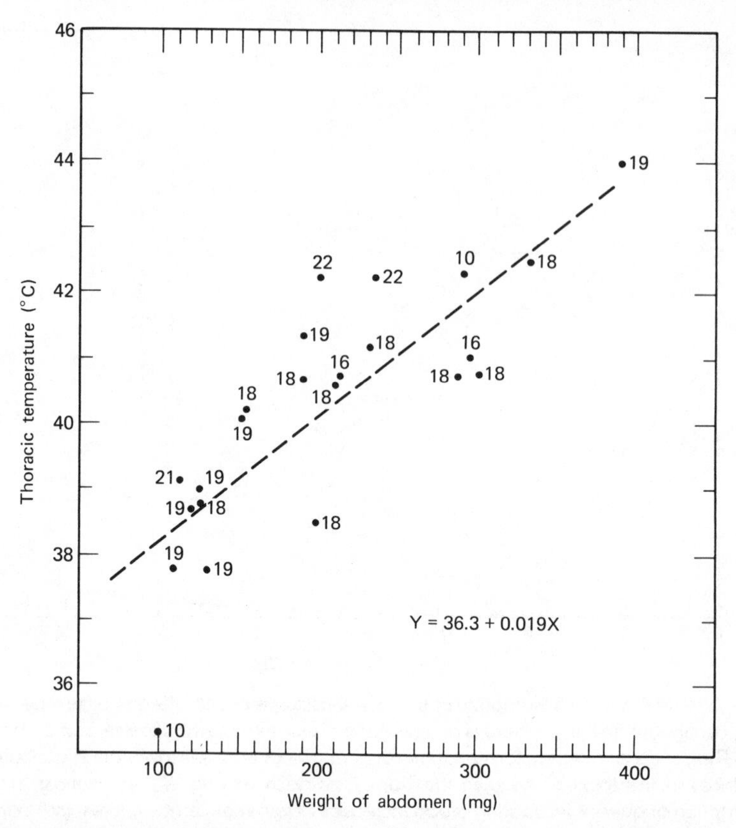

Fig. 9 Thoracic temperature of B. edwardsii queens increases with load lifted in free flight. [The weight of the abdomen reflects the weight of sugar syrup carried in the honey stomach; thoracic weights were similar (mean 0.143 g, range 0.106–0.167 g, N = 40).] The increase in T_{Th} is independent of ambient temperature (given by small numerals next to the data points) (Heinrich, 1975).

is complicated by several factors, including insulation, relative proportion of the body mass devoted to flight muscles, specialization of the flight mechanism (e.g., neurogenic and myogenic flight rhythms), and the pattern of excitation of the flight muscles. Only the latter will be considered here (for further discussion see Kammer and Heinrich, 1978). The influence of motor pattern—in this case, muscle potentials per wingbeat cycle—can be illustrated by comparing sphingid and saturniid moths. Some species of these two families have comparably high thoracic temperatures during flight (35–40°C at T_A = 7°C and 15–17°C), although they differ in wing loading [i.e., about 40 mg/cm² for saturniids versus 80–120

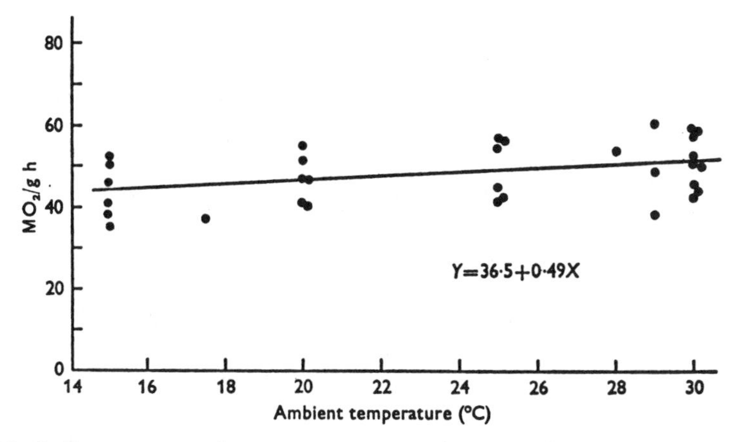

Fig. 10 In *M. sexta*, rates of oxygen consumption during free flight in a 10.2-liter jar are independent of ambient temperatures. (From B. Heinrich, *J. Exp. Biol.* 154, 1971.)

mg/cm² for sphingids (selected values from Fig. 8 in Bartholomew and Heinrich, 1973)]. These moths also differ in wingbeat frequency (saturniids, 5–12 beats/sec, versus sphingids, 24–50 beats/sec) and motor pattern (Fig. 12A and C). If oxygen consumption depends on average action potential (spike) frequency, as it does in bumblebees (Kammer and Heinrich, 1974), then a low wingbeat frequency (e.g., 8 beats/sec) and a burst of four to six impulses per cycle result in an average spike frequency (40 spikes/sec) and a rate of oxygen consumption equivalent to that of a sphingid moth (1–2 impulses/cycle, 30 beats/sec, average about 45 spikes/sec). This example suggests that morphological measurements such as wing loading provide a useful index to expected metabolic rates only when considered in conjunction with physiological specializations.

During free flight, temperature regulation does not depend on varying rates of heat production. The metabolic rate in free flight does not vary systematically with ambient temperature in *M. sexta* (Fig. 10) (Heinrich, 1971a), other sphingid moths (Heinrich and Casey, 1973), *Bombus* (Fig. 8; Heinrich, 1975), and honeybees (Heinrich, 1980). If metabolic rate were adjusted as part of temperature regulation, a higher rate of oxygen consumption would be expected at lower T_A in order to compensate for the greater rate of heat loss, and this is not observed. It is important that such measurements be made on freely flying insects, because of the effects of flight effort on metabolic rate (Heinrich, 1974a). In tethered insects wingbeat frequency may be low, wingbeat cycles omitted, or the force per wingbeat less than that required for free flight. Calculations of metabolic rate based on temperature excess and passive cooling curves are also questionable because, as discussed previously, the rate of cooling is varied physiologically. In addition, the rates of passive heat loss of a stationary and a flying insect are not identical. Because the insects cited above were measured in free flight, it can be concluded that, as Heinrich has suggested (1974a), the flight muscles are dedicated to supplying power for flight and are not simultaneously controlled to supply heat for temperature regulation.

Theoretically, it is possible that insects lacking the well-developed circulatory control and the effective abdominal radiators of large moths and bumblebees vary the rate of heat production as part of their thermoregulatory repertoire during flight. Insects can control the power output of the flight muscles by varying the excitation of the flight muscles, and it has been suggested that heat production during flight is increased by changing the flight behavior (Heath et al., 1971). However, the available data argue against this hypothesis. Nevertheless, the diversity of insects should not be forgotten.

On-off mechanisms are also possible, if an insect could alternate flight with quiescent periods at high ambient temperatures, or alternate flight and warm-up at low ambient temperatures. For example, by switching between flight and warm-up, foraging bumblebee workers remained active at air temperatures less than 10°C, although continuous flight was not possible (cf. Fig. 5) (Heinrich, 1975). Similar results were obtained with honeybees, which returned to the hive at ambient temperatures of 7°C although they could fly continuously only at T_A greater than 10°C (Heinrich, 1979a). Presumably the reduced rate of cooling by convection in stationary bees allowed them to conserve the heat produced by shivering. At high ambient temperatures insects that glide can reduce heat production by alternating flying and gliding. For example, various dragonflies spend most of their time gliding on hot, sunny days but beat their wings continuously in the early morning or evening (Corbet, 1963; May, 1976).

3.2 Preactivity Warm-Up

A marked increase in the rate of heat production independent of locomotion occurs as a precursor of in-flight temperature regulation. Large insects that regulate at a high thoracic temperature during flight—large moths, bumblebees, aeshnid dragonflies, and beetles—cannot produce enough lift to fly with body temperatures near the ambient temperatures usually encountered. They warm up prior to flight by activating the flight muscles with characteristic motor patterns (Figs. 11 and 12). Similarly, warm-up precedes singing in katydids such as *Neoconocephalus robustus*, which produces sound by rubbing its forewings at extraordinarily

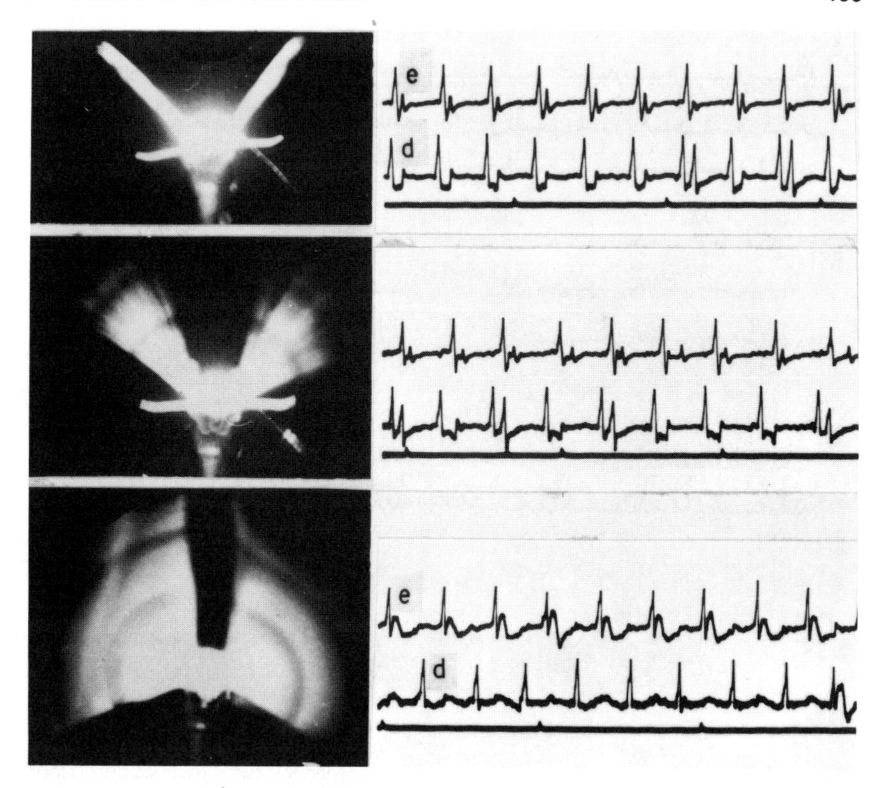

Fig. 11 Wing movements and motor patterns during warm-up and flight of a hawkmoth, *Mimas tiliae*. The moth was waxed to a support and fine wires were inserted to record the electrical activity of the flight muscles. e, Action potential from a wing elevator muscle, the tergosternal; d, potential from a wing depressor, the subalar; time mark 100 msec. Early in warm-up (top) the amplitude of wing movements was small and antagonistic muscles were excited synchronously. Later in the warm-up, that is, shortly before takeoff (middle), the amplitude of wing movements was larger and the phase relationships of motor units were altered. In flight (bottom) the wing strokes were large, and antagonistic muscles were excited alternately.

high frequencies of 150–200/sec (Heath and Josephson, 1970; Josephson and Halverson, 1971; Stevens and Josephson, 1977). The warm-up motor patterns from these various insects have in common the feature that muscles antagonistic in flight (wing elevator and wing depressor muscles) are activated at approximately the same time. Hence the isometrically contracting muscles produce heat but relatively little wing movement.

In moths and dragonflies, wing vibrations of small amplitude are apparent, and the synchrony among the motor units that drive the wings appears to be sufficient to account for the small wing vibrations in moths

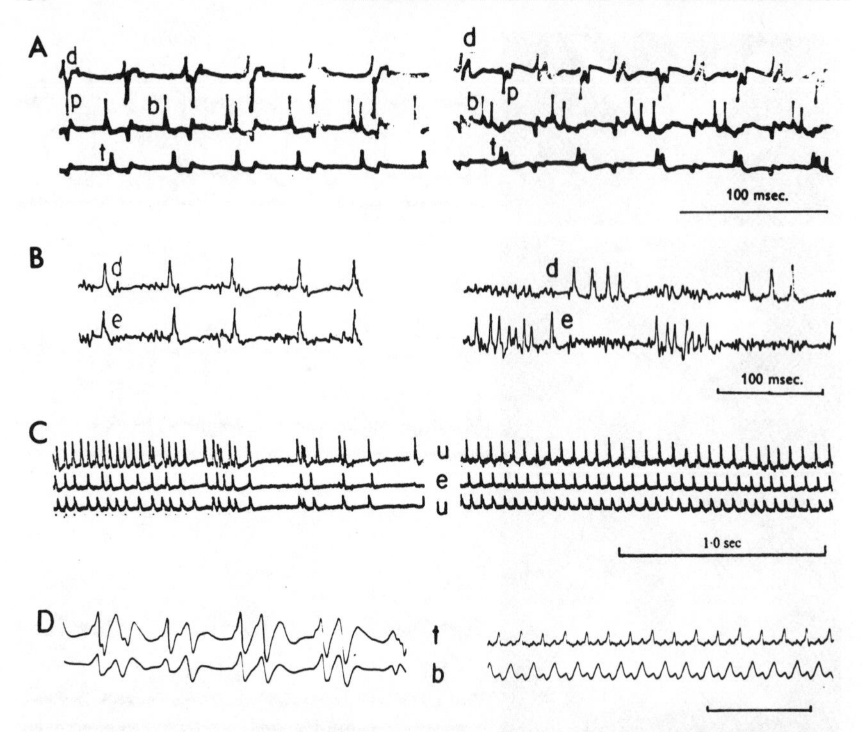

Fig. 12 Muscle potentials recorded during warm-up (left column) and flight (right, A–C) or singing (right, D), showing different motor patterns involving the same muscles during these behaviors. (A) Hawkmoth, **M. sexta.** During warm-up sets of antagonistic muscles are activated simultaneously, but some sets are not synchronous. Muscles: d, dorsal longitudinal (indirect depressor of wing); p, posterior tergocoxal (wing elevator); b, coxobasalar (direct depressor and pronator of wing); t, tergotrochanteral (wing elevator and leg depressor) (Kammer, 1970b). (B) Saturniid moth, **Samia cynthia.** In addition to the phase differences, the motor patterns differ in the number of action potentials per burst and in the time between bursts. Muscles: d, dorsal longitudinal; e, elevator. (From A. Kammer, **J. Exp. Biol.** 48, 1968.) (C) Bumblebee, **B. vosnesenskii.** The fibrillar muscles are activated simultaneously during warm-up but show no preferred phase during flight. Muscles: u, unidentified, probably dorsal longitudinal; e, left dorsoventral (elevator). (From B. Heinrich and A. Kammer, **J. Exp. Biol.** 58, 1973.) (D) Katydid, **N. robustus.** Fast, nonfibrillar muscles are excited simultaneously during warm-up; during singing the action potentials (large upward deflections) of antagonistic muscles alternate. Muscles: t, tergocoxal; b, basalar; time mark 25 msec. (From Josephson and Halverson, 1971.)

(Kammer, 1968, 1970b; Pond, 1973). Thoracic vibrations have been observed in large beetles, suggesting that the flight muscles are activated (Leston et al., 1965; Bartholomew and Casey, 1977a; Bartholomew and Heinrich, 1978), but the motor patterns have not been described. In katydids there was "little obvious movement of the thorax or wings" (Heath and Josephson, 1970), and the motor pattern consisted of syn-

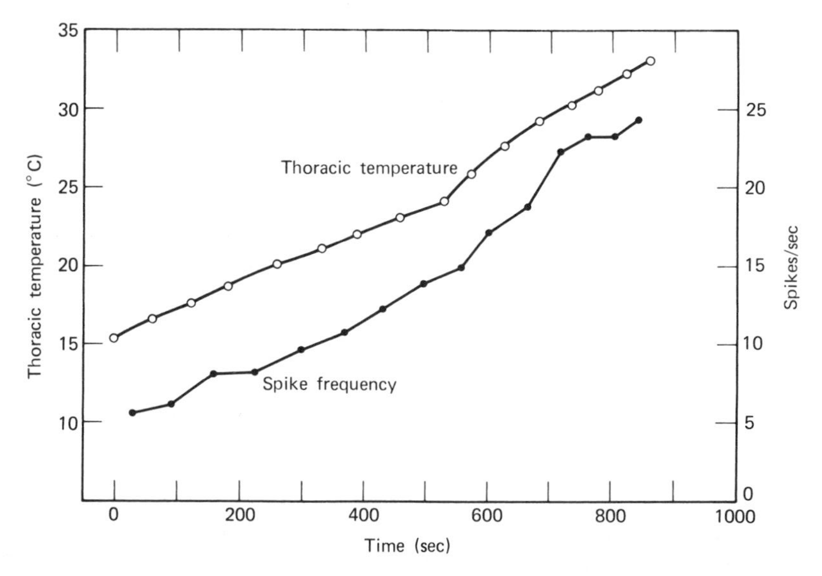

Fig. 13 Action potential (spike) frequency and thoracic temperature during warm-up in B. vosnesenskii at T_A = 11°C. (From B. Heinrich and A. Kammer, J. Exp. Biol. 58, 1973.)

chronous muscle potentials usually in bursts (Josephson and Halverson, 1971); the lack of movement could be explained by the synchrony and by the folded position of the wings. In honeybees and bumblebees, which shiver with wings folded dorsally above the abdomen, no vibrations were seen or heard (Esch, 1964; Kammer and Heinrich, 1972; Heinrich and Kammer, 1973). In syrphid flies, on the other hand, thoracic vibrations appeared and disappeared with no apparent change in the motor pattern recorded from the fibrillar muscles (Heinrich and Pantle, 1975), and only 60% of the action potentials from antagonistic muscles occurred synchronously (phase of 0.8–1.0 and 0.0–0.2; calculated from Fig. 11 in Heinrich and Pantle, 1975). For these Hymenoptera and Diptera with "asynchronous" or "fibrillar" flight muscle, a mechanism other than or in addition to synchronous contraction of antagonistic muscles must be responsible for the lack of wing movement during shivering. Perhaps the wings are mechanically uncoupled from the thorax so that they do not move (Leston et al., 1965), or perhaps the thoracic box is held sufficiently rigid by accessory muscles that the specialized fibrillar muscles cannot start the cycle of stretch and release needed for shortening.

During warm-up the rate of thoracic temperature rise is usually linear (Figs. 4B and 13). Therefore the rate of heat production must increase exponentially, both to account for the heat storage that gives the rise in T_{Th} and also to offset the increased rate of heat loss consequent to an

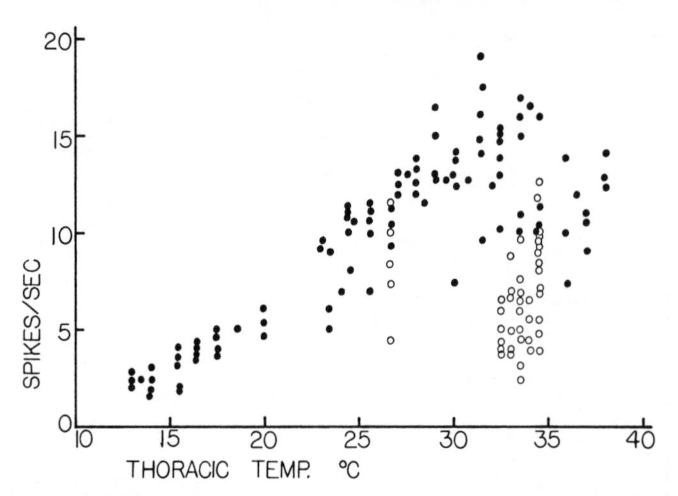

Fig. 14 Action potential (spike) frequency in relation to thoracic temperature during warm-up (●) and during the stabilization of thoracic temperature (○). Data from one bumblebee (**B. vosnesenskii**) at several ambient temperatures. (From B. Heinrich and A. Kammer, **J. Exp. Biol.** 58, 1973.)

increased difference between T_{Th} and T_A. The increased rate of heat production results primarily from the usual acceleratory effect of higher temperatures on muscle metabolism and on central nervous system activity. As a consequence of the response of the neural control system to higher temperatures, the frequency of action potentials in the muscles, hence of muscle contractions, increases as the thoracic temperature rises (Figs. 13–15) (*Manduca*, McCrea and Heath, 1971; Heinrich and Bartholomew, 1971; *Bombus*, Heinrich and Kammer, 1973). McCrea and Heath (1971) calculated for *Manduca* that the change in action potential frequency (measured as wing vibrations) could account for approximately 75% of the increased heat production. In *Hyalophora cecropia,* however, the Q_{10} for stroke frequency during warm-up from 25 to 37°C was only 1.2 (Fig. 15) (Hanegan and Heath, 1970a), whereas a Q_{10} of approximately 2 is expected. The apparent insensitivity of the neural pattern generator to temperature is unexplained, as is the mechanism by which heat production increases as thoracic temperature rises linearly. In *N. robustus* also action potential frequency varied irregularly or changed only slightly as thoracic temperature increased (Heath and Josephson, 1970; Josephson and Halverson, 1971). Further investigation of the temperature sensitivity of the central nervous system and muscles in these insects is needed.

Because the preflight warm-up only sets the stage for subsequent behavior, one would expect that as little time and energy as possible would

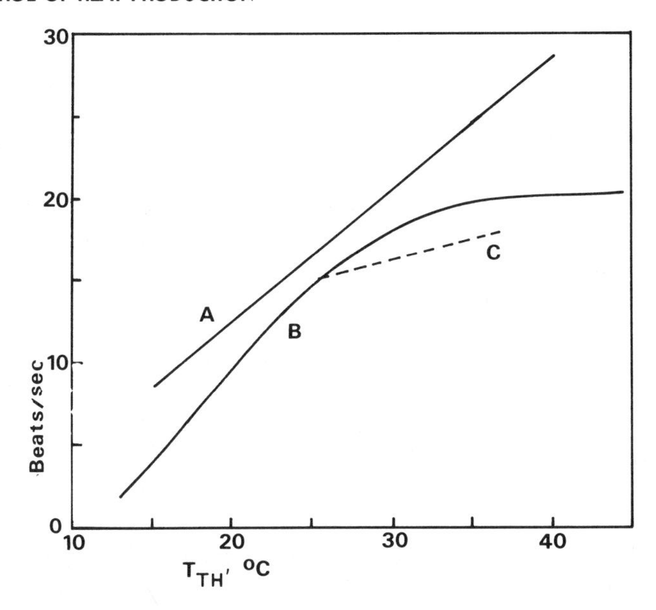

Fig. 15 Effect of thoracic temperature (T_{Th}) on frequency of wing movements. (A) *Manduca sexta*, warm-up. Regression line from Heinrich and Bartholomew (1971). (B) *M. sexta*, large-amplitude movements (as in flight) forced by tactile stimulation. Line drawn by eye through data in Heinrich and Bartholomew (1971). (C) *Hyalophora cecropia*, warm-up. Data from Hanegan and Heath (1970a).

be expended in warm-up, and thus as much as possible of the energy would go into heat storage, increasing thoracic temperature, rather than be dissipated to the environment. In accord with this expectation, abdominal temperatures remain within a few degrees of ambient while the thorax is warmed (Figs. 4B and 18) (*Manduca,* Heinrich and Bartholomew, 1971; *Neoconocephalus,* Heath and Josephson, 1970; *Bombus*, Heinrich, 1975; *Anax*, May, 1976; beetles (to some extent), Bartholomew and Casey, 1977a; Bartholomew and Heinrich, 1978), thus saving for the thorax energy that would otherwise be expended in warming the abdomen and also sparing heat loss from this poorly insulated region. One would also expect that the rate of heat production would be the highest possible at any thoracic temperature and independent of ambient temperature. In other words, one would expect that the heat-producing mechanism would be turned on completely (within limits set by the thoracic temperature) and not varied to produce heat at various rates. These expectations are for the most part fulfilled.

In various insects the warm-up rate increases directly with ambient temperature, indicating that the rate of heat production is not adjusted according to the rate of heat loss (various sphingids, Dorsett, 1962; *M. sexta*, Heinrich and Bartholomew, 1971; *H. cecropia*, Bartholomew and Casey, 1973; monarch butterfly, *D. plexippus*, Kammer, 1970a; *Bombus*, Heinrich, 1975; syrphid flies, Heinrich and Pantle, 1975; some dragonflies, May 1976). For earlier claims to the contrary, see Heath and Adams (1967), Hanegan and Heath (1970b), and McCrea and Heath (1971). A more recent exception to the conclusion that warm-up rate depends strongly on ambient temperature has been reported by May (1976): Male *Anax* dragonflies warmed up at the same rate at ambient temperatures of 15, 20, and 25°C.

Warm-up rates are not always uniformly high, and heat production is not simply switched on at a rate dependent only on thoracic temperature. In moths, the rate of warming typically decreases as the takeoff temperature is approached (Dorsett, 1962; McCrea and Heath, 1971; Heinrich and Bartholomew, 1971). In dragonflies, thoracic temperature may rise linearly until flight begins (Fig. 4B), or the rate may decline shortly before takeoff (May, 1976). Over a short period of time one individual may warm up at different rates (McCrea and Heath, 1971; Heinrich and Pantle, 1975), and different individuals warm up at different rates at a given ambient temperature. For example, in *Manduca* at T_A = 30°C, the maximum warm-up rate exceeds the minimum rate by 4.5°C/min (Heinrich and Bartholomew, 1971). The katydid *N. robustus* provides a further example (Heath and Josephson, 1970). Warm-up rates varied from 0.94°C/min to 2.20°C/min (whether these rates varied with ambient temperature, which ranged from 22.5 to 29°C, was not stated). In one individual the frequency of muscle action potentials varied irregularly and repeatedly from 0 up to 100/sec during the first 2 min of a warm-up and during the fourth minute stayed almost constant although the thoracic temperature rose 3°C. Thus the rate of heat production was not the maximal rate possible at a given T_{Th}. In *Neoconocephalus* this period of elevated thoracic temperatures in nonsinging individuals may have behavioral correlates not yet known. Some of the variability in rates of heat production may result from the laboratory situation, particularly the restraining thermocouple leads, and some may be due to physiological conditions such as age or "central excitatory state."

From these studies on warm-up, the following conclusions can be drawn. Since the warm-up rate varies with ambient temperature and with individual unknown factors, there does not appear to be a physiological set point for regulation of warm-up rate, hence there is no reason to think that the rate of heat production is regulated during warm-up. In addition, under some as yet poorly defined circumstances the rate of heat production is not the maximum possible at a given thoracic temperature.

3.3 Endothermy and Temperature Regulation during Terrestrial Activity

Terrestrial activity, in comparison with flight, does not require the *continuous* output of metabolically expensive aerodynamic work, and thus the rate of heat production can be reduced and controlled over a wide range in order to regulate body temperature. However, among insects there are only a few instances in which the benefits of temperature regulation, or even endothermy without regulation, compensate for the energy costs. Heat production depends primarily on activation of the flight muscles, which are specialized asynchronous or fibrillar muscles in insects known to execute this behavior. Walking or standing depends on the activity of different muscles which are presumably ordinary synchronous muscles. Since different groups of muscles are employed, these insects can shiver and walk concurrently.

Bumblebees brooding eggs, larvae, and pupae in the nest maintain elevated thoracic temperatures for hours (Heinrich, 1972c). In contrast with flight, heat production (measured as oxygen consumption) during brooding increases as ambient temperature decreases (Fig. 8) (Heinrich, 1974b). Although direct observations have not been made on brooding bees, heat production is probably controlled by neural excitation of the flight muscles at different frequencies. Variation in action potential frequency has been observed during stabilization of thoracic temperature in restrained bees (Figs. 14 and 16) (Heinrich and Kammer, 1973), and oxygen consumption has been well correlated with action potential frequency (Kammer and Heinrich, 1974). Heat production by activation of the flight muscles can also account for the continuous elevation of thoracic temperatures in foraging bumblebees making frequent, brief visits to scattered but nectar-rich flowers (Heinrich, 1972a,b). On certain flowers at low ambient temperatures, the thoracic temperatures of workers foraging for pollen were about 38°C, 2°C higher than the thoracic temperature of workers collecting nectar (Heinrich, 1972b). Pollen gatherers usually vibrate their wings after landing and thereby shake pollen loose from the anthers; this activity of the flight muscles necessarily produces heat rapidly.

The mechanisms of endothermy described for bumblebees are presumably also employed by honeybees during foraging and regulation of hive temperature, although for the latter additional refinements have been added (Seeley and Heinrich, this volume). Honeybees produce elevated thoracic temperatures by activation of the flight muscles (Esch, 1964; Esch and Bastian, 1968; Bastian and Esch, 1970), but the details of the motor patterns involved are not known.

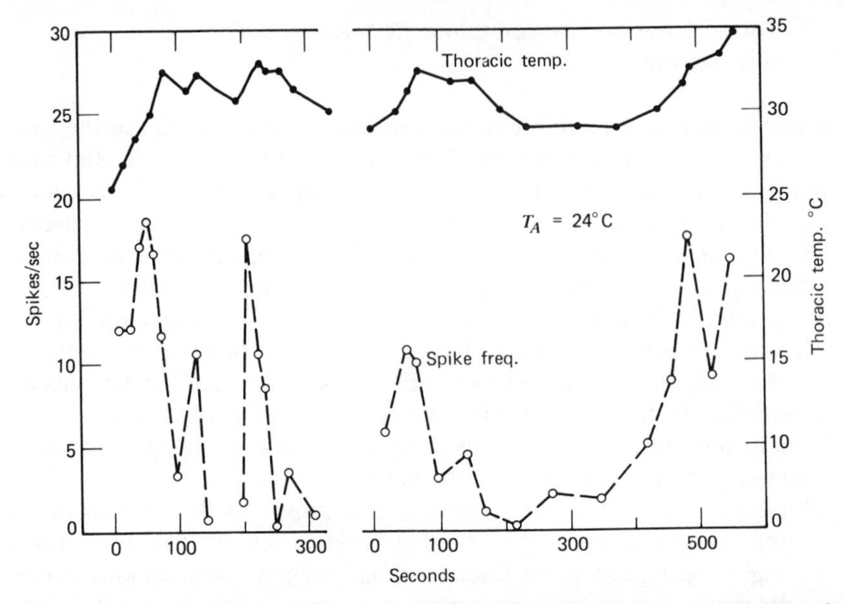

Fig. 16 Thoracic temperature and spike frequency at the end of warm-up and during the stabilization of thoracic temperature in stationary bumblebees, B. vosnesenskii. Note variations in spike frequency and corresponding changes in T_{Th}. (From B. Heinrich and A. Kammer, J. Exp. Biol., 58, 1973.)

Large beetles may have elevated metathoracic temperatures while walking or while stationary (Figs. 17 and 18) (Bartholomew and Casey, 1977a,b; Bartholomew and Heinrich, 1978). The tropical American beetles *Strategus aloeus* and *Stenodontes molarium* maintained metathoracic temperatures 5–7°C or more above ambient temperature for several hours (Bartholomew and Casey, 1977a). Thoracic temperature was not stable but oscillated over a few degrees (Fig. 18). The increased thoracic temperature and aerobic scope may provide a competitive advantage in activities that involve walking, but the function of endothermy in these beetles has not been established.

In African dung beetles, metathoracic temperatures recorded during walking were sometimes near ambient and sometimes substantially elevated (Fig. 17B) (Heinrich and Bartholomew, 1978). Metathoracic temperatures of diurnal species making or rolling dung balls were 3–7°C above ambient but (depending on the species) 2.5–13°C less than during flight. In the nocturnal *Scarabaeus laevistriatus*, which makes dung balls in a fraction of the time required by diurnal species of similar size, the mean metathoracic temperature excess during ball rolling was only 3°C

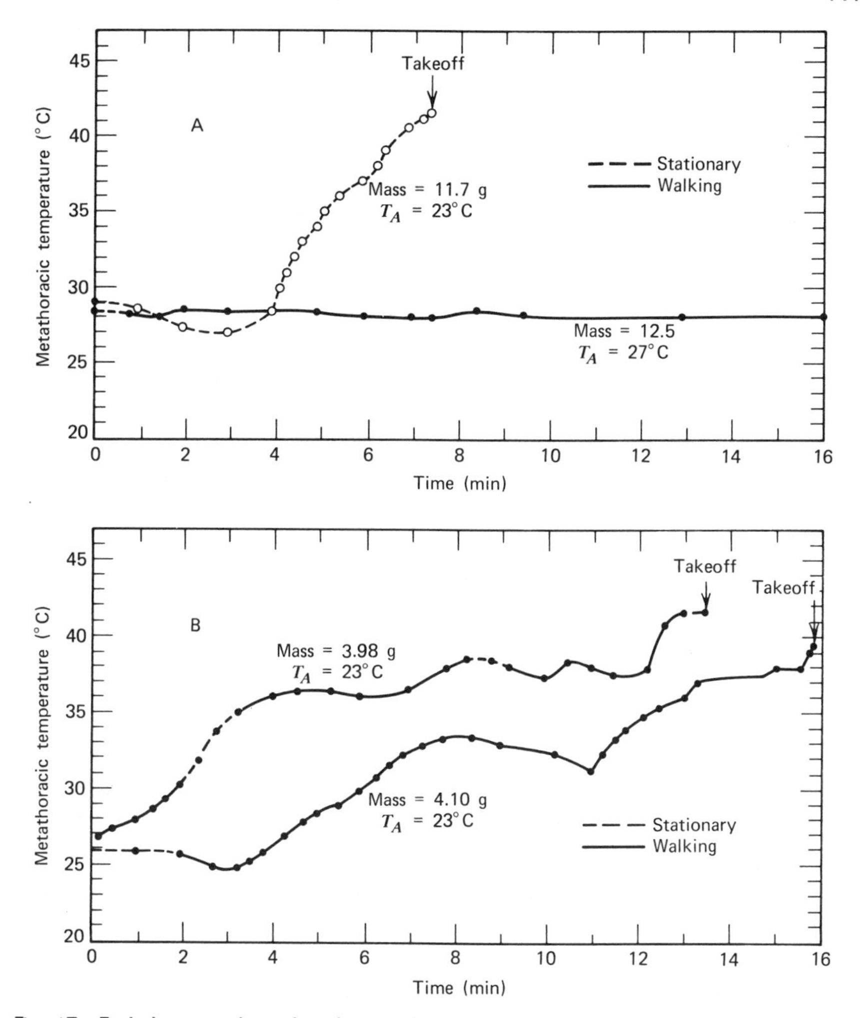

Fig. 17 Endothermy in large beetles is independent of locomotion. (A) **Heliocopris dilloni** walking with T_{Th} near T_A; in another case, a rapid warm-up while stationary. (B) Elevated but variable metathoracic temperatures during walking in **Scarabaeus laevistriatus.** (From G. H. Bartholomew and B. Heinrich, **J. Exp. Biol.** 73, 1978.)

less than the mean temperature excess during flight. Furthermore, metathoracic temperature during ball rolling in some species varied with the work done, "work" meaning pushing the dung ball rather than riding on it, or rolling the ball at greater velocities. For example, body temperatures of *S. laevistriatus* rose from 28 to 43°C as ball velocity quadrupled (Bartholomew and Heinrich, 1978). In these beetles, as in other insects,

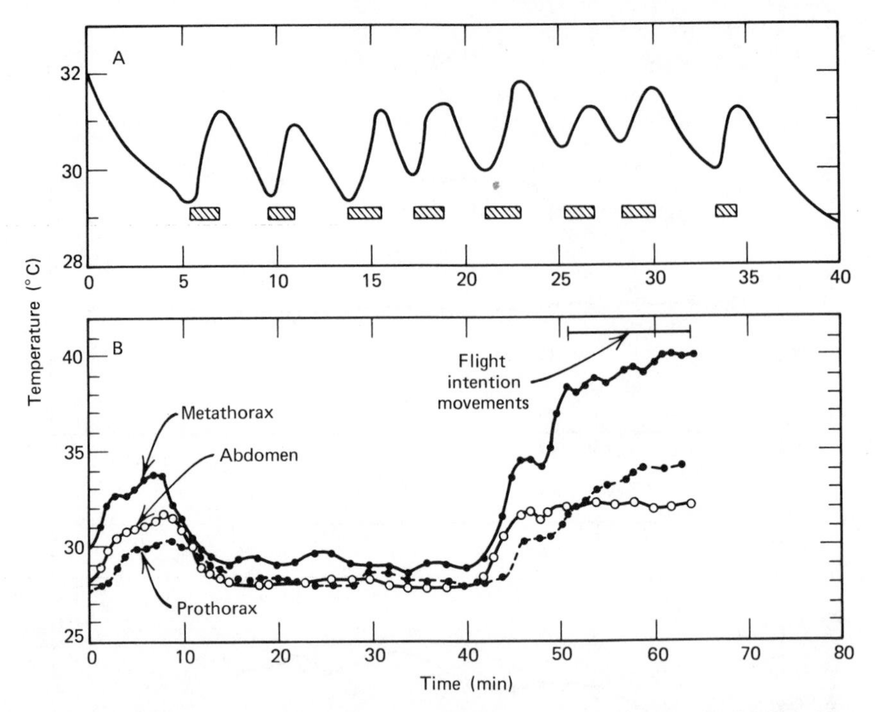

Fig. 18 Endothermy in large beetles. (A) Metathoracic temperature was elevated but variable during a 5-hr period of almost continuous terrestrial activity in the beetle **Strategus aloeus** (T_A = 23.0 – 23.5°C). During warming, abdominal respiratory pumping rates (horizontal blocks) were high. (B) Metathoracic temperature was higher than that of the prothorax or the abdomen during a period of sustained endothermy accompanied by walking and a preflight warm-up in **Stenodontes molarium** (T_A = 23°C). (From Bartholomew and Casey, 1977a.) Copyright 1977 by the American Association for the Advancement of Science.

the main source of heat appears to be the flight muscles, as suggested by three observations: (1) the temperature of the metathorax, which also provides most of the power for flight, was higher than that of the prothorax (Fig. 18*b*); (2) vibrations of the metathorax (but not of the wings or elytra) were observed; (3) elevated temperatures (15°C greater than ambient) were produced while beetles were walking, stationary, and rolling dung balls. Walking beetles may experience only a slight increase (1.0–1.5°C) in thoracic temperature (Fig. 17A), suggesting that activity in the leg muscles alone is insufficient for substantial heat production. Dung beetles such as *S. laevistriatus* compete for an essential but limited food resource. Elevated thoracic temperatures allow faster leg movements, hence more rapid ball building and rolling. Endothermy provides a selective advantage in both inter- and intraspecific competition for fresh dung (Bartholomew and Heinrich, 1978; Heinrich and Bartholomew, 1979).

4 NEURAL CONTROL MECHANISMS

In contrast to the recent advances in our understanding of the physiological mechanisms for altering heat production and heat loss in insects, relatively little progress has been made in analyzing the neural control of these mechanisms. For example, it is not known if regions of thermal sensitivity involved in temperature regulation are distributed along the ventral nerve cord or are restricted to one location. Possible roles of temperature sensors in other parts of the body also have not been investigated. It is not known if the insect brain contains a coordinating center analogous to the vertebrate hypothalamus, or if the thoracic ganglia function like both the brain and spinal cord.

4.1 Central Nervous System (CNS) Control

The meager available evidence suggests that the thoracic ganglia are temperature-sensitive and important in controlling the effectors essential for temperature regulation. In *H. cecropia*, localized heating of the thoracic ganglia with a thermode caused quiescent moths to execute flight movements (17/20 moths responded). Moths already warming up began to produce the flight motor pattern when heated (3/9), and flight activity stopped when the ganglia were cooled (4/4) (Hanegan and Heath, 1970a). In this experiment the moths usually switched between quiescence and the flight motor pattern, rather than warming up, perhaps because the ganglia were heated too fast for the full sequence to occur (responses occurred in 10–60 sec, whereas normal warm-up from 25 to 37°C took about 3 min). The results suggest that the temperature of the thoracic ganglia influences the transitions between quiescence, warm-up, and flight, but additional unknown factors are also involved. Other experiments that demonstrate the involvement of the thoracic ganglia employed electrical stimulation. Moran and Ewer (1966) found that moderate dc stimulation of the prothoracic ganglion of moths elicited warm-up movements, and stronger stimulation elicited flight movements. Stimulation of the combined meso- and metathoracic ganglia resulted in flight, regardless of stimulus intensity. Hanegan (1972) reported that ac stimulation of the thoracic ganglia of *H. cecropia* elicited warm-up unless the moth was already flying; dc stimulation elicited brief flight.

Gentle mechanical stimulation or a change in light intensity can also initiate preflight warm-up in moths and other insects, and stronger mechanical stimulation can result in flight movements although the thoracic temperature, hence the wingbeat frequency, is insufficient for free flight. These sensory inputs presumably interact with signals from a coordinating center in the brain to determine the motor output, in a system analogous to that controlling cricket song (Huber, 1960; Otto, 1971; Bentley, 1977). Thus both temperature and excitation (or release from inhibition) influence the production of warm-up or flight motor patterns by the thoracic ganglia.

In moths the thoracic ganglia (and perhaps the brain) also control the pulsations of the abdominal heart and thereby the rate of heat loss, as shown by transecting the ventral nerve cord between thorax and abdomen (*Manduca*, Heinrich, 1970b) or by localized heating of the thoracic ganglia (*H. cecropia,* Hanegan, 1973). In addition, the thoracic ganglia alone, severed from the brain and abdominal ganglia, could coordinate movements away from an artificial heat source that caused the thoracic temperature to rise to the maximum ''voluntarily'' tolerated (*Manduca*, McCrea and Heath, 1971).

Almost nothing is known about the contribution of the brain and the possible role of temperature sensors in the head, except in honeybees. Recently Heinrich (1979b, 1980), prompted in part by a question raised at the symposium from which this volume originated, measured the temperature of the head of honeybees, *Apis mellifera*. The head with a droplet of fluid serves as a heat disperser at high ambient temperatures, as discussed previously. Responses important in regulation are produced in response to head temperature, as shown by Heinrich's (1979b) experiments. Heating the head with a focused lamp resulted in extrusion of a nectar droplet, large-amplitude pulsations of the aorta, and thereby a rapid decline in head and thoracic temperatures. Heating the thorax resulted in aortic pulsations in 9 out of 21 bees, but only when head temperature, which is closely coupled to thoracic temperature, exceeded 45°C. Therefore in honeybees the brain functions as a temperature-control center, rather than the thorax. There are no data on head temperatures during free or tethered flight in other insects, nor is there information about the possible influence of head temperatures on insects that, in contrast to honeybees, regulate by dumping heat via the abdomen.

4.2 Temperature Sensitivity of the CNS

Understanding neural mechanisms used in the control of temperature is complicated by the fact that neuronal processes are temperature-sensitive. Temperature affects the membrane potentials of excitable cells and alters such properties as threshold, conduction velocity, and pacemaker activity (Sperelakis, 1970). Temperature also affects neuromuscular transmission in various animals (Jensen, 1972; White, 1976; Florey and Hoyle, 1976; Harri and Florey, 1977; Bennett et al., 1977) and presumably central synaptic transmission. In some cases, however, two opposing membrane processes may be accelerated by a rise in temperature, so that the overall behavior of the neuron appears to be independent of temperature (Burkhardt, 1959; Partridge and Conner, 1978). Unfortunately, the effects of temperature on insect neurons and CNSs have been explored only rarely. One would expect the functional capabilities of the CNS, like those of muscle (Josephson, this volume), to be influenced by and adapted to the temperatures prevalent under its working conditions.

The effects of temperature on the nervous system of insects that physiologically regulate their thoracic temperature are apparent in the increased rates of wing vibration or muscle potential frequency as thoracic temperature rises during warm-up (Dorsett, 1962; Kammer, 1968; McCrea and Heath, 1971; Henrich and Bartholomew, 1971; Heinrich and Kammer, 1973) or during forced flight (Kammer, 1970a; Heinrich and Bartholomew, 1971) (Figs. 13 and 15). This coordinated motor activity at different temperatures suggests that the neurons controlling the muscle activity have similar Q_{10}s. Although these neurons have not been investigated, several motor neurons in the metathoracic ganglia of *Schistocerca nitens* have been shown to have similar responses to temperature changes. These motor neurons exhibited a transient increase in action potential threshold followed by a steady-state decrease in spike threshold with heating (Fig. 19) (Heitler, Goodman and Rowell, 1977). The action potential frequencies of tonically active neurons also changed with temperature (Fig. 20) in a manner explicable by decreases in spike threshold at higher temperatures. However, neurons within the same ganglion may differ in their responses to temperature. In a certain clone of *Schistocerca* some motor neurons responded with a decrease in spike threshold at higher temperatures, but in other neurons the threshold increased with increasing temperature (Goodman and Heitler, 1977). Individuals of this clone were sluggish and less likely to jump than normal locusts at the same temperature. Differences in the responses of the individual central neurons to temperature changes could be part of the mechanisms by which behavior is changed in response to thoracic temperature, for example, the switch from warm-up to flight, or the movement out of warm areas when the maximum tolerated temperature is reached.

In spite of this influence of temperature on neurons and pattern-generating mechanisms, the CNS exhibits some independence of temperature. Three examples have been mentioned: (1) the variation in warm-up

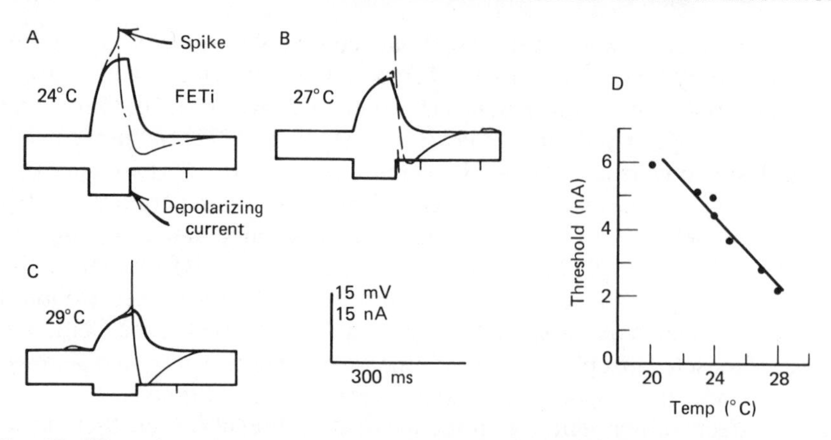

Fig. 19 Effect of temperature on insect neurons. The threshold of the fast motor neuron supplying the extensor tibialis muscle (FETi) of a locust, *Schistocerca nitens*, decreases with increasing temperature. Membrane potential (upper trace in A–C) was recorded with one microelectrode, and 100-ms current pulses (monitored in the lower trace) were injected via a second microelectrode in the soma. Threshold was measured in terms of the amount of current sufficient to elicit a spike. In each record two responses are shown, one just subthreshold, the other just above threshold. Temperature of the ganglion was changed by changing the temperature of the saline flowing over the preparation. The results are summarized in D. (From W. J. Heitler et al., *J. Comp. Physiol.*, A117, 1977.)

rates and spike frequencies during warm-up, (2) the different responses of warm-up and flight to increases in thoracic temperature (Fig. 15), and (3) changes in spike frequency during stabilization of thoracic temperature in bumblebees (Figs. 14 and 16). Presumably these different neural outputs at the same temperature reflect different neural inputs to the pattern-generating mechanism.

4.3 Interactions between Neural Control of Body Temperature and Behavior

Understanding neural mechanisms for controlling body temperature is further complicated by interactions between these mechanisms and the neural systems controlling behavior. These interactions are suggested by two kinds of observations: (1) the responses to elevated thoracic temperatures in restrained versus freely behaving insects and (2) regulation of different thoracic temperatures for different behavioral acts.

Thermal stimulation alone may not elicit regulatory responses unless the insect is executing the relevant behavior. For example, in externally heated *H. cecropia*, heart rate was not correlated with either thoracic or abdominal temperature; instead it appeared to be correlated with flight ac-

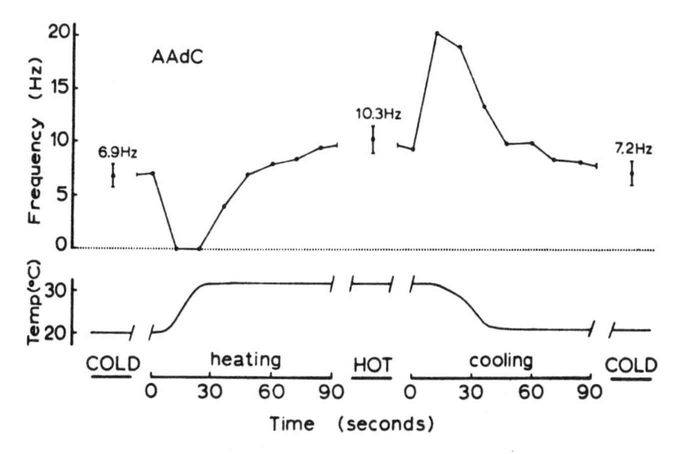

Fig. 20 Effect of temperature on the spike frequency of a tonically active motor neuron, the anterior coxal adductor motor neuron (AAdC) of a locust, S. nitens. Spike frequency increased by about 40% after a temperature increase of 12°C. Larger changes in frequency occurred during temperature changes. (From W. J. Heitler et al., J. Comp. Physiol., A117, 1977.)

tivity (Hanegan, 1973). Restrained insects exogenously heated may, however, show regulatory responses. Aeshnid dragonflies so treated transferred more heat to the abdomen at higher T_{Th} (May, 1976; Heinrich and Casey, 1978). The heart rate of restrained *Manduca* (Heinrich, 1970b, 1971b) or bumblebees (Heinrich, 1976) increased as thoracic temperatures rose. It is not clear how the neural control mechanisms active under these circumstances relate to the control mechanisms active during free flight. Bumblebee queens regulate at 39–42°C during free flight, but in exogenously heated bees the low-frequency, large-amplitude heartbeats that facilitate heat transfer to the abdomen were observed at thoracic temperatures greater than 42–44°C (Heinrich, 1976). On the other hand, in restrained *Manduca* the abdominal heart rate started to increase at a thoracic temperature of 34°C, although moths in free flight regulate at above 41°C. Possibly neural signals activating or participating in the control of the flight muscles provide excitation to the neurons controlling heart rate and amplitude. As a consequence, regulatory responses of the heart would be initiated at different thoracic temperatures during free flight and during the exogenous heating of a quiescent insect.

Interactions between neural mechanisms controlling behavior and those controlling body temperature are more clearly shown by observations that in some insects the regulated thoracic temperature is different for different activities. For example, some species of insects tolerate higher thoracic temperatures when they are flying than when they are quiescent or basking and heated by an external source (Fig. 3; McCrea and Heath, 1971; May, 1976, 1977); that is, the regulated thoracic temperature is altered along with activation of the flight-control system.

Body temperature may be regulated at different levels depending on the work load. The thoracic temperature at which sphingid moths completed warm-up and began to fly increased with wing loading. An increase of 50 mg/cm^2 was correlated with an increase of 5.75°C in thoracic temperature at takeoff (Dorsett, 1962). If and/or how the CNS measures wing loading before the moths begin to fly is unknown. Regulation at different temperatures during flight was demonstrated by studying bumblebees that had loaded their honey crops with different amounts of sugar syrup (Heinrich, 1975). Thoracic temperature rose as the load lifted increased, even at temperatures that posed no impediment to cooling (Fig. 9). For example, a bumblebee with an abdomen weighing 300 mg generated a temperature excess ($T_{Th} - T_A$) of 32°C, whereas lighter bees (abdomen 130 mg) generated a temperature excess of only 20°C. One interpretation of these observations is that the receptors monitoring flight speed and direction provide input to both the flight pattern generator, which controls lift, and to the neural mechanism controlling blood circulation which, when additional power is needed, allows thoracic temperature to rise. Observations relating body temperature to exercise have also been made on humans. Rectal temperatures increased linearly from 36.8 to 39.1°C as the work rate (measured as oxygen uptake in liters per minute) increased, and this change was independent of ambient temperature (Nielsen, 1938). [Parenthetically it may be noted that this classic study on human work physiology provoked one of the first studies on insect warm-up (Krogh and Zeuthen, 1941).] Since these changes in body temperature are independent of ambient temperature, it can be argued that they reflect not a failure of the temperature-control system but a finely tuned regulation. However, the idea of a changed set point during exercise has been challenged vigorously (Cabanac, 1975; Stitt, 1979). In Hymenoptera it seems likely that the regulated thoracic temperature is changed in accord with the aerodynamic work performed. The thoracic muscles of bumblebees function at high rates only at temperatures greater than 30°C—witness their inability to fly and sluggish walking movements at low temperatures. Some muscular efficiency presumably is gained by a higher temperature and a restricted range of optimal temperatures (Kammer and Heinrich, 1974; Heinrich, 1977). Within this range increasing temperatures increase metabolic rate and the capacity to do work. Thus the thoracic temperature is finely adjusted to the work load. The neural mechanisms are not known, but the receptors that provide feedback to help control flight may also provide input to the temperature-control system.

Temperature regulation is similarly complex in insects that regulate during different kinds of behavior. In bumblebees foraging by making brief visits to rich, distributed sources, thoracic temperature was regulated at 32–33°C, a few degrees above the minimum required for flight (Heinrich, 1972a). When the richness and viscosity of the nectar was artificially supplemented, the bees remained longer on the flowers and regulated at 34.8°C, a statistically significantly higher thoracic temperature (*Bombus vagans*, T_A = 9–24°C). Similar-sized *Bombus ternarius* had comparably high temperatures when flying to or from their nest (T_{Th} = 34.5 and 34.7°C, respectively at T_A = 11–15°C). Higher thoracic temperatures are possible; *Bombus edwardsii* queens maintained T_{Th} between 36 and 45°C during free flight at T_A = 2–35°C (Heinrich, 1975). While foraging for nectar on adjacent florets that could be reached by walking, *Bombus terricola* allowed its temperature to drop but, when foraging for pollen from the same flowers, thoracic temperature was regulated (Heinrich, 1972b). Thus thermoregulatory mechanisms are turned on and off in conjunction with specific kinds of foraging behavior. The data also suggest that thoracic temperature is regulated at different levels during different behaviors, but this interpretation needs further documentation.

Complexity of temperature control similar to that in bumblebees has also been observed in honeybees, although the data from different studies are not entirely consistent. Honeybees, *A. mellifera*, can regulate as a consequence of endothermy when not in flight. Worker bees restrained in a small cage regulated thoracic temperature at 34–36.5°C over T_A = 15–35°C, and oxygen consumption increased linearly with decreasing T_A (Cahill and Lustick, 1976). During continuous flight at ambient temperatures of 10–26°C in the laboratory, thoracic temperature was not regulated but was 14–15°C higher than ambient (Heinrich, 1979b). Abdominal temperature also varied closely in parallel with T_A. At the extreme ambient temperature of 46°C, continuous flight was possible and thoracic temperature was only about 0.5°C above ambient (Heinrich, 1979b). As discussed previously, regulation was accomplished by evaporative cooling. Under field conditions bees sometimes regulated over a wider range of ambient temperature than in the laboratory and sometimes did not. At T_A less than 25°C, the thoracic temperatures of bees arriving at a feeding station varied with ambient temperature and with distance between the station and the hive (Esch, 1960). In a different study, however, it was found that bees traveling in the opposite direction, that is, returning to a hive, had thoracic temperatures of 30°C, independent of ambient temperature (Heinrich, 1979a). Bees attacking a predator (the investigator) had higher temperatures than bees that were foraging (Table 1). Since the maximum temperature difference between thorax and environment was

Table 1 Thoracic temperatures of *Apis mellifera adonsonii* during different behaviors[a]

Behavior	T_{Th} (°C)	T_A (°C)
Attacking[b]	37.5	8–20
Leaving hive[b]	37.2	8–17
On brood	32.4	—
Foraging[b]	31.9	11–22
Returning to hive[b]	30.3	10–17

[a] Heinrich, 1979a.
[b] T_{Th} was independent of T_A, hence was regulated. (Minimum T_{Th} for flight was 27°C.)

14–15°C, regulation of thoracic temperature at low ambient temperatures (10–15°C) may be accomplished under some circumstances by interrupting flights with bouts of warm-up (Esch, 1976; Heinrich, 1979a). To regulate at the higher ambient temperature, bees under natural conditions may commence evaporative cooling at lower T_{Th} than in the laboratory. If so, then factors other than body temperature influence activation of the cooling mechanism. It may be postulated that, in honeybees and also in other insects discussed above, the neural system initiating and maintaining a behavior provides input to the neural mechanism controlling body temperature.

4.4 Components of a Temperature-Control System

In mammals, neural control of body temperature involves coordination of various autonomic and behavioral responses. Several regions of the CNS participate in this process, as shown by thermal stimulation of the spinal cord, preoptic area of the hypothalamus, and other areas of the brain stem (Satinoff, 1978; Banet et al., 1978). Recently it has been proposed that, rather than a single integrator with multiple inputs and outputs, there is a hierarchy of integrators, each with its own input and controlled response (Satinoff, 1978). These several integrators interact, inhibiting or facilitating activity at different levels in the CNS, and thereby produce the coordinated responses of the normal animal.

In contrast to this complexity, insect thermoregulatory responses may appear simpler, and there may be only one coordinating center for these regulatory responses. However, the neural control mechanism must be sufficiently complex to account for heterothermy, variations in warm-up rate, and differences in the regulated temperature associated with different behaviors and work loads. The components necessary to explain the neural control of temperature include (1) a higher integrative center that initiates a certain activity, (2) a motor-pattern generator for flight, walking, or other movements, (3) mechanisms for controlling the muscles that produce the thermoregulatory response (shivering, change in heartbeat, etc.), (4) temperature sensors (neurons involved in the previous mechanisms or additional neurons or both), (5) an integrator of information about current thoracic temperature and T_{Th} required for the behavior commanded by the higher center, (6) receptors that monitor the movements performed, and (7) receptors that respond to environmental stimuli and provide input to excite certain behaviors (escape, attack, foraging, brooding, etc.). Moving beyond this list of hypothetical components to identifying the actual neural elements represents a substantial challenge for future work.

5 DEVELOPMENT

Development of physiological mechanisms for controlling body temperature appears to parallel development of the flight muscles, as would be expected, because these mechanisms depend on heat produced by the flight muscles. On the other hand, behavioral mechanisms that do not depend on fully developed muscles can be employed by the larvae of endopterygote insects that regulate as adults, for example, sphingid moths (Casey, 1976a), or by both immature and adult exopterygote insects that regulate only behaviorally, for example, locusts (Uvarov, 1977). Whether exopterygote insects that shiver before they stridulate (the katydids discussed previously and certain crickets) also shiver when they are immature but motile nymphs remains to be investigated.

Larvae and pupae of the social Hymenoptera may have temperatures maintained above ambient because nest temperature is regulated or the brood is incubated by thermoregulating adults. The brood or pupae alone do not produce sufficient heat to raise their body temperature significantly above ambient (Ishnay, 1973; Heinrich, 1974b).

Development of flight muscle sometimes continues for a few days after the final molt into the adult stage (review, Kammer and Heinrich, 1978). Parallel improvement in ability to temperature-regulate might be expected. Young worker honeybees apparently have poorer regulatory ability than older workers (Allen, 1959); workers 1–2 days after emergence had temperatures only 1.4°C above ambient, whereas workers more than 20 days old had temperatures 12–13°C above ambient when foraging or guarding the hive entrance (Himmer, 1932). Whether these differences reflect differences in ability to produce heat or differences related to behavior, which changes with age, remains an open question.

Parallel development of flight muscles and temperature regulation dependent on endothermy is well illustrated by sphingid moths. Larvae of these insects may regulate to some extent by appropriate behavior (Casey, this volume). The flight muscles develop during the pupal stage. During the last few days of this stage both flight and warm-up motor patterns are produced (Kammer and Rheuben, 1976). Preliminary evidence suggests that the motor patterns cannot be initiated by external stimuli; cooling a pharate moth does not elicit the warm-up motor pattern. Furthermore, the motor patterns recorded from the developing flight muscles represent postsynaptic potentials that are usually below threshold for initiation of action potentials. Hence the muscles do not contract, and there is neither movement nor sufficient heat production to elevate thoracic temperature above ambient (Kammer and Kinnamon, 1979). The motor patterns are spontaneously produced by the pharate moth, for whatever developmental reason, without the energy expenditure required for muscle activity and temperature regulation.

Comparison of developing and mature physiological mechanisms for regulating body temperature provides one approach to the challenge, mentioned earlier, of understanding the neural bases of this process.

6 SUMMARY

Significant advances in our understanding of physiological mechanisms of insect temperature regulation have come in the last decade with the introduction of an emphasis on the control of heat loss. It is now clear that, in moths, bees, and beetles, stabilization of thoracic temperature during continuous flight over a range of ambient temperatures is accomplished by controlling the rate of heat loss, not the rate of heat production. This conclusion follows from several observations that metabolic rates in free flight vary with aerodynamic work and are necessarily continuously high, but metabolic rates do not vary as would be expected if heat production were altered for temperature regulation.

Maintaining a constant thoracic temperature during continuous flight therefore depends on maximizing the rate of heat loss at high T_A and minimizing heat loss at low T_A. In moths, bumblebees, and dragonflies that regulate T_{Th} at high T_A circulating hemolymph transfers excess heat from the thorax to the abdomen, from which heat is readily lost to the environment. At low T_A heat transfer to the abdomen is reduced. In *Manduca* this is accomplished by a reduction in heartbeat rate and amplitude. In *Bombus* there are changes in the activity of both the heart and ventral diaphragm such that countercurrent flow is possible between hemolymph leaving the thorax and that returning to it; the countercurrent mechanism would retain heat within the thorax. In honeybees, on the other hand, the arrangement of the aorta in the petiole suggests an efficient countercurrent mechanism for retention of heat in the thorax, but without a mechanism for altering the flow of heat to the abdomen at high T_A. Under such circumstances honeybees cool by evaporation of a droplet of regurgitated fluid held on their tongue.

During behavior other than continuous flight, the rate of heat production may be adjusted to maintain body temperature within adaptive limits. At ambient temperatures outside the range over which continuous flight is possible, flight may be intermittent and heat production may be regulated by on-off behavior. During pauses at high T_A, flight muscles are inactive and heat production is reduced, whereas at low T_A heat production is increased by shivering, that is, isometric contraction of the flight muscles.

Increased heat production occurs during preflight warm-up in several large insects and also as a prelude to singing in certain katydids. The higher rates of heat production result from increased metabolic activity of the flight muscles. Motor patterns observed during warm-up differ from those characteristic of flight, primarily in the synchronous excitation of muscles that are antagonistic in flight and also, in some species, in the frequency of action potentials. Muscle activity is not controlled to maintain a constant rate of heat production or to compensate for higher rates of heat loss as thoracic temperature rises above ambient.

Endothermy during strictly terrestrial activity not preparatory for flight has been demonstrated to date in conjunction with three kinds of behavior, the production of elevated brood or nest temperatures by bees, singing at high wingbeat frequencies by some katydids, and competition for food among African dung beetles. Since the metabolic rate of bumblebees incubating brood and of small groups of honeybees increases as ambient temperature decreases, these insects regulate heat production. The available evidence suggests that regulation of heat production depends on neural control of flight muscles, that is, on changes in action potential frequency.

The neural mechanisms coordinating the machinery that effects heat production or heat loss are poorly understood. Temperature-sensitive neurons that activate thermoregulatory responses are located in the thoracic ganglia of moths and in the head of honeybees.

Although neuronal activity, like any biological process, is influenced by temperature, temperature is not the only determinant of the neural output coordinating thermoregulatory responses. These responses may appear at the appropriate body temperature only when the insect is behaving normally, rather than restrained under experimental conditions. Furthermore, the regulated temperature may be adjusted according to the behavior; for example, bees flying with a heavy load have a higher thoracic temperature than bees flying with a light load, and the T_{Th} of attacking bees is higher than the T_{Th} of foraging bees. Thus the neural mechanisms controlling a given behavior interact with the neural mechanisms controlling body temperature.

Development of physiological mechanisms for controlling body temperature parallels development of the related behavior and the necessary machinery, particularly the flight muscles. In moths both flight and warm-up motor patterns are produced by pharate adults enclosed within the pupal cuticle, but neither movement nor increased heat production results. In honeybees thermoregulatory performance improves with age. However, relatively little is known about this aspect of insect regulatory physiology.

ACKNOWLEDGMENTS

This chapter was written while I was on sabbatical leave at the University of California, Los Angeles, and supported in part by NSF grant BNS 75-18569. I thank Dr. S. Hagiwara for the hospitality and stimulating environment of his laboratory. I also thank Dr. Bernd Heinrich for his helpful comments on the manuscript, Gloria Caffey, Sue Kinnamon, and Dr. John Kinnamon for their assistance in preparing the figures, and Sharon Edgmon and Coleen Beck for typing.

REFERENCES

Adams, P. A. and Heath, J. E. (1964a). Temperature regulation in the sphinx moth, *Celerio lineata*. *Nature* **201,** 20–22.

Adams, P. A. and Heath, J. E. (1964b). An evaporative cooling mechanism in *Pholus achemon* (Sphingidae). *J. Res. Lepid.* **3,** 69–72.

Allen, M. D. (1959). Respiration rates of worker honeybees of different ages and at different temperatures. *J. Exp. Biol.* **36,** 92–101.

Banet, M., Hensel, H. and Liebermann, H. (1978). The central control of shivering and non-shivering thermogenesis in the rat. *J. Physiol.* **283,** 569–584.

Bartholomew, G. A. and Casey, T. M. (1973). Effects of ambient temperature on warm-up in the moth *Hyalophora cecropia*. *J. Exp. Biol.* **58,** 503–507.

Bartholomew, G. A. and Casey, T. M. (1977a). Endothermy during terrestrial activity in large beetles. *Science* **195,** 882–883.

Bartholomew, G. A. and Casey, T. M. (1977b). Body temperature and oxygen consumption during rest and activity in relation to body size in some tropical beetles. *J. Therm. Biol.* **2,** 173–176.

Bartholomew, G. A. and Heinrich, B. (1973). A field study of flight temperatures in moths in relation to body weight and wing loading. *J. Exp. Biol.* **58,** 123–135.

Bartholomew, G. A. and Heinrich, B. (1978). Endothermy in African dung beetles during flight, ball making, and ball rolling. *J. Exp. Biol.* **73,** 65–83.

Bastian, J. and Esch, H. (1970). The nervous control of the indirect flight muscles of the honeybee. *Z. Vergl. Physiol.* **67,** 307–324.

Bennett, M. R., Fisher, C., Florin, T., Quine, M., and Robinson, J. (1977). The effect of calcium ions and temperature on the binomial parameters that control acetylcholine release by a nerve impulse at amphibian neuromuscular synapses. *J. Physiol.* **271,** 641–672.

Bentley, D. (1977). Control of cricket song patterns by descending interneurons. *J. Comp. Physiol.* **116,** 19–38.

Burkhardt, D. (1959). Die Erregungsvorgänge sensibler Ganglienzellen in Abhängigkeit von der Temperatur. *Biol. Zentralbl.* **78,** 22–62.

Cabanac, M. (1975). Temperature regulation. *Ann. Rev. Physiol.* **37,** 415–439.

Cahill, K. and Lustick, S. (1976). Oxygen consumption and thermoregulation in *Apis mellifera* workers and drones. *Comp. Biochem. Physiol.* **55A,** 355–357.

Casey, T. M. (1976a). Activity patterns, body temperature and thermal ecology in two desert caterpillars (Lepidoptera: Sphingidae). *Ecology*, **57,** 485–497.

Casey, T. M. (1976b). Flight energetics of sphinx moths: Heat production and heat loss in *Hyles lineatra* during free flight. *J. Exp. Biol.* **64,** 545–560.

Church, N. S. (1960a). Heat loss and the body temperature of flying insects. I. Heat loss by evaporation of water from the body. *J. Exp. Biol.* **37,** 171–185.

Church, N. S. (1960b). Heat loss and the body temperature of flying insects. II. Heat conduction within the body and its loss by radiation and convection. *J. Exp. Biol.* **37,** 186–212.

Corbet, P. S. (1963). *A Biology of Dragonflies*. Quadrangle: Chicago.

Dorsett, D. A. (1962). Preparation for flight by hawk-moths. *J. Exp. Biol.* **39,** 579–588.

Douglas, M. M. (1978). Thermal niche partitioning in the Sphingidae. *Am. Zool.* **18,** 573.

Edney, E. B. and Barrass, R. (1962). The body temperature of the tsetse fly, *Glossina morsitans* Westwood (Diptera, Muscidae). *J. Insect Physiol.* **8,** 469–481.

Esch, H. (1960). Über die Körpertemperaturen und den Wärmehaushalt von *Apis mellifica*. *Z. Vergl. Physiol.* **43,** 305–335.

Esch, H. (1964). Über den Zusammenhang zwischen Temperatur, Aktionspotentialen und Thoraxbewegungen bei der Honigbiene (*Apis mellifica* L.). *Z. Vergl. Physiol.* **48,** 547–551.

Esch, H. (1976). Body temperature and flight performance of honeybees in a servo-mechanically controlled wind tunnel. *J. Comp. Physiol.* **A109,** 265–277.

Esch, H. and Bastian, J. (1968). Mechanical and electrical activity in the indirect flight muscles of the honeybee. *Z. Vergl. Physiol.* **58,** 429–440.

Florey, E. and Hoyle, G. (1976). Effects of temperature on a nerve-muscle system of the Hawaiian ghost crab, *Ocypode ceratophthalma* (Pallas). *J. Comp. Physiol.* **A110,** 51–64.

Goodman, C. S. and Heitler, W. J. (1977). Isogenic locusts and genetic variability in the effects of temperature on neuronal threshold. *J. Comp. Physiol.* **117,** 183–207.

Hanegan, J. L. (1972). Pattern generators of the moth flight motor. *Comp. Biochem. Physiol.* **41A,** 105–113.

Hanegan, J. L. (1973). Control of heart rate in cecropia moths; response to thermal stimulation. *J. Exp. Biol.* **59,** 67–76.

Hanegan, J. L. and Heath, J. E. (1970a). Temperature dependence of the neural control of the moth flight system. *J. Exp. Biol.* **53,** 629–639.

Hanegan, J. L. and Heath, J. E. (1970b). Mechanisms for the control of body temperature in the moth, *Hyalophora cecropia. J. Exp. Biol.* **53,** 349–362.

Harri, M. and Florey, E. (1977). Effects of temperature on a neuromuscular system of the crayfish, *Astacus leptodactylus. J. Comp. Physiol.* **A117,** 47–61.

Heath, J. E. and Adams, P. A. (1965). Temperature regulation in the sphinx moth during flight. *Nature* **205,** 309–310.

Heath, J. E. and Adams, P. A. (1967). Regulation of heat production by large moths. *J. Exp. Biol.* **47,** 21–33.

Heath, J. E., Hanegan, J. L., Wilkin, P. J., and Heath, M. S. (1971). Adaptation of the thermal responses of insects. *Am. Zool.* **11,** 147–158.

Heath, J. E. and Josephson, R. K. (1970). Body temperature and singing in the katydid. *Neoconocephalus robustus* (Orthoptera, Tettigoniidae). *Biol. Bull.* **138,** 272–285.

Heinrich, B. (1970a). Thoracic temperature stabilization by blood circulation in a free-flying moth. *Science*, **168,** 580–582.

Heinrich, B. (1970b). Nervous control of the heart during thoracic temperature regulation in a sphinx moth. *Science*, **169,** 606–607.

Heinrich, B. (1971a). Temperature regulation of the sphinx moth, *Manduca sexta*. I. Flight energetics and body temperature during free and tethered flight. *J. Exp. Biol.* **54,** 141–152.

Heinrich, B. (1971b). Temperature regulation of the sphinx moth, *Manduca sexta*. II. Regulation of heat loss by control of blood circulation. *J. Exp. Biol.* **54,** 153–166.

Heinrich, B. (1972a). Temperature regulation in the bumblebee *Bombus vagans:* A field study. *Science*, **175,** 185–187.

Heinrich, B. (1972b). Energetics of temperature regulation and foraging in a bumblebee *Bombus terricola. J. Comp. Physiol.* **77,** 49–64.

Heinrich, B. (1972c). Patterns of endothermy in bumblebee queens, drones and workers. *J. Comp. Physiol.* **77,** 65–79.

Heinrich, B. (1974a). Thermoregulation in endothermic insects. *Science*, **185,** 747–756.

Heinrich, B. (1974b). Thermoregulation in bumblebees: I. Brood incubation by *Bombus vosnesenskii* queens. *J. Comp. Physiol.* **88,** 129–140.

Heinrich, B. (1975). Thermoregulation in bumblebees. II. Energetics of warm-up and free flight. *J. Comp. Physiol.* **B96,** 155–166.

Heinrich, B. (1976). Heat exchange in relation to blood flow between thorax and abdomen in bumblebees. *J. Exp. Biol.* **64,** 561–585.

Heinrich, B. (1977). Why have some animals evolved to regulate a high body temperature? *Am. Nat.* **111,** 623–640.

Heinrich, B. (1979a). Thermoregulation of African and European honeybees during foraging, attack, and hive exits and returns. *J. Exp. Biol.* **80,** 217–229.

Heinrich, B. (1979b). Keeping a cool head by honeybees, *Apis mellifera*. *Science* **205,** 1269–1271.

Heinrich, B. (1980). Mechanisms of body-temperature regulation in honeybees, *Apis mellifera*. I. Regulation of head temperature. *J. Exp. Biol.* **85,** 61–87.

Heinrich, B. and Bartholomew, G. A. (1971). An analysis of pre-flight warm-up in the sphinx moth, *Manduca sexta*. *J. Exp. Biol.* **55,** 223–239.

Heinrich, B. and Bartholomew, G. A. (1979). Roles of endothermy and size in inter- and intraspecific competition for elephant dung in an African dung beetle, *Scarabaeus laevistriatus*. *Physiol. Zool.* **52,** 484–496.

Heinrich, B. and Casey, T. M. (1973). Metabolic rate and endothermy in sphinx moths. *J. Comp. Physiol.* **82,** 195–206.

Heinrich, B. and Casey, T. M. (1978). Heat transfer in dragonflies: "Fliers" and "perchers." *J. Exp. Biol.* **74,** 17–36.

Heinrich, B. and Kammer, A. E. (1973). Activation of the fibrillar muscles in the bumblebee during warm-up, stabilization of thoracic temperature and flight. *J. Exp. Biol.* **58,** 677–688.

Heinrich, B. and Pantle, C. (1975). Thermoregulation in small flies (*Syrphus sp.*): Basking and shivering. *J. Exp. Biol.* **62,** 599–610.

Heitler, W. J., Goodman, C. S., and Rowell, C. H. F. (1977). The effects of temperature on the threshold of the identified neurons in the locust. *J. Comp. Physiol.* **A117,** 163–182.

Himmer, A. (1932). Die Temperaturverhältnisse bei den socialen Hymenopteren. *Biol. Rev.* **7,** 224–253.

Huber, F. (1960). Untersuchungen über die Funktion des Zentralnervensystems und ins besondere des Gehirnes bei der Fortbewegung und der Lauterzeugung der Grillen. *Z. Vergl. Physiol.* **44,** 60–132.

Ishay, J. (1973). Thermoregulation by social wasps: Behavior and pheromones. *Trans. N.Y. Acad. Sci.* **35,** 447–462.

Josephson, R. K. and Halverson. R. C. (1971). High frequency muscles used in sound production by a katydid. I. Organization of the motor system. *Biol. Bull.* **141,** 411–433.

Jensen, D. W. (1972). The effect of temperature on transmission at the neuromuscular junction of the sartorious muscle of *Rana pipiens*. *Comp. Biochem. Physiol.* **41A,** 685–695.

Kammer, A. E. (1968). Motor patterns during flight and warm-up in Lepidoptera. *J. Exp. Biol.* **48,** 89–109.

Kammer, A. E. (1970a). Thoracic temperature, shivering and flight in the monarch butterfly, *Danaus plexippus* (L.). *Z. Vergl. Physiol.* **68,** 334–344.

Kammer, A. E. (1970b). A comparative study of motor patterns during preflight warm-up in hawkmoths. *Z. Vergl. Physiol.* **70,** 45–56.

Kammer, A. E. (1971). Influence of acclimation temperature on the shivering behavior of the butterfly *Danaus plexippus* (L.). *Z. Vergl. Physiol.* **72,** 364–369.

Kammer, A. E. and Heinrich, B. (1972). Neural control of bumblebee fibrillar muscle during shivering. *J. Comp. Physiol.* **78,** 337–345.

Kammer, A. E. and Heinrich, B. (1974). Metabolic rates related to muscle activity in bumblebees. *J. Exp. Biol.* **61,** 219–227.

Kammer, A. E. and Heinrich, B. (1978). Insect flight metabolism. *Adv. Insect Physiol.* **13,** 133–228.

Kammer, A. E. and Kinnamon, S. C. (1979). Maturation of the flight motor pattern without movement in *Manduca sexta*. *J. Comp. Physiol.* **130,** 29–37.

Kammer, A. E. and Rheuben, M. B. (1976). Adult motor patterns produced by moth pupae during development. *J. Exp. Biol.* **65,** 65–84.

Krogh, A. and Zeuthen, E. (1941). The mechanism of flight preparation in some insects. *J. Exp. Biol.* **18,** 1–9.

Leston, D., Pringle, J. W. S., and White, D. C. S. (1965). Muscle activity during preparation for flight in a beetle. *J. Exp. Biol.* **42,** 409–414.

McCrea, M. J. and Heath, J. E. (1971). Dependence of flight on temperature regulation in the moth, *Manduca sexta*. *J. Exp. Biol.* **54,** 415–435.

May, M. L. (1976). Thermoregulation and adaptation to temperature in dragonflies. *Ecol. Monog.* **46,** 1–32.

May, M. L. (1977). Thermoregulation and reproductive activity in tropical dragonflies of the genus *Micrathyria*. *Ecology*, **58,** 787–798.

May, M. L. (1979). Insect thermoregulation. *Ann. Rev. Entomol.* **24,** 313–349.

Moran, V. C. and Ewer, D. C. (1966). Observations on certain characteristics of the flight motor of sphingid and saturniid moths. *J. Insect Physiol.* **12,** 457–463.

Nielsen, M. (1938). Die Regulation der Korpertemperatur bei Muskelarbeit. *Skan. Arch. Physiol.* **79,** 193–230.

Otto, D. (1971). Untersuchungen zur zentralnervösen Kontrolle der Lauterzeugung von Grillen. *Z. Vergl. Physiol.* **74,** 227–271.

Partridge, L. D. and Conner, J. A. (1978). A mechanism for minimizing temperature effects on repetitive firing frequency. *Am. J. Physiol.* **234,** C155–C161.

Pond, C. M. (1973). Initiation of flight and pre-flight behavior of anisopterous dragonflies *Aeschna* spp. *J. Insect Physiol.* **19,** 2225–2229.

Porter, W. P. and Gates, D. M. (1969). Thermodynamic equilibria of animals with environment. *Ecol. Monogr.* **39,** 227–244.

Satinoff, E. (1978). Neural organization and evolution of thermal regulation in mammals. *Science*, **201,** 16–22.

Sperelakis, N. (1970). Effects of temperature on membrane potentials of excitable cells. In *Physiological and Behavioral Temperature Regulation*, J. D. Hardy, A. P. Gagge, and J. A. J. Stolwijk, Eds. Thomas: Springfield, Ill.

Stevens, E. D. and Josephson, R. K. (1977). Metabolic rate and body temperature in singing katydids. *Physiol. Zool.* **50,** 31–42.

Stitt, J. T. (1979). Fever versus hyperthermia. *Fed. Proc.* **38,** 39–43.

Uvarov, B. (1977). *Grasshoppers and Locusts*, vol. 2. Centre for Overseas Pest Research: London.

White, R. (1976). Effects of high temperature and low calcium on neuromuscular transmission in frog. *J. Therm. Biol.* **1,** 227–232.

Wigglesworth, V. B. (1972). *The Principles of Insect Physiology*, 7th ed. Chapman Hall: London.

6

Regulation of Temperature in the Nests of Social Insects

THOMAS SEELEY AND BERND HEINRICH

1 INTRODUCTION

The ultimate adaptation to the physical environment is control of the environment, and this is the level of adaptation the social insects—social wasps, ants, social bees, and termites—have repeatedly achieved with respect to the thermal environment inside their nests.

Since the last comprehensive review of thermoregulation in insect societies, by Himmer (1932), there has been relatively little work that has altered the pattern of occurrence of temperature control already recognized at that time. However, there has been an impressive growth in experimental studies on the physiological and behavioral bases of thermal homeostasis in insect colonies. We here review the older natural history literature together with the more recent physiologically oriented studies.

2 THE TERMITES

Termites stand premier among soil- and wood-dwelling animals in their ability to construct nests by tunneling through and manipulating soil and organic matter. Although they present a great array of nest forms, from crude burrows in damp wood to regularly formed mounds and towers, their nests are invariably enclosed except when occasionally opened for nest expansion, foraging activities, or release of reproductives. Nest enclosure means that the microenvironment inside a termitary is more or less isolated from the vast external environment. It also means that a termite nest's internal environment can be relatively responsive to any climate control activities undertaken by its inhabitants.

The elaboration of colony thermoregulation processes in termites appears to have been paced by the evolution of nesting habits. The primitive, wood-dwelling termites in the families Kalotermitidae and Hodotermitidae rely primarily on the insulating properties of their nest substrate and on their ability to move from one nest chamber to another as local conditions demand. The more advanced termites—some genera of the Hodotermitidae, the Rhinotermitidae, and the Termitidae—exercise a wider and more precise choice of nest sites in wood or soil. They often construct elaborate and well-insulated nests, and in some cases they elevate the nest's internal temperature through the clustering of colony members.

In most termites that exhibit colony thermoregulation, temperature control rarely exceeds damping of temperature fluctuations, avoidance of excessively warm locations, and possibly some elevation of the average nest temperature. There are very few known cases of regulation at or near a specific set point. The following examples illustrate the range of sophistication found among termites in nest temperature control.

Cowles (1930) monitored the temperature at the center of a *Trinervitermes trinervoides* mound and of the surrounding air. Measurements were made just before sunrise and again between 1300 and 1400 hr, every day for 22 days. The mean nest temperatures at these times were 23.8 and 28.2°C, respectively, while those of the air were 14.1 and 34.9°C. Thus the nest temperature varied considerably within a day, though much less than that of the surrounding air. This pattern of temperature variation is probably typical of most termite nests. A more finely tuned system of colony thermoregulation occurs in the species *Nasutitermes exitiosus*. Holdaway and Gay (1948) measured hourly for 56 consecutive hours the temperature of the nursery region of an *N. exitiosus* nest mound and of the surrounding air. They found that the central nursery temperature varied between 23.8 and 27.2°C, while that of the air ranged from 1.2 to 13.5°C. Long-term observations over a year revealed that the nursery temperature remained relatively stable over short periods, as demonstrated by the records from the 56-hr period, but that it changed gradually with the seasons in a course approximately paralleling the seasonal changes in average air temperature. On a still higher level, Greaves (1964), working with the tree-dwelling *Coptotermes acinaciformis*, described what may be nest temperature regulation at a set point. Winter and summer temperatures for one colony's nest showed a fairly narrow range, from 33 to 38°C. And temperature comparisons between the nest and the center of a control tree showed differences of 18–20°C and 12–13°C for winter and summer, respectively. Thus the termites were able to compensate for lower ambient temperatures during winter.

Perhaps the ultimate in temperature homeostasis within a termite nest occurs in the nurseries of the immense, castle-shaped nests of *Macrotermes bellicosus* (formerly *M. natalensis*) (Lüscher, 1951, 1955, 1961). Continuous records from one nest for 25 days showed that the nursery temperature ranged between 28.4 and 31.5°C, averaging 29.8°C, while the air temperature outside the nest varied between 15 and 38.0°C, averaging 22.5°C (Ruelle, 1964). Year-round observations indicated that the daily average temperature in the nursery deviated by less than 0.5° from 30.0°C, and that temperature fluctuations in the nursery rarely exceeded 4°C within a day.

The range of competence in temperature control in termite nests, from crude damping of external fluctuations to precise temperature stabilization at an elevated level, is due to a wide array of mechanisms.

2.1 Nest Site Selection

Nest site selection can be a powerful first step toward nest temperature homeostasis if it places the nest in a microenvironment where temperatures are more favorable than those found in the general environment. Most termites nest partially or wholly in subterranean galleries and so receive a high degree of temperature buffering from their nest substrate. For example, while studying the thermal conditions of subterranean *Trinervitermes geminatus* nests, Josens (1971) found that, whereas the soil surface temperature varied up to 40°C over a day, the range of temperatures 40 cm underground was just 2–3°C. Similarly, Greaves (1964) found when studying *C. acinaciformis* colonies in tree trunks that a 22°C fluctuation in a tree's bark was damped to just 1°C at the tree's center (tree diameter about 65 cm). But in addition to these basic forms of habitat selection, some termites show additional, more precise, choices of nest sites.

Grassé (1949) states that in the Sahara Desert, where daily temperature oscillations can be extreme, species of *Anacanthotermes, Psammotermes,* and *Amitermes* sink their subterranean nests deeply, where exposure to temperature fluctuations are minimized and/or a high nest humidity is maximized. Another pattern in termite nest site selection in hot climates is the construction of nest mounds in the shade of trees or shrubs. Kemp (1955) found that the most prevalent termites in the dry bushland areas in Tanzania were underground nesters, and that the relatively few surface-nesting (mound-building) species, such as members of the genus *Cubitermes,* were restricted to shaded thickets. Presumably nest site choice and not differential colony mortality produced this distribution pattern. Similarly, in Australia, the desert termite *Tumulitermes tumuli* almost invariably builds its spirelike nest mounds in the shade of a tree or bush (Gay and Calaby, 1970). The large mounds of *Nasutitermes longipennis* observed by Lee and Wood (1971) in the Northern Territory were almost always built against the base of living *Eucalyptus* trees. In northern Nigeria, 75% of the *Trinervitermes oeconomus* and *T. occidentalis* mounds were shaded by trees, while 90% of the *T. geminatus* mounds were fully exposed to the sun (Sands, 1969). The strikingly different distributions for these three cooccurring species suggest that they exercise nest site selection for or against shaded trees.

However, some caution is needed in interpreting these nest shade patterns, since factors other than shade are associated with plants. *Tumulitermes tumuli, N. longipennis, T. oeconomus,* and *T. occidentalis* are all grass- and litter-feeding species and therefore probably do not live beside trees to be near a food supply, but this is always a possibility for other species. Furthermore, the true reason for associations between soil-dwelling termites and nearby trees may be quite obscure, such as the influence of soil texture. Greaves and Florence (1966) report that the subterranean gallery systems of *C. acinaciformis* are restricted to tree roots in loose, sandy soil in which tunneling is impossible. Bouillon (1970) reviews the diverse symbioses between termites and plant roots in the Ethiopian region.

In cooler climates some termites apparently select sunny sites. For instance, in Australia, in dry forests east of Adelaide and in the Canberra district, the mounds of *N. exitiosus* occur primarily on sunny, north-facing slopes with low vegetation and are rare on adjacent, shaded southerly slopes with tall trees. Aggregations of *Heterotermes ferox, Amitermes neogermanus, A. xylophagus,* and *Microcerotermes* sp. are commonly found under stones in southern Australia. All these species live primarily under stones exposed to the sun, apparently as the result of a preference for warm sites (Lee and Wood, 1971).

2.2 Nest Architecture

More than any of the other social insects, termites rely upon and have reached great heights in adapting nest architecture to the service of nest temperature control. However, the effectiveness of termites' formed nests in helping regulate nest temperature varies enormously. Representative of one extreme are the nests of *Amitermes evuncifer* and *Thoracotermes macrothorax* constructed on the soil surface and with thin, cardboardlike nest walls. Their internal temperatures closely track the temperature fluctuations in the surrounding environment. In striking contrast are the nests of species such as *Cephalotermes rectangularis* and *M. bellicosus,* which are also constructed on the soil surface but with thick earth or carton walls. Their interior temperatures show little variation, generally less than 3°C, within a day (Lüscher, 1961; Noirot, 1970).

One general complication in studying the relationship between nest form and colony temperature homeostasis is the ever-present interaction among three important properties of a termite nest's microclimate: temperature, humidity, and carbon dioxide. Usually all three are simultaneously influenced by nest design, and it is often impossible to tell, given our current knowledge of termite physiology and micrometeorology, which property is actually being responded to at excessively high or low levels—and which are simply being incidentally affected. Careful experiments in which one atmospheric property is varied while all others are held constant are needed to disentangle the stimulus–response relationships underlying termite nest-building habits.

The basic termite nest is a series of galleries in wood or soil, but in higher termites the making of aboveground mounds has been added to the termite's architectural repertoire. The adaptive significance of nest mounds seems to lie in one of three directions: (1) providing a refuge during flooding, (2) increasing nest ventilation, or (3) perhaps, as in ants (see Section 3.3), providing a warm, solar-heated nest space. For example, *Cubitermes glebae* builds subterranean nests on the dry Masai plateau but switches to mound nests up to 30 cm high in wetter parts of Tanzania. *Microcerotermes parvulus,* another African species whose range encompasses both dry and swampy grassland, likewise shows a transition from strictly underground to mound nests as it extends into wetter and wetter areas (Bouillon, 1970). The role of mounds in nest ventilation is suggested by the ontogeny of nests built by species whose colony populations can grow into the millions, such as *M. bellicosus* and *N. exitiosus.* Nests of these species start small and underground and expand upward over several years like miniature volcanoes as the colonies' populations grow. A mature colony's mound may rise to two or more meters in height.

These mounds lack the thermal protection of surrounding soil, but they exhibit architectural innovations in nest thermoregulation. The relatively high sensitivity of termite mounds to ambient temperatures is demonstrated by the variation in percentage of nest occurring as mound in wide-ranging species such as *N. exitiosus.* In northern Australia, where winters are mild, this species' nests are almost wholly aboveground, while near Canberra, where winters are more severe, about half the nest is subterranean and mounds are correspondingly smaller (Lee and Wood, 1971).

One means of increasing a mound interior's isolation from the outside thermal environment is by increasing the insulation effects of its walls. However, increased wall insulation probably also affects a colony's ventilation needs. Another factor affecting nest insulation may be the presence of an airspace between the inner nest region and the armorlike outer covering wall. Both *C. acinaciformis* (Lee and Wood, 1971) and *M. bellicosus* (Nye, 1955) employ this architectural feature. Several of the grass-storing species in Australia—*Amitermes vitiosus, Drepanotermes rubriceps, Nasutitermes triodiae, Tumulitermes hastilis,* and *T. pastinator*—fill the outer galleries of their mounds with loosely packed grass fragments, and in *Nasutitermes magnus* the harvested grass is often concentrated in a zone of galleries surrounding the central nursery chamber (Lee and Wood, 1971). Although providing a food reserve is probably the primary function of this stored grass, its orderly arrangement about the nest periphery may also be an adaptation for insulation. En-

Fig. 1 Mound structure of the nests of compass termites, **A. meridionalis**, in Australia. Their narrow ends face north and south, while their broad sides face east and west. The relatively weak morning and evening sun cannot greatly heat the nest, and during the hot midday only a narrow nest end intercepts the sun's rays. (Adapted from von Frisch, 1974.)

hanced insulation is apparently not the only modification of the nest mound for temperature homeostasis. On very hot days *N. triodiae* clear temporary openings in the roofs of their mounds (Lee and Wood, 1971). This may let hot air escape from inside their nests.

A conspicuous adaptation of mound architecture to colony thermoregulation is the shape of the "compass" or "magnetic" nests of *Amitermes meridionalis* (Fig. 1). Nests of this northern Australian species are strongly wedge-shaped with their axis of elongation in an approximately north-south direction. Grigg and Underwood (1977) measured the orientation of 248 mounds of *Amitermes laurensis,* which sometimes also constructs wedgelike nests, and found their north-south orientation was well developed with all nests aligned between 349° and 30° from true north. The function of this orientation, hypothesized by Hill (1942) and Gay and Calaby (1970), and experimentally confirmed by Grigg (1973), is minimization of heat gain under the strong midday sun by having just one narrow nest end facing the sun. Grigg's experimental study involved comparing the temperature dynamics in a nest of *A. laurensis* that was first observed in its normal orientation and then sawed off at its base and pivoted 90° to an east-west orientation where the sun, which is somewhat to the north, could strike the flat surface of the mound. In the normal north-south orientation, the core temperature plateaued each day between 1000 and 1730 hr at 33–35°C, but in the experimental east-west orientation the daily plateau in temperature disappeared. Instead, there appeared a pattern of steady rise and fall in temperature with peaks at midday between 40 and 42°C.

Full understanding of the adaptive significance of the wedge-shaped nests requires consideration of the mound's role as a refuge during flooding. *Amitermes meridionalis* generally occupies swampy floodplains, and its nests are invariably elongated in a north-south direction. On the other hand, *A. laurensis* and *A. vitiosus* inhabit both dry and swampy habitats and construct wedge-shaped, north-south mounds only in poorly drained areas (Gay and Calaby, 1970). Thus this nest form, and its consequent reduction of the mound's maximum internal temperature, appear to be special adaptations for situations in which the termites are forced into their mounds by flooding and cannot escape mound overheating by retreating underground. One nest-building behavior that may be responsible, at least in part, for the elongated nest form is responding to "hot spots" within the nest by intense building activity in the overheated regions. Stuart (1977) has observed this behavior pattern in laboratory experiments with a Panamanian termite, *Nasutitermes corniger*.

Another architectural feature in some termite's mounds that may influence nest temperature is a system of air ducts. These apparently facilitate mound ventilation and so simultaneously influence nest temperature, humidity, and carbon dioxide. In all cases it is unclear which microclimate variable(s) presents a problem for termite colonies and thereby underlies the evolution of these air ducts. Our feeling is that carbon dioxide elimination is probably the function of most of these duct systems but, as the observations described below demonstrate, the air passages can also strongly influence nest temperature and humidity.

The simplest such ventilation system is the set of chimneys—well-like depressions in a mound's surface—found in certain nest mounds, notably those of *Odontotermes* spp., *Protermes* spp., and *Ancistrotermes cavithorax*. These chimneys, which were originally thought to connect with a nest's inhabited galleries underground (Doflein, 1906; Escherich, 1911), are actually dead-end tunnels and so cannot be the site of regular air currents (Grassé, 1944; 1945). However, they probably still aid nest ventilation by facilitating air exchange by diffusion between the nest interior and the outside air. The walls separating the chimneys from the nest galleries are often thin and porous. Also supporting this functional interpretation is the observation by Grassé and Noirot (1958) that species in the genus *Protermes* that construct chimneys nest more deeply than species lacking them.

A more sophisticated system of air ducts, one that clearly involves airflow, is found in the mounds described by Weir (1973) of *Macrotermes subhyalinus* on the Serengeti Plains in Tanzania. These mounds bear two general types of openings on the surface: rimless pits near the mound base and well-rimmed funnels atop the mound. Both kinds of openings are 10 cm or more in diameter. Small mounds (0.5–1.0 m high) have 2–3 openings, while large mounds (2.0–3.0 m high) can bear 10–18 openings. These openings lead into tunnels which connect underground, forming a network of ducts which converge in a space beneath the compact nursery region of the nest. Air flows in through the basal pits and passes out the mound-top funnels. To explain this airflow, Weir suggests that air is drawn through the nest by venturi forces generated by wind blowing over the rimmed, mound-top funnels. With winds of 9–10 km/hr, Weir measured an airflow of about 1100 liters/min through one small mound. Airflows through large mounds ranged up to about 12,000 liters/min. The cooling effect of this ventilation was demonstrated by stoppering all the pits and funnels in one small mound, whereupon the temperature deep in one funnel rose to 10°C above that in nearby control mounds. But even more important may be the regulation of nest humidity provided by the air circulation. Within 48 hr of stoppering the mound, its walls were damp and soggy throughout and crumbled when one experimenter stepped on the previously firm mound.

The apex in the evolution of termite nest ventilation systems is probably the air-conditioned mound of *M. bellicosus,* whose principles of operation, first elucidated in full by Lüscher (1955, 1961), are shown in Fig. 2. The large mounds of mature colonies of this species are heavily fluted with projecting ribs down their sides. These enclose part of an air passage system, which begins above the nursery and fungus garden region as a central chimney, divide in the nest attic into a dozen or so radial canals about 10 cm in diameter and then further divides while descending through the outer ribs into smaller channels 2–3 cm in diameter. These peripheral galleries rejoin below ground into ducts opening into the nest's cellar cavity. Openings in the cellar's ceiling lead into the nursery and

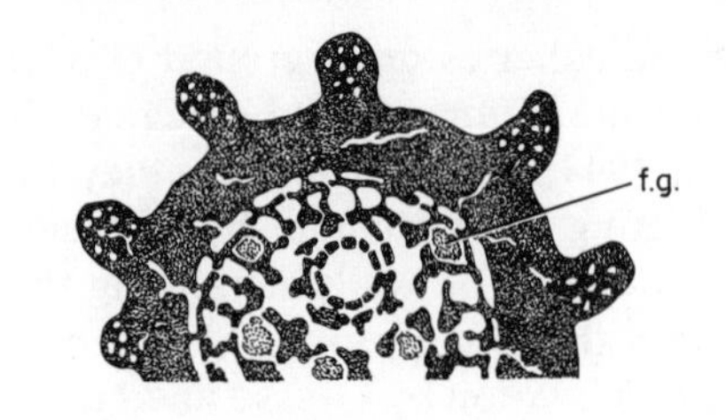

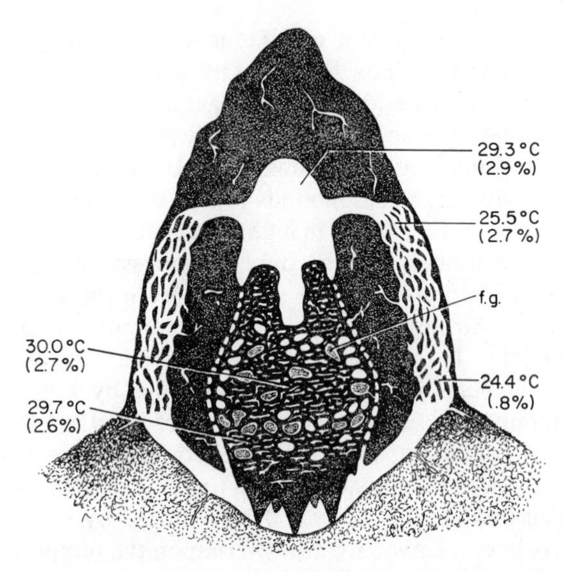

Fig. 2 Airflow and microclimate regulation in a nest of **M. bellicosus** in Africa. The horizontal (top) and vertical cross sections of a nest are shown here with its labyrinthine network of air ducts greatly simplified. Air circulation is established as the air is heated in the nest core by the colony members and their fungus gardens (f. g.). The air cools, and gas exchange takes place in the peripheral, capillary-like network of passageways in the flutings on the sides of the nest. Changes in temperature and carbon dioxide levels during circulation are shown. (Modified from Lüscher, 1961.)

thus complete the circle of air ducts around the nest. Air can circulate through this system as a thermosiphon, being heated by the million or more inhabitants and their fungus gardens and being cooled in the outer ribs. Carbon dioxide and oxygen are also exchanged through the ribs' thin walls.

The full circularity of the airways and the nature of the air currents in *M. bellicosus* nests, which actually were just inferred by Lüscher from his temperature measurements, have since been confirmed and elaborated upon by Ruelle (1964) and Loos (1964). Ruelle made a cement cast of a nest's duct system by pouring 400 kg of cement into a nest and then washing away its soil walls; the cast showed that complete circulation of air in a nest was possible. Ruelle also repeated Lüscher's temperature measurements and found a similar pattern of temperatures—for example, 29.8, 26.4, and 23.4°C in the nest nursery, attic, and cellar, respectively. His temperature measurements over a year indicated that the nursery was always the warmest nest region.

Loos (1964) inserted a periscope-like microanemometer into a nest to measure air currents directly and thus test the thermosiphon idea. On mornings, when the outside air was calm and relatively cool, the air flowed *upward* at 20–35 cm/min in the nest attic and *downward* at about 7 cm/min in the peripheral galleries and at 30–50 cm/min in the basal conduits leading to the nest cellar. But in the afternoon, with the outside air still calm but warmer, and with the nest's walls heated by the sun, the air currents inside the nest flowed *upward* in the peripheral galleries and basal conduits at 15–50 and 7–20 cm/min, respectively. The warm air may have percolated out through thin portions of the nest wall near the nest top. Thus it appears that the thermosiphon operates sometimes (when the nest's walls are cool relative to the nest's center) but not always. When the external air becomes windy, steady-state airflow disappears. Loos registered rapidly reversing air currents in the nest's outer galleries at velocities of 80–270 cm/min when the outside air gusted between 0 and 20 km/hr. All of Loos's measurements were made during the wet season when the nest's walls were being rebuilt and so were quite porous. During the dry season, when nest walls are more thoroughly sealed, steady-state airflows within the nest may prove to be the rule.

As already stated, the temperature stability achieved in *M. bellicosus* nests is truly impressive. When Ruelle (1964) continuously monitored the nursery temperature in a large nest for 25 days, he found that it averaged 29.8°C, about 5°C above the average temperature in nearby uninhabited nests, with a range of 28.4–31.5°C. Meanwhile the outside air temperature ranged between 15.0 and 38.0°C and averaged 22.5°C. The daily span of nursery temperatures reached 4°C under alternating cooling rains and solar heating but was usually about 1°C. The thermal inertia of the massive nest mound appears to protect the colony from overheating on very hot days. For example, Ebner (1926), in Sudan, measured a temperature of 27°C inside a 1-m-high nest when the outside shade temperature was 43°C. The high level of microclimate control may be the critical factor behind this species' unusually broad geographical and ecological ranges, from Chad to the Cape of Good Hope in Africa, and from the dry steppes in Nigeria, through the swampy savannahs in Gabon, to even the forest habitats in various locations (Noirot, 1958).

Whether nest cooling or carbon dioxide removal is the true function of the *M. bellicosus* ventilation system, or whether it should be viewed as a solution to two equally demanding problems of nest microclimate, is not clear from our current knowledge of this species. Based on respiration rate measurements for *Kalotermes flavicollis* and *Zootermopsis nevadensis* (about 500 μl oxygen/g·hr and a colony population estimate of 2 million inhabitants, Lüscher (1955) has calculated that a colony consumes 240 liters oxygen/day. Thus colonies apparently require high rates of respiratory gas exchange. But could *M. bellicosus* colonies achieve this exchange by simple diffusion of oxygen and carbon dioxide through solid nest walls, or is the elaborate labyrinth of air passages and associated air currents essential to the exchange?

2.3 Colony Metabolism

Besides choosing a favorable microenvironment in which to nest and further improving it by constructing a well-insulated and/or ventilated nest, a termite colony with a large population can still further enhance the microenvironment inside its nest by aggregating its members to combine their metabolic heat and thus raise the nest temperature. Also, the fungus gardens of certain termites provide them with an additional, important heat source. However, Noirot (1970) is probably correct in considering that any thermoregulatory effect of fungus gardens is incidental to their role in aiding termites in digesting lignin-cellulose materials (Sands, 1969; Martin and Martin, 1978).

Strong evidence that a colony's metabolic heat can raise its nest temperature comes from studies by Holdaway and Gay (1948) on *N. exitiosus* in Australia. Colonies of this mound-building species contain up to 2.5 million inhabitants (Holdaway et al., 1948; Gay and Wetherly, 1970). Hourly temperature readings for a nest's nursery region, nearby soil, and air, over 56 hr in late winter, showed that the nursery temperature ranged from 23.8 to 27.2° and thus was relatively high and stable compared with those of the soil (6.7–9.5°C) and air (1.2–13.5°C). They further tested the hypothesis of nest warming by the inhabitants by performing "colony-kill" experiments. Two fairly large colonies, closely matched in mound size and location, were monitored for 4 weeks before one colony was killed with white arsenic. Prior to the poisoning, the two colonies' nursery temperatures differed by 5.1°C on average, and after killing the cooler colony the temperature difference rose to 12.6–13.7°C. Thus the temperature elevation produced by the living termites was at least 7–8°C, since any heat from decaying termites would produce an underestimate of the live termites' heat production. The colony not killed contained approximately 800,000 termites.

Greaves (1964) also reported significant temperature elevations in nests of *C. acinaciformis,* an Australian termite that occupies tree trunks and produces colonies with more than a million members (Greaves, 1967). In one nest whose temperature was periodically measured throughout the year, the temperature at the nest center ranged from 33 to 38°C, which was 12–20°C above that at the center of a nearby control tree of similar size.

Turning to the fungus-growing termites, we have already noted how the nursery in *M. bellicosus* nests, which may house a million or more termites, is the warmest portion of the nest and is about 5°C warmer in live than in dead colonies (Ruelle, 1964). Lüscher (1951) demonstrated the possibility that the fungi in these nests contribute to their heating by placing about 3 liters of fungus gardens in an enclosed container and noting that its temperature rose up to 19.5°C above ambient temperature. Also, Geyer (1951) made comparative measurements over 12 months on the temperature in fungus gardens of *Termes badius* and in adjacent soil. The monthly averages for fungus gardens ranged from 12.9 to 23.8°C, and for the soil from 11.8 to 22.5°C. The fungus garden temperature tracked that of the surrounding soil but was generally 1–4°C warmer. Similar measurements for the fungus gardens of *Odontotermes badius* in India (Cheema et al., 1960) detected a mean yearly temperature elevation of 3.2°C relative to the surrounding soil.

The importance of nestmate aggregation in raising nest temperature is suggested by Josens' (1971) observation of no temperature difference between the nest and surrounding soil for *T. geminatus*. Although the two colonies studied had respectable populations, approximately 95,000 and 170,000 individuals, their inhabitants were relatively loosely dispersed throughout the nest dome and adjoining galleries. In this respect they differ strikingly from the species of *Nasutitermes, Coptotermes,* and *Macrotermes* mentioned above whose colony populations are concentrated in compact nests and can apparently warm their nests.

2.4 Mass Movements in the Nest

Large-scale migrations in nests commonly occur over diurnal and seasonal time scales as termites shift to locations offering the best available conditions of temperature, humidity, and perhaps also carbon dioxide concentration. An example of a daily movement is given by Hill (1942) for *A.*

meridionalis inside their north-south-aligned, wedge-shaped nests. In the winter, these termites show a marked daily movement to the warmest part of the nest—to the eastern side in the morning and to the western side in the afternoon. In the summer these termites tend to move vertically, retreating to the mound base or even underground on very hot days. Skaife (1955) and Coaton (1948) report similar patterns of movement for *Amitermes hastatus* and *Trinervitermes* spp., respectively, with the termites moving to the nest periphery for warmth on cool but sunny days and returning to the nest interior when the temperature becomes excessive or perhaps the humidity falls to dangerously low levels. *Reticulitermes flavipes,* subterranean termites in eastern North America, show clear seasonal migrations in their nests (Snyder, 1926). In summer they are active near the soil surface, but in winter, in areas of cold climate, they burrow below the frost line. Greaves (1967) experimentally induced retreats to the nursery in a nest of tree-nesting *C. acinaciformis* by artificially cooling the tree trunk to 4°C above and below the nursery temperature with belts of dry ice wrapped around the tree.

Sometimes these mass population movements may represent active regulation of the nest microclimate instead of a passive response by shifting to a more favorable location. For example, Holdaway and Gay (1948) observed during the summer in a nest of *N. exitiosus* that, when the ambient temperature dropped over 24 hr from 24 to 18°C, the nursery temperature rose from 29 to 34°C, apparently a reflection of termites returning to the mound. Also, for this species, it is clear that the colony members aggregate much more in the nest in winter than in summer (Holdaway et al., 1935). However, no experiments have been performed to determine the precise effect on nest temperature of this variable intensity of aggregation.

The observations by Greaves (1964) on nests of *C. acinaciformis* living in trees strongly suggest that variation in the density of aggregation by termites has important thermoregulatory consequences. Comparative measurements on a summer and a winter day, when the air temperatures were approximately 25 and 15°C, respectively, showed that summer and winter nursery temperatures were both above about 30°C. However, the temperature elevation above ambient was largely confined to the nursery in winter, whereas it extended for about 1 m above the nursery in summer. Greaves (1964) also reported his observations from a "natural experiment." After several weeks of drought during which the termites of a colony were largely confined to their nest nursery, rain fell and the nursery temperature dropped 7°C when many termites left the nursery for construction sites. Meanwhile the temperature outside the nursery remained relatively constant.

3 THE ANTS

Like termites, ants are principally soil- and wood-dwelling insects. Thus they too benefit in nest temperature control from the circumstance that, at depths below several centimeters in soil and wood, temperature fluctuations are small compared with those in the aerial environment. Unlike termites, ants thrive in cold temperature regions of the world as well as in the wet and dry tropics, but across their wide distribution they show general shifts in nest site which represent, at least in part, first-level adaptations for proper nest temperatures.

In tropical rain forests the largest number of ant species nest in small pieces of rotting wood on the ground, while true soil-dwelling species are relatively scarce (Wilson, 1959). Apparently, aboveground nest sites in the tropical environment with its relatively stable air temperatures offer favorable temperature and humidity conditions. In contrast, in the desert with its enormous temperature fluctuations, most species nest deep underground where temperature variations are strongly moderated. For example, in the Sahara Desert of Algeria a variety of formicine and myrmicine ants nest about 50 cm underground. At this depth the temperature ranges annually between just 15 and 38°C, while the surface temperatures span 0 and 70°C (Délye, 1967). In warm temperate regions, on the other hand, the shallow soil habitat appears to be nearly ideal for ants. Most species here build nests at or in the soil surface beneath the covering layer of leaf litter and humus.

In cold temperate areas, the largest number of ant species nest at sites that can be warmed by the sun. These include spaces beneath flat and shallowly set rocks, under the bark of dead logs or stumps, and inside mounds of soil. The tendency to inhabit sunlit sites is especially marked in the boreal coniferous forests, which are the northernmost limit for the distribution of ants as a whole (Wilson, 1971), and shading of such nests by growing vegetation is often lethal to the ant colony (Brian, 1956).

A further generalization about nest temperature control by ants is that the brood is moved about within the nest chambers to spots best suited for development. Usually a colony's brood is placed in the warmest chambers available. The upper temperature limit varies somewhere within 20–40°C, depending on the species (Fielde, 1904; Brian, 1973; Ceusters, 1977). Workers in *Lasius flavus* colonies, for example, begin evacuating their brood from beneath stones when the temperature approaches 28°C; members of the species *Formica fusca* delay until the brood chamber warms to about 36°C (Steiner, 1929). The fungus-growing ant *Atta cephalotes* will even reposition its fungus gardens to maintain them at about 25°C (Quinlan and Cherrett, 1978). When selecting a proper microenvironment in a nest's network of galleries and chambers, ants commonly distinguish among those appropriate for eggs and larvae, pupae, or adults. Pupal chambers are generally the warmest (Wilson, 1971). In some cases the choice of brood chamber temperature may be even more precise, as in *Formica polyctena* where no less than four distinct temperature preferenda—one for eggs and small larvae, one for medium larvae, one for large larvae, and one for pupae—are sought by nurse ants (Ceusters, 1977). Identification of the different stages of immatures apparently involves antennal contact for detection of surface pheromones (Walsh and Tschinkel, 1974; Bigley and Vinson, 1975; Brian, 1975), and perhaps tactile cues such as size and hairiness of the larvae.

The observations of Steiner (1929) on *L. flavus* provide a clear example of the hour-by-hour shifting of brood about a nest for temperature control. From 1900 to 2400 hr a colony's brood lay 10 cm deep in its small mound; from 2400 and 0730 hr, as the night cooled, at 20–30 cm; from 0730 to 1200 hr back up to the warmth at 2 cm; and from 1200 to 1330 hr, during the peak in daytime temperature, back down to 5 cm. Seasonal adjustments in nest location can be even more striking. During the summer, European wood ants (*Formica rufa*) are active high in their meter or so tall nests, but during the winter they retreat into their subterranean galleries as deep as 2 m beneath ground level (Otto, 1962).

3.1 "Fugitive" Ants

Colonies of most ant species can extend brood transport inside nests to movement of the entire colony to a new nest site when temperature or other conditions at the old site become intolerable. The degree of reliance upon this crude thermoregulatory technique varies widely among species. For most it is a last resort. In general, the more elaborate the species' nest architecture, the less mobile the colonies. Mound-building ants of the genus *Formica,* harvesting ants (*Pogonomyrmex*), and fungus-growing ants (*Atta*), for example, emigrate only when a nest site has seriously deteriorated.

At the opposite extreme are a few fugitive species such as *Monomorium pharaonis, Tapinoma erraticum, Leptothorax rugatulus, Iridomyrmex humilis,* and *Paratrechina longicornis,* which specialize in building simple nests at flimsy sites—inside a twig or acorn, under a small stone, between a few fallen leaves—and depend upon frequent colony emigrations to maintain a favorable nest environment. Captive colonies of these species in the laboratory appear "nervous." The slightest disturbance causes them to pick up brood and walk around (Wilson, 1971). In nature, nest emigration begins with scouting for a new nest site by a small fraction of a colony's workers. A successful scout ant then guides additional nestmates to a new-found site using diverse recruitment techniques, including depositing chemical trails, performing jerky invitation displays, and direct piloting of nestmates to the site via tandem running (Wilson, 1971; Abraham and Pasteels, 1977; Meudec, 1977; Möglich, 1978). Finally, when a sizable fraction of the colony has learned the location of the new nest site, the remaining workers, queen, and brood are quickly carried there by the informed workers. The entire process may take only an hour or so for fugitive species, whereas for other ants colony migration is usually a slow process requiring many hours or even a few days.

Intermediate between fugitive species and species with highly stable nests are the weaver ant *Oecophylla longinodis* and another ant with silken nests, *Polyrachis simplex*. Both species reposition their nests seasonally, apparently as a thermoregulatory measure. *Oecophylla* nests are strictly arboreal and are formed by the ants drawing together adjacent tree leaves to produce an enclosure and then lashing them in place with strands of larval silk. Weaver ant colonies located in Zanzibar (5° S), and thus near the equator, move alternately with the seasons between the northern and southern sides of their nest trees. By shifting their position just after each equinox, the colonies keep their nests well exposed to the sun (Vanderplank, 1960). Colonies of *P. simplex* along stream banks in Israel migrate seasonally between cavities in streamside cliffs which they inhabit during the winter, and vegetation along streams occupied during the summer. During the winter the cliff surfaces average in the daytime 17.5–24.0°C, but by May they average close to 30°C. Migration to the streamside vegetation enables the ants to avoid severe heating. The wintertime return to the cliffs may be to avoid floodwaters (Ofer, 1970).

3.2 Stable Nesting Ants without Mounds

For the vast majority of ant species, colony thermoregulation is achieved by nesting in a generally favorable, long-lived microhabitat and then fine-tuning the nest temperature by moving the colony short distances within the boundaries of the nest. *Camponotus acvapimensis,* for example, a soil-dwelling ant in the Ivory Coast, nests about 25 cm underground and there encounters a nearly steady and ideal nest temperature of 27 ± 2°C. At most, this ant's colonies shift up or down a few centimeters in the soil as they enter the cool and hot seasons, respectively (Levieux, 1972).

In some cases, special nest site selection behaviors may be important in ensuring that a colony occupies a favorable microhabitat. This is especially true for ants living in the cold temperate zones, which prefer to occupy sites warmed by sunlight during cool weather. Queens of the species *F. fusca* and *Myrmica ruginodus* in Britain often aggregate beneath the bark on the warm southern and eastern sides of sunlit stumps. The role of temperature in shaping this pattern was demonstrated by placing colony-founding queens of both species in a glass dish and warming one side to 30°C (ambient temperature 17°C). The queens invariably settled on the warmest side of the dish (Brian, 1952). Similarly, Pontin (1960) observed in southern Britain that queens of *Lasius niger* and *L. flavus* alighted predominantly on patches of bare, sunlit soil before shedding their wings and burrowing into the ground to start their colonies. Adjacent, 18-m^2 plots of garden, one bare earth and the other planted with *Brassica,* yielded 67 and 14 *Lasius* queens, respectively. As a further example of this predilection by ants for sunny nest sites, Benois (1972) reports that the nests of *Camponotus vagus* inside pine trees on the French Riviera are nearly absent from north-facing slopes. However, this pattern could reflect differential colony mortality as well as nest site choice by founding queens.

Perhaps the most common example of a nest site adaptation for increased nest warmth is the habit of living beneath stones lying on the soil surface. In the Swiss Alps at 1400–2200 m many species of ants (*M. ruginodis, M. rubra-ruginodis, F. fusca, F. exsecta, L. niger, L. flavus,* and others) live under flat (1- to 20-cm-thick) stones. At these elevations frosts occur regularly at night in the summer and winds are often strong, but stones provide a temporary source of heat during the day as the sun shines on them (Fig. 3). Because of their low specific heat, dry stones are particularly effective in heating up rapidly. The damp ground, on the other hand, is slow to increase in temperature because of the high heat capacity of the water and the large number of calories dissipated during water evaporation.

The ants excavate cavities directly under the flat stones, but these cavities constitute only the upper portion of their nests. Intercommunicating passageways lead downward into the ground beneath the stones where temperatures remain below the optimum for growth and development of the brood. The brood is transported to areas of the nest where the most suitable temperatures are available. Usually it is transported into the cavities directly under the stone in the late morning and back into the lower portion of the nest in the evening when the temperature under the stone declines. Steiner (1926, 1929) further examined the brood-carrying behavior of the ants by heating the stone with a flatiron or shading it from the sun; brood transport was related directly to temperature changes.

The importance of sunlight in creating favorable microhabitats for ants at central and northern European latitudes has been documented in sev-

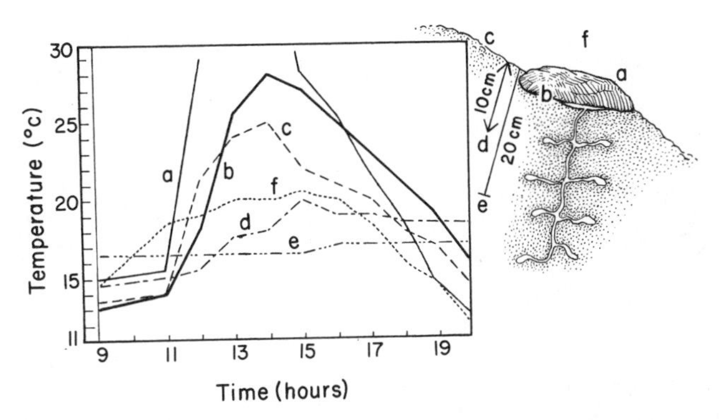

Fig. 3 Temperatures and brood transport by a colony of L. flavus nesting beneath a stone in the Alps. Temperatures shown over the daily cycle at left were taken at locations shown in diagram at right. Brood transport to space b, beneath the stone: 0900–1100 hr, no brood, few workers; 1200 hr, no brood, many workers; 1300 hr, brood transport in galleries beneath the stone; 1400 hr, 1 pupa; 1500–1600 hr, brood chamber b filled with pupae; 1830 hr, 2 pupae; 1900 hr, no brood, some workers. (Adapted from Steiner, 1929.)

eral ways. Pontin (1960) experimentally seeded founding queens of *L. flavus* and *L. niger* in glass tube nests placed at open sites and in shaded grassland and woodland locations. After one summer the broods of shaded queens had not advanced beyond the larval stage, while in comparison the broods of sunlit queens averaged over 60% pupae. Also, when Brian (1956) followed over 4 years a population of colonies within a shrinking glade in a spruce-pine plantation, he observed that the ant community dwindled from 45 nests of 4 species to 9 nests of 3 species. Meanwhile, the glade shrank from 38.2 to 20.7 m^2. The most detailed study of the effect of insolation on ant populations is that of Brian and Brian (1951), in which they compared *M. ruginodis* colonies of open and wooded habitats in western Scotland. They discovered that the latter relative to the former had brood retarded in development by about 1 month, contained less brood, possessed smaller workers, and were smaller in population. Most importantly, however, the shaded colonies were never found rearing reproductives. Thus in an ultimate, genetic sense, the shade was lethal to the colonies.

3.3 Mound-Building Ants

A more advanced means of nest temperature control has been attained by a small minority of ant species that build mounds. These mounds consist of excavated soil often mixed with organic materials—twigs, pine nee-

Fig. 4 Large mound of the wood ant **F. polyctena** in Finland. The mound is heaped against the southern side of a tree. Pine needles, bits of leaves, small twigs, and other organic debris provide thatching for the mound. (Photograph courtesy of B. Hölldobler.)

dles, leaf fragments, bits of lichen, grass stems—and are riddled with a dense network of galleries and chambers which provide living quarters for the ants. Mounds range in size from only a few centimeters tall to the impressive hills of *F. polyctena* (Fig. 4), which may rise a meter or more high and extend two or more meters across at the base. The soil beneath the mounds contains an extensive labyrinth of galleries and thus provides a refuge for the ants when mound temperatures become too extreme. Mound-building species are found in the myrmicine genera *Atta, Acromyrmex, Myrmica, Pogonomyrmex,* and *Solenopsis* in the tropics and warm temperate regions of the world, in the dolichoderine genus *Iridomyrmex* in Australia, and in the formicine genera *Formica* and *Lasius* in Europe, Asia, and North America (Wilson, 1971).

True mounds occur in a wide variety of habitats, but they are most frequently encountered in habitats subject to extremes of temperature, such as stream banks, temperate zone woodlands, and deserts. A second indicator of the mound's role in nest thermoregulation is their predominance in sunny clearings and along forest margins with southern and eastern exposures (Sudd et al., 1977; Dreyer, 1942). A survey of *Formica* mounds in four forest areas in Germany gave values of 2.4, 37.1, 28.6, and 5.9 mounds/10 km of forest edge with northern, eastern, southern, and western exposures, respectively (Wellenstein, 1967). In a similar pattern, Elton (1932) observed that *F. rufa* in Norway heaped its nests only against the southern sides of trees.

As long as 170 years ago Pierre Huber (1810) suggested that the primary function of mounds was to act as solar collectors to help provide warmth during cool weather. This hypothesis, later crystallized by Forel in his "Théorie des Dômes" (1874), has since been tested by a long line of European, American, and Japanese investigators. The most detailed analyses are those of Andrews (1927) on *F. exsectoides*; Wellenstein (1928) and Zahn (1958) on *F. rufa*; Steiner (1924, 1926, 1929) on members of the *F. rufa* group of species, *F. exsecta* and *Lasius* spp. (see Fig. 5); Kâto (1939) on *F. truncorum*; Raignier (1948) and Kneitz (1964) on *F. polyctena*; and Scherba (1962) on *F. ulkei*.

All these studies have confirmed that mound structures provide higher and often more stable temperatures from spring to fall than the adjacent soil or surrounding air. For example, Raignier's (1948) temperature measurements from 10 mounds of *F. polyctena* in Belgium spaced over May 17–August 26, with readings taken in all kinds of weather but usually between the hours of 0730 and 1600, yielded a mean, overall range, and mean daily range for temperatures in the mounds (30 cm deep) of 26.9, 18–33, and 3.1°C, respectively. The same measurements for the air were 19.7, 7.5–26.5, and 5.0°C. The soil temperature at a 10-cm depth averaged just 13.3°C. However, the monthly average temperatures for the mound at a 30-cm depth and the soil at a 10-cm depth differed strongly (by more than 12°C) only during May, June, July, and August. In the winter months of November, December, January, and February there was no significant difference, and the spring and fall months of March and April and September and October were times of transition with the average temperatures of the mound and soil differing by less than 8°C. Scherba's (1962) measurements on 20 colonies of *F. ulkei* in Illinois revealed a similar pattern of annual mound and soil temperature differences.

The precise pattern of mound heating in sunshine and cooling at sundown is of course highly variable, for it depends upon many independent factors, such as a mound's insulation, its sun exposure, and the wind speed and air temperature. But what is probably a typical pattern of temperature dynamics for a large mound of *F. rufa* or *F. polyctena* was recorded by Raignier (1948). Under steady solar illumination, the following temperature rises were recorded between 0730 and 1130: air, 17–21.7°C; mound surface, 22.3–54°C; and mound at a depth of 30 cm, 29.0–32.2°C. For the rest of the daylight hours the sky was overcast, and

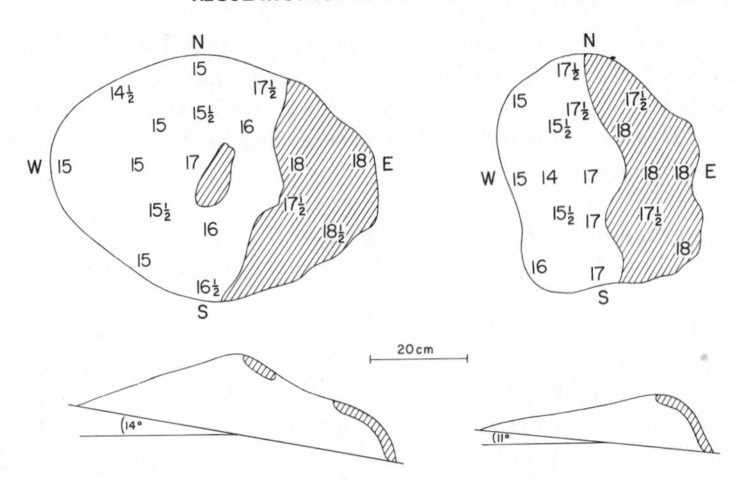

Fig. 5 Two nest mounds of L. flavus, showing temperatures (°C) at a depth of 2 cm at 1000–1130 hr. The lined portions of the top profiles and cross sections below them indicate the regions where the brood was located at the time the measurements were made. Air temperature = 13–14°C. (Adapted from Steiner, 1929.)

the mound's surface temperature dropped in 2 hr to about 25°C while those of the air and mound interior (30 cm) remained steady at about 22.5 and 31.5°C, respectively. Finally, during the night the mound interior slowly dropped to about 29°C. Thus it took the mound 20 hr to lose by slow conduction and radiation the heat it had taken just 4 hr to accumulate under intense heating by the sun.

Such extensive heat storage is probably limited to large mounds of mature colonies of species like *F. polyctena* and *F. rufa*. The small mounds, often just 10–20 cm high, of young colonies of these species, or of mature colonies of *L. niger, L. flavus* (Fig. 5), and *F. fusca* also show considerable warming by the sun. But because they have little mass for heat storage and are less thoroughly insulated with layers of organic matter, they cool down rapidly when they are no longer heated by the sun's rays. The daytime temperature averages for points 2 cm deep in an 8-cm-tall *F. fusca* mound and in the adjacent soil were 24.02 and 20.15°C, respectively, on one sunny day. The nighttime averages were reversed in order, however, 12.25 and 14.25°C, respectively (Steiner, 1929). Thus these ants must shift their brood between soil and mound on a daily basis. But through the use of the mound, the ants can rear their brood in an environment that is warmer on the average than that provided by the underground chambers alone.

The principal *raison d'être* of mounds appears to be their ability to be warmed more rapidly by the sun than simpler subterranean nests. The precise details of this property are still not clear, since controlled experimentation with such variables of ant mounds as their shape, composition, and internal structure remain to be performed. But at least in part their temperature response to solar illumination must be a function of their shape. Steiner (1929) has calculated how much more sunlight a hemispherical mound at the latitude of Bern, Switzerland, intercepts than does the circular ground surface it covers. The difference is especially large when the sun is low in the sky (and the air is cool), as occurs in the spring and fall, and in early mornings and late afternoons. For example, the ratio of illumination for a hemisphere relative to a horizontal circle with the same diameter is 3.33 and 1.23 at 07.00 hr and noon, respectively, on March 21, but only 1.05 at noon on June 21.

Other features of raised mounds probably also contribute to their thermoregulatory character. The outer layer of organic material, or thatching, found on many but not all species' mounds, may help insulate the mound by creating trapped air spaces. Also, the mounds of most species have faces (Figs. 4 and 5) sloping to the south and southeast, which increase the amount of sunlight intercepted still more (Forel, 1874; Andrews, 1927; Linder, 1908; Scherba, 1958; Hubbard and Cunningham, 1977). A mound's inhabitants often apparently foster its sun exposure by "weeding" vegetation off the mound's south-facing surfaces (Dreyer, 1942; Linder, 1908). These surfaces often consist of loose, sandy, sparsely vegetated soil but are invaded by vegetation when the colony inside dies. Still another means of fostering mound exposure, for which there is both descriptive and experimental evidence, is by increasing the vertical elongation of mounds. For example, shaded mounds of *F. rufa* and *F. polyctena* are usually taller than those that are highly exposed (Raignier, 1948). In laboratory experiments Lange (1959) found that decreasing the intensity of lamp illumination on nests of these two species induced the ants to rebuild their mounds in fairly pointed forms.

That the sun's illumination of mounds is important to the ants inside is suggested by the repeated observation of increased shading of mounds in habitats undergoing succession being accompanied by extinction of the ants (Waloff and Blacklith, 1962; Pontin, 1963). As one example, Pontin (1963) followed for 5 years the mound-building colonies of *L. flavus* in a grassland in Berkshire. The major vegetational change over the observation period was the spread of *Brachypodium pinnatum,* a grass that makes very dense stands and completely shades the underlying soil. Where the grass completely surrounded *L. flavus* colonies they died out, probably because of shading.

Another possible aspect of the relationship between ant mounds and sunlight is the *Wärmeträgertheorie* of Zahn (1958). In the spring, a common sight on the surface of *F. rufa* and *F. polyctena* mounds is hundreds of ants sitting where the sun's rays strike the mound. Zahn observed that these individuals often basked in the sun for 15–45 min and then ran inside the nest. He suggests that these individuals help to warm the mound interior by becoming heated outside and then dashing inside while still warm and radiating heat to the mound interior. However, the amount of heat flux generated by the movements of these workers is still unclear.

Besides the sun, the metabolic heat of the colony may provide significant heat in large colonies of *F. rufa* and *F. polyctena,* whose populations may attain 100,000 or more members (Yung, 1900). The evidence for this comes from colony-kill experiments similar to the one described above (see Section 2.3) for termites. Raignier (1948) monitored the temperatures on the surface and at various depths inside two similar-sized (no dimensions given) colonies of *F. polyctena* 30 m apart for 1 month and found no significant temperature differences. Then one colony was killed on August 9 by pouring 3.5 liters of carbon disulfide over its mound. During the second half of August, the average temperatures at 15- and 30-cm depths in the dead mound were lower than the comparable average temperatures in the live mound by 7 and 9°, respectively. Moreover, the surface temperature of the dead nest averaged 3°C *above* that of the live nest. Thus either the metabolic heat of the living ants, or some other effect of living ants, such as their maintenance of mound insulation or their regulation of openings on the mound, causes live mounds of this species to be warmer than dead mounds. Steiner (1924) has reported that workers of *F. rufa* colonies create openings on the mound surface when it is warm and close them at night, but the effect of this adjustment on mound temperature is not clear.

Another observation suggesting significant metabolic heat production in these mounds is the repeated (Steiner, 1924; Raignier, 1948) observation of increases of about 3°C during the night at depths of 15–30 cm in mounds of *F. rufa* and *F. polyctena*. For example, one night between 19.00 and 22.00 hr Raignier (1948) detected a temperature rise from 25.6 to 27.4°C at a point 15 cm beneath the apex of a *F. polyctena* mound. Raignier simultaneously monitored the rates of ant departure and return to this nest along a foraging trail and found no marked retreat to the nest associated with the temperature rise. However, even without increasing their number, the ants might create local temperature elevations by clustering together. Kneitz (1964) measured the metabolic rate of *F. polyctena* workers at 20°C to be about 700 μl oxygen/g·hr. Given 1 million inhabitants in a nest (probably a generous estimate of a colony's population), a respiratory quotient of 1.00, and a nest atmosphere with about 95% RH,

Kneitz calculates that a colony generates heat at a rate of 4.3 kcal/hr. Without good estimates of the heat capacity of the mound material, mound insulation, and ant dispersion inside the mound, it is presently impossible to know how important this amount of metabolic heat is for nest warming. Perhaps the simplest way to determine this is by monitoring the effects on mound temperature of artificial heat sources of various strengths and spatial arrangements implanted in abandoned mounds.

There is no solid evidence supporting Wasmann's (1915) hypothesis that decomposition of organic materials provides a third heat source in mounds (Kneitz, 1964). However, this absence of evidence may merely reflect the difficulty of distinguishing heat from ant metabolism from that of mound decomposition. Even finding elevated temperatures or carbon dioxide levels inside shaded, uninhabited mounds would only suggest that decomposition occurs in occupied mounds. Living ants can inhibit the growth of fungi and microorganisms in their nests with the acid secretion of their metapleural glands (Maschwitz et al., 1970). And Raignier (1948) observed that the pH of organic materials dropped from about 8 to about 5 when they became incorporated into *F. polyctena* nests.

3.4 Army Ants

The surface-dwelling army ants of the genus *Eciton,* whose colonies can contain hundreds of thousands of workers, present the clearest case among ants where metabolic heat significantly affects colony temperature. They are also distinctive among ants in their nesting habits. Each nest, or bivouac, consists of a colony's hundreds of thousands of workers hooked together by their tarsal claws into a single mass. This mass is suspended from a log or tree root near the rain forest floor. Colonies of *Eciton burchelli* and *E. hamatum,* the two species whose colony thermoregulation properties have been studied (Schneirla et al., 1954; Jackson, 1957), continuously alternate between a nomadic phase roughly 15 days long and an approximately 20-day-long statary phase. During the nomadic phase a colony's brood is all in the larval stage and the ants change their nest site nightly following the completion of each day's swarm raid. During the statary phase a colony remains at the same bivouac site while the larval brood undergoes metamorphosis to the adult form and new eggs are laid.

The temperature inside the massed ant cluster with its eggs, larvae, and pupae is consistently 2–5°C higher than in the surrounding air and much more stable (Fig. 6). Temperature elevation and high stability inside the ant mass is in part achieved by the trapping of metabolic heat within the air spaces created by the intertwined bodies of the workers.

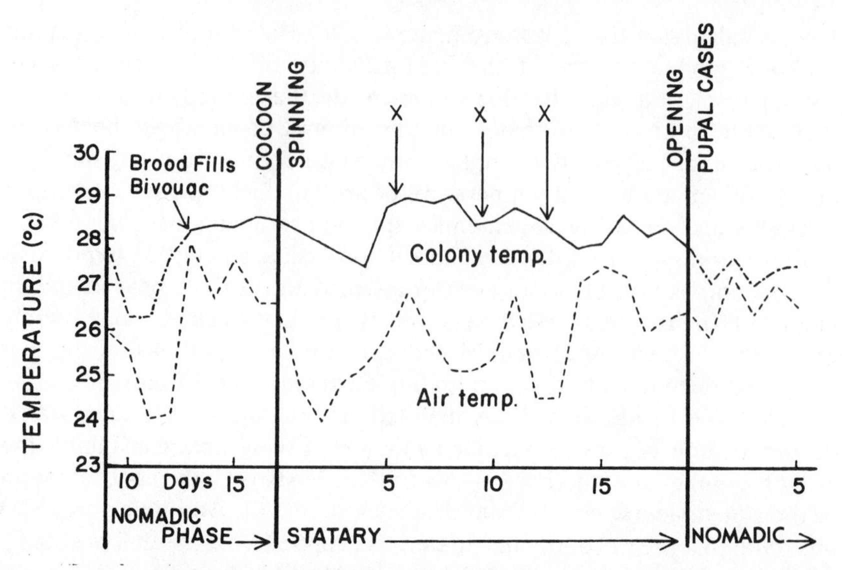

Fig. 6 Temperatures maintained in the clusters of army ants, **E. hamatum**, in a Central American rain forest during different stages of the colony cycle. The colony temperatures refer only to the clusters, or so-called bivouacs, of hundreds of thousands of workers entwined together while suspended from a log or from tree roots. x, Bivouac moved to a different location; solid line, brood present; dashed line, brood absent in portion of cluster measured. (From Schneirla et al., 1954.)

The settling of bivouacs probably does not involve deliberate site selection. Instead, it seems to follow from other characteristics of raiding behavior. For example, bivouacs are established along daytime raid trails which tend to be within cover, since raiding ants avoid sunlit areas when advancing. Bivouacs also tend to become established at division points along the raid trails, where food gathered during the raid is deposited, and these too are set up in spots not disturbingly bright, hot, or dry. Finally, bivouacs form where hanging strands of workers can readily form, such as spaces beneath logs, vine masses, and roots. When establishing statary bivouacs, the workers appear unusually sensitive to air movements, sun flecks, and other disturbances, and so only quite sheltered sites come to serve as nesting sites during the statary phase.

4 THE SOCIAL WASPS

Current knowledge of thermoregulation in wasp societies is based upon studies with just two kinds of social wasps, the socially primitive paper wasps (genus *Polistes*) and the more socially advanced yellow jackets and hornets (subfamily Vespinae). Wasps of the former group are characterized by small colonies, usually with fewer than 30 adult members, living upon a single, naked comb constructed from paper and suspended from an aerial support. Yellow jackets and hornets form much larger colonies containing up to several thousand individuals. Their nests are also built of paper but contain several tiers of combs enclosed within a multilayered jacket. These nests may be found in tree cavities (*Vespa crabro*), in underground holes (*Paravespula vulgaris, P. germanica,* and *V. orientalis*), or aboveground suspended from tree branches (*Dolichovespula arenaria, D. maculata*).

4.1 Polistes Paper Wasps

Polistes wasps are primarily tropical in distribution, but many species also occur in the temperate regions. The thermoregulatory strategy of *Polistes* wasp colonies consists primarily of inhabiting microenvironments that are warm and sheltered and cooling the nest when overheating threatens. Weyrauch (1928) and Steiner (1930) report finding *Polistes gallicus* and *P. biglumis* in Europe along southern and eastern sides of barns, chalets, and walls; north- and west-facing surfaces are apparently avoided. Because their paper nests are built of dark-grey carton, they warm rapidly when illuminated by the sun, and they are highly susceptible to overheating.

How do wasps prevent overheating of the nest? First, they begin fanning when the nest temperature reaches 31–36°C. If the nest temperature rises to 34–37.5°C, they begin transporting water, applying it as droplets inside cells, atop capped pupal cells, and on the back of the nest comb. This, coupled with fanning, results in evaporative cooling. Usually the queen performs over half the water collection flights, even when she may have up to about 20 other wasps with her on the nest (Steiner, 1932). The other wasps, however, participate actively in fanning. A single water collection flight may take only about 20–30 sec, as wasps gather water from nearby sources. When a queen is alone during the early stages of colony growth and nest overheating threatens, she may devote herself exclusively to nest cooling, making eight or more water collection flights every 15 min. Water distribution and bouts of fanning fill the intervals between collection flights. Although adult wasps can receive fluid secretions from their larval brood (Morimoto, 1960), there is no report that they use the brood as a quick source of cooling fluid as is done in some yellow jacket and hornet societies (see below).

Steiner (1930) demonstrated the efficacy of nest-cooling behavior in two ways. First, he experimentally caged a solitary queen wasp who was

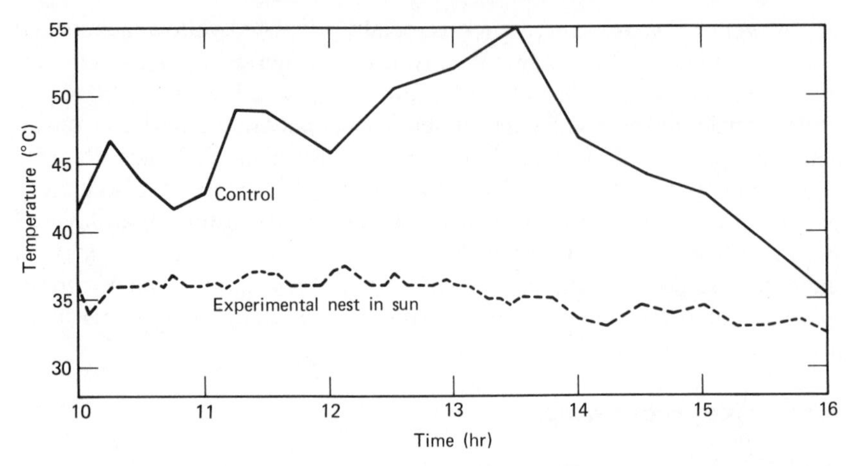

Fig. 7 Temperatures of two P. gallica nests in sunshine for 6 hr on July 31. One nest (solid line) was uninhabited, and the other (dashed line) had an attending queen that brought water for evaporative cooling and fanned. (From Steiner, 1930.)

actively cooling her nest. Earlier in the day, the temperature of the sunlit control nest had risen from 35°C at 1015 hr to 55°C at 1313 hr; meanwhile the temperature of the queen's nest had not exceeded 38°C (Fig. 7). Throughout this time the queen wasp had been making 6–11 water collection flights every 15 min. When she was confined for 10 min, at 13.30 hr, her nest's temperature quickly rose from 36 to 39.5°C, but it dropped down to 36°C after she was released and again became heavily engaged in water transport.

The second demonstration involved calculating the cooling power of the water spread over a nest. By collecting regurgitated water with a syringe, Steiner determined that a queen's crop volume was 30–35 μl. The maximum frequency of water collection trips observed was 52/hr. Therefore the maximum volume of water a queen presumably transported in 1 hr was about 1.8 ml. The evaporation of this volume of water would dissipate 1050 cal at 35°C. Thus the temperature of a nest with brood weighing 15.4 g (Steiner's estimate for a typical nest) would plummet by about 68°C if all this water evaporated instantaneously and if all the calories for the dissipation of this 1.8 ml were derived from the nest and brood. However, the evaporative cooling is extended over the hour, and some of the water cools the environment and not just the nest. Nevertheless, Steiner's calculations indicate that nest cooling by water evaporation is physically feasible. Weyrauch (1936) also observed water droplets on hot days atop the nests of several European species of *Polistes* (*P.*

Fig. 8 Vertical cross section of a mature nest of a colony of vespine paper wasps. Brood is reared in downward-projecting cells assembled in tiers of comb. A multi-layered, paper jacket encloses the nest, providing protection for colony defense and temperature regulation.

gallicus, P. biglumis, and *P. foederata*), thus evaporative cooling for nest temperature control may be a widespread practice in this genus.

Incubation of brood has not been reported for *Polistes,* and the brood does not produce enough heat to elevate nest temperature significantly (Steiner, 1930).

4.2 Yellow Jackets and Hornets

Yellow jacket and hornet societies are widespread throughout the Northern Hemisphere, and they have attained a more advanced level of microclimate control than is found in colonies of *Polistes* wasps. In addition to nest cooling, vespine wasps can warm their colonies by the production of metabolic heat. Because their nests are enclosed, either by a thick paper jacket (Fig. 8) or the walls of a tree or soil cavity, heat generated inside their nests can accumulate to produce significant warmth. This possession of coupled cooling and heating capacities means that wasp colonies can tune their nests' temperatures with fair precision. Based on daily measurements taken every 2–3 hr between 0700 and 2100 hr, over the period July 24–August 31, Himmer (1927) found that the temperature inside a large *P. vulgaris* colony averaged 30.7°C, range 26–36°C (average daily range 2.5°C), whereas the ambient temperature averaged just 18.4°C, with a range of 9–34°C (Fig. 9).

Such precision in colony thermoregulation, which comes close to that observed in honeybees, occurs only in the middle portion of the colony cycle when colonies have large contingents of workers. Both before this period, during the founding stage in late spring when colonies are still small, and afterward, when colonies have begun releasing their crops of

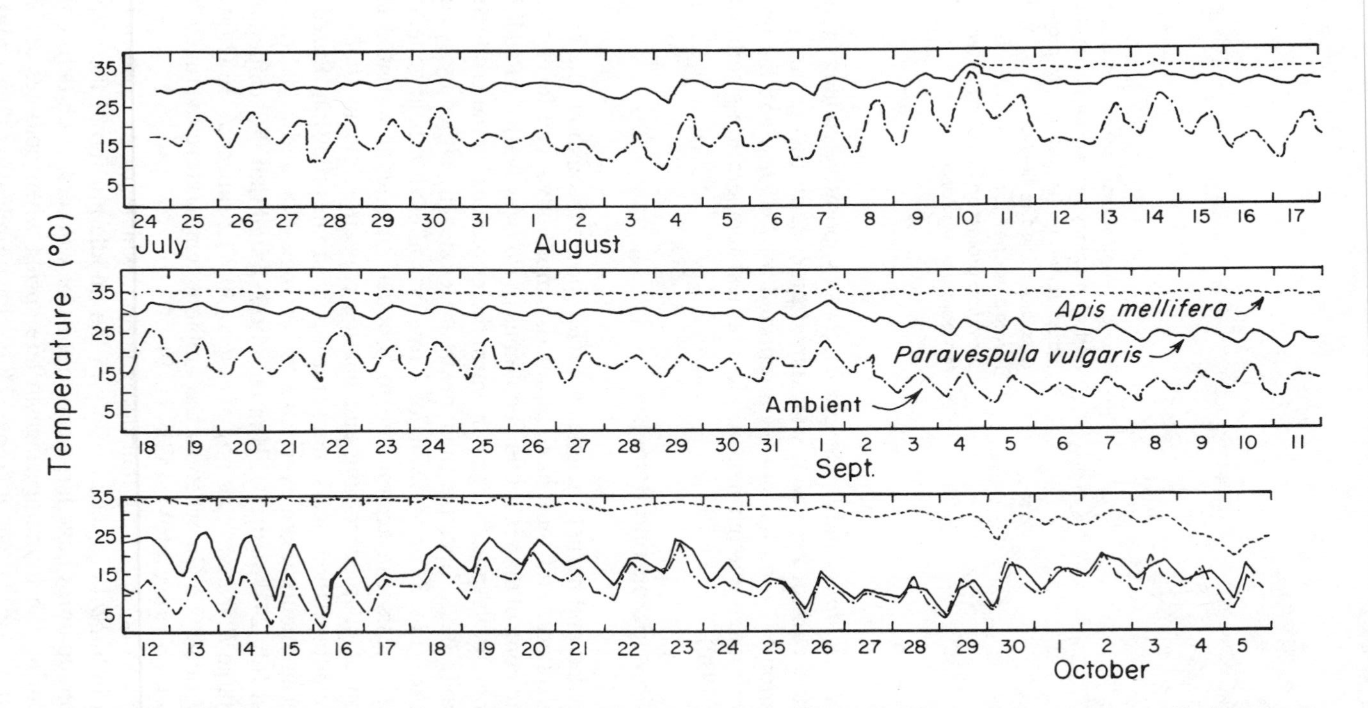

Fig. 9 Nest temperature records for a colony of yellow jacket wasps, **P. vulgaris**, in midsummer in comparison with the common honeybee, **Apis mellifera.** (From Himmer, 1927.)

reproductives and have begun disintegrating, the effectiveness of temperature control is far smaller.

The sequence of development, maturation, and senescence in nest temperature control has been documented by Gibo et al. (1974, 1977) in their studies on the bald-faced hornet, *D. maculata* in southern Canada. Colonies of this wasp species are founded in late May by solitary queens. Colonies grow in population by rearing workers until mid-August, when production of each colony's crop of males and queens begins. A colony's worker force then soon declines, and by October most colonies are dead. On June 2–4, a newly founded nest about 5 cm in diameter, with two layers in the nest envelope and containing a queen with about 20 larvae, could maintain itself up to 4°C above ambient temperature T_A. Nest temperature T_N ranged from 11 to 31°C, while T_A varied between 8 and 28°C. Two weeks later, when the colony had produced five adult workers and added two more layers of nest envelope, T_N ranged from 18 to 36°C over a T_A span of 12–36°C. By early July the nest had grown to about 10 cm in diameter with eight complete envelope layers. The thermoregulatory capacity of colonies was tested by placing them once a week in a 5°C environmental chamber for 4 hr and measuring T_N at the end. A typical hornet colony showed increasing ability in heat production throughout late July and August, emerging from the stress test on August 2, 19, and 30 with T_N of 6, 13, and 19°C, respectively. In early September new reproductives appeared in the colony, and its heating capacity quickly dwindled. On September 6, T_N following the test was just 11°C, and by September 20 it was 5°C, thus in equilibrium with T_A.

Heat production within vespine wasp societies operates as a complex interplay between adults and larvae, for although the adult wasps are the principal heat producers, the larvae also contribute by generating some heat and by providing a colony food reserve.

Adult wasps probably produce heat through microvibrations of their large flight muscles, without wing movements, as do bumblebees and honeybees (see Kammer, this volume). However, this has not been studied in detail in wasps. Himmer (1927) found that thorax temperatures of "active" *P. vulgaris* workers exceeded T_A by about 7.5°C on average (T_A and wasp activity pattern were not specified), and the power of this heat production was demonstrated by giving a small group of workers a comb containing brood. The adults clustered upon the brood and boosted its temperature from 20–22°C (T_A) to 30–32°C in 5–7 min (Ishay, 1972). As with honeybees (see Section 5.4.4), Stussi (1972) found that workers of the genus *Paravespula* raised their metabolic rate as T_A decreased, presumably to stabilize their body temperature. For example, they increased their metabolic rate from 1000 to 4700 μl oxygen per wasp per hour as T_A dropped from 30 to 15°C. It is unclear if individual wasps strive to warm the general nest interior, if they only attempt to regulate their own body temperature, and/or if they respond to the temperature of the brood.

Incubation of brood among the Vespinae (Ishay and Ruttner, 1971; Ishay, 1973) is directed against the pupae only and is perhaps performed only when colonies are under severe cold stress. Ishay (1973) has demonstrated that when an adult wasp huddles over a pupal cell, or occupies a vacant cell adjacent to the pupa, it presses its abdomen tightly against it and begins rapid pumping of the abdomen. The pupa can be warmed 10°C or more to restore it to 30–32°C. When the number of pupae exceeds the number of brooding workers, each worker spreads her efforts by spending 15–40 min warming one pupa and then moving on to another. The cues by which workers orient their incubation behavior include a pheromone released from pupal cuticle (Ishay, 1972, 1973), as well as aspects of the pupa's physical characteristics. For *V. crabro,* the pheromone is *cis*-9-pentacosene (Veith and Koeniger, 1978), and the physical characteristics of a pupa inside a cell can be mimicked by a 350- to 400-mg pebble slipped into a cell (Koeniger, 1975; Fabritius, 1976). The biological importance of pupal brooding was demonstrated by experimentally maintaining *V. crabro* pupae at 4, 20, or 32°C. No adults emerged from the 4°C group. About 65% of the 20°C group yielded an adult, but 25% of these wasps possessed malformed wings, legs, and/or sting apparatus. Nearly all (87%) of the 32°C group emerged, and of these only a very few (2–4%) bore a deformity (Ishay, 1973).

The inattention of incubating adults to larvae may ultimately be related to the lower temperature sensitivity of the larvae. It might also reflect the capacity of larvae to warm themselves. For example, when Ishay and Ruttner (1971) removed the adults from a *V. crabro* comb containing about 200 larvae, they observed that the control comb's temperature decreased from 25 to 22.4°C but that the experimental comb's temperature dropped only 1.5°C from 30°C. Thus the larvae maintained themselves 6.1°C above ambient after 18 hr without adults.

A distinctive larval behavior appears to be part of their heat production process. Under cool conditions a larva swings its downward-hanging body like a circular pendulum with its dorsal side scraping over the cell's walls. This round-and-round behavior is easily distinguished from the up-and-down movements made by larvae scraping their mandibles against a cell wall to signal their hunger to adult wasps (Ishay and Ruttner, 1971).

The larvae of vespine wasps are a store of food reserves for the colony, and they can thus make a second, less obvious but perhaps more important, contribution to colony thermoregulation. When a *Dolichovespula*

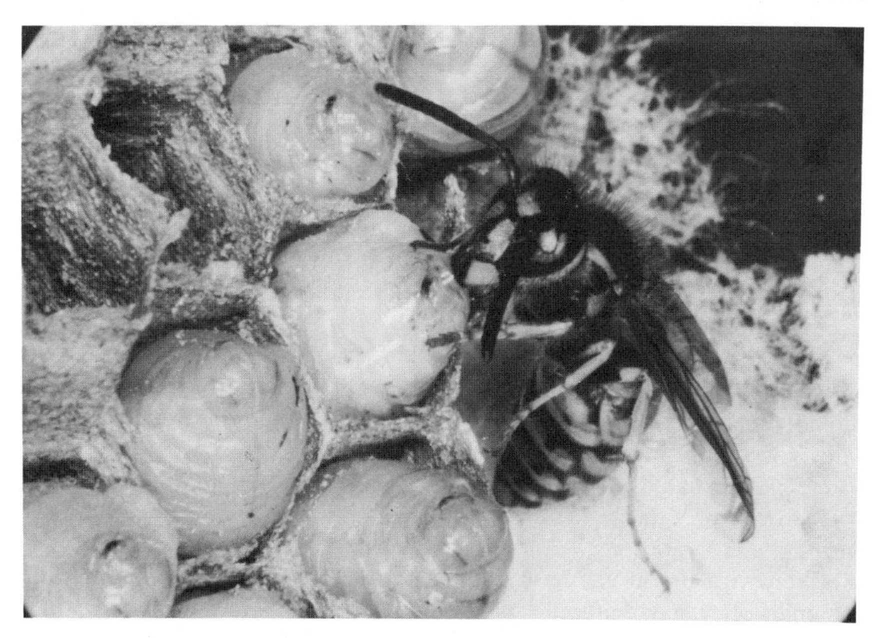

Fig. 10 A worker of *P. vulgaris* has tapped with her mandibles on the mouthparts of a large larva and is sucking up the sugary secretion released by the larva. (Photograph courtesy of U. Maschwitz.)

arenaria queen is incubating her newly founded nest, she can generate over 120 cal heat/hr. If she uses nectar with 40% sugar to fuel this heating, she will metabolize about 70 mg of nectar each hour, and this is 30% of her body weight (Gibo et al., 1977). Obviously such intense metabolism cannot be maintained for long without feeding. During days with good weather incubating wasps can refuel by foraging, but at night or in inclement weather they rely upon sugary secretions from the larvae. Food exchange between adult wasps and their larvae was described in 1742 by Réaumur and in 1895 by Janet. Both observers reported that a light touch on a larva's mouthparts by either an adult wasp or the tip of an experimenter's pencil triggered it to lean backward, spread its mandibles, and disgorge a droplet of clear liquid (Fig. 10). The observation that male wasps in the fall when they are refused food by workers consume little else besides larval secretions, suggested to Janet (1895) and Montagner (1964, 1966) that the larval secretion was nutritive. Chemical analysis of the larval secretions of *P. vulgaris*, *P. germanica* (Maschwitz, 1966), and *V. orientalis* (Ishay and Ikan, 1968) indicated that they contained approximately 9% trehalose and glucose, or about four times the carbohydrate concentration found in larval hemolymph. Amino acids and proteins are also present in the secretion, but at only one-fifth the hemolymph concentration. The secretion is produced by the strikingly large salivary glands of the larvae. Maschwitz (1966) measured the metabolic rate of resting *P. vulgaris* workers to be approximately 50 μg glucose/hr. Thus 1 μl of the larval secretion, a 9% sugar solution, could sustain a wasp about 1.8 hr, and the sugar released in a single "milking" (about 7 μl) of a larva could suffice a worker for half a day. Of course, incubating workers, with their much higher metabolic rates, would require more frequent feedings from the larvae.

More recently, Ishay and Ikan (1968) have added another chapter to the story on wasp larvae as a food reserve. They have discovered that larvae do not merely sequester sugars from their food in the secretion fed back to the workers. They also synthesize sugars from the proteins of their principally meat diet. Within 3 hr of feeding *V. orientalis* larvae with ^{14}C-labeled protein, radioactive glucose and other sugars were found in the salivary secretions.

Even with the larvae as a colony food reserve, it appears that wasp colonies often have difficulty maintaining nest temperatures at an optimal level of 28–32°C throughout cool evenings. When wasp colonies are continuously monitored for temperature, a sudden drop in nest temperature late at night or early in the morning is often reported (Gaul, 1952; Roland, 1969). Apparently colonies run low on fuel at these times, and they must "turn down their thermostats." For example, an experimental colony of *P. vulgaris* monitored by Roland (1969) maintained a nest temperature of 28–30°C during the daytime, but each night, a little after midnight, the temperature fell off rapidly to 18–20°C. But when Roland reprovisioned the wasps with honey, their nest temperature again rose to 28–30°C.

The nest-cooling behaviors found in vespine wasp societies closely resemble those already described for colonies of *Polistes* wasps. Fanning begins at low levels of overheating, and water collection starts if simple ventilation is inadequate. Water evaporation for nest cooling has been reported for *P. vulgaris* and *P. germanica* (Himmer, 1933), *D. silvestris* (Himmer, 1933), *D. arenaria* and *D. maculata* (Gaul, 1952), and *V. orientalis* (Ishay et al., 1967). Although this behavior appears to be widespread among Vespinae, it is perhaps not universal. When a colony of *V. crabro* (which had built a nest inside an empty beehive) was heated from below with an electrical heating pad, vigorous fanning was observed when the nest temperature reached 37°C. However, this ventilation had little cooling effect, and the nest continued to warm to 39.2°C, at which point

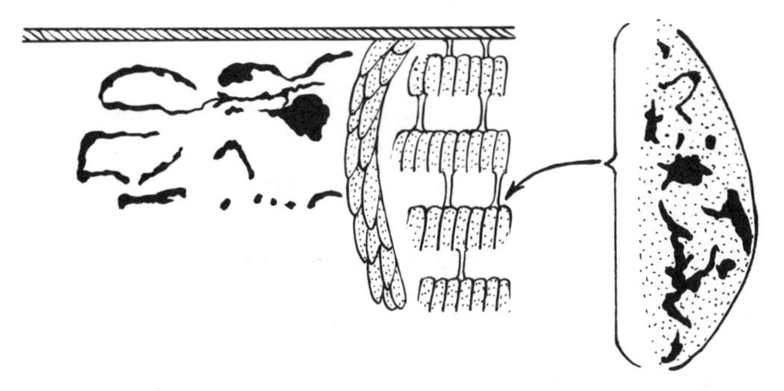

Fig. 11 Shape of fluid droplets deposited by wasps, P. vulgaris L., on the top surface of their combs in the nest (right) and on the wall next to the nest (left) during nest overheating. (From Weyrauch, 1936.)

the hornets deserted it. At no time was water collection observed (Himmer, 1931). In the field *V. crabro* colonies generally occupy hollow tree trunks where they are probably well insulated from high ambient temperatures. Thus it is perhaps not surprising that they lack the responses associated with nest cooling by water evaporation.

Species that employ evaporative cooling may sometimes turn to larvae as a source of water. Weyrauch (1936) observed workers in overheated colonies of *P. vulgaris* and *P. germanica* proceed down rows of larvae with each adult wasp visiting eight or more larvae and so collecting a large volume of larval secretion. These workers then applied the secretion as droplets and smears about the nest (Fig. 11). It may be that larval secretions provide an immediately available fluid source which wasps rely on only when water collection from the field is inadequate.

5 THE SOCIAL BEES

Of all the major groups of social insects, social bees present the greatest diversity in levels of sociality. And this broad span in social evolution is matched by the wide range of temperature-control mechanisms found among bee societies. Moreover, bumblebees and honeybees have been more intensively studied with respect to nest temperature control than any other social insects, and they provide insights into the physiology, behavior, and sociology of temperature control that are unavailable elsewhere.

5.1 Sweat Bees

The little sweat bees of the worldwide subfamily Halictinae are most notable for possessing various levels of social organization, from solitary species to primitively social species with colonies containing up to several hundred individuals. In all species females generally dig a simple shaft into the soil and excavate small cavities or cells laterally from it. Each cell is provisioned with a ball of mixed honey and pollen onto which the female oviposits a single egg before sealing the cell (Michener, 1974).

Because the brood cells are fixed in place their location is of prime importance in ensuring conditions favorable for development. Proper cell positioning involves two features for halictines, selection of the general nest habitat and choice of soil depth within this habitat. Most halictines avoid ground with thick plant cover, preferring instead bare soil surfaces that are warmed by the sun. One indicator of these bees' preference for exposed ground is the richness of the bee fauna in certain scattered microhabitats created by humans, such as pathways, parks, and road and railway cuts (Sakagami and Michener, 1962). Another measure of this preference for sunlit areas is the difference in nest density on roadbanks on opposite sides of east-west roads. Michener and collaborators (1958) in Brazil found on average 5.35 bee (primarily Halictinae) nests/m^2 on banks facing north, and thus toward the midday sun, but only 0.87 nests/m^2 on average on south-facing banks. Soil samples from a 20-cm depth taken from opposite north and south banks along a road did not differ noticeably in humidity; both contained about 14–20% water, so soil temperature was probably the important difference. In their study on *Lasioglossum duplex* nesting in the Botanical Garden of the University of Hokkaido, Sakagami and Hayashida (1961) found the nests distributed in areas that received morning sun. Areas receiving comparable amounts of afternoon illumination were nearly devoid of nests. Thus warming of the soil on cool mornings may be the most important feature of sunlit nest sites.

To more finely tune the temperature of cells, halictines apparently dig their burrows to depths providing the desired temperature characteristics. The evidence for this choice of nest depth is the progressive deepening of newly constructed cells during the progressively hotter weather of summer. For example, in *Lasioglossum imitatum* in Kansas, the mean depth of the uppermost inhabited cell dropped from 7.7 cm (range 3.5–11 cm) in the cool spring (April–May) soil to 41.0 cm (range 7–63 cm) in the hot August and September soil (Michener and Wille, 1961). However, soil humidity probably varies with the seasons as does soil temperature, and this increase in nest depth may be more a response to desiccated rather than overheated soil.

5.2 Bumblebees

The large, hairy bumblebees (genus *Bombus*) are notable among the social insects for their adaptation to colder climates. Most species live within the temperate zones of North America and Eurasia, and within these continents many species occur only in the northern regions and above the timberline on high mountains. Several are found near the Artic Circle.

As in the vast majority of social bees and wasps endemic to the cold temperate zones, the life cycle of bumblebee colonies is annual (Free and Butler, 1959; Alford, 1975). Understanding the stages in colony life throughout a year is basic to understanding thermoregulation in bumblebee societies. As colonies proceed through the different phases of the life cycle, their social structure changes markedly, and these social changes in turn influence nest temperature control (Heinrich, 1979a).

Bumblebee societies begin in the spring, as queen bees leave their hibernation burrows in soil or beneath matted vegetation and found colonies individually. Each queen takes up residence in an abandoned rodent nest, hummock of grass, or the like, and rears a small brood of worker bees. Because the queen is at first alone in her nest, she must initially perform all the colony operations, including nest construction, foraging for food, and brood incubation. The first workers emerge about 1 month after the start of egg laying. The workers aid the queen in all colony activities except egg laying, and the queen spends little time outside the nest following their appearance. The production of workers lasts for from one to several broods and is followed in the mid to late summer by the production of reproductives—queens and males. The colony life cycle is complete when the newly fertilized queens move into sheltered sites for hibernation.

5.2.1 Colony Growth and Thermoregulation Pattern

The ability of bumblebee colonies to control nest temperature changes in a regular pattern as they pass through the successive stages of the colony cycle—founding, growth, and reproduction. In reference to colony thermoregulation, Hasselrot (1960) has termed these stages the periods of "upbuilding," "equilibrium," and "decline." During the period of colony founding when each colony's queen works alone, nest temperatures show large variations. Each queen must leave her nest daily at frequent intervals to gather food, and nest temperature variations are closely related to a queen's presence or absence. Richards (1973) monitored nest temperature continuously in founding colonies of *Bombus polaris* at Lake Hazen (81° N) on Ellesmere Island, Canada. When a queen was on her nest, the

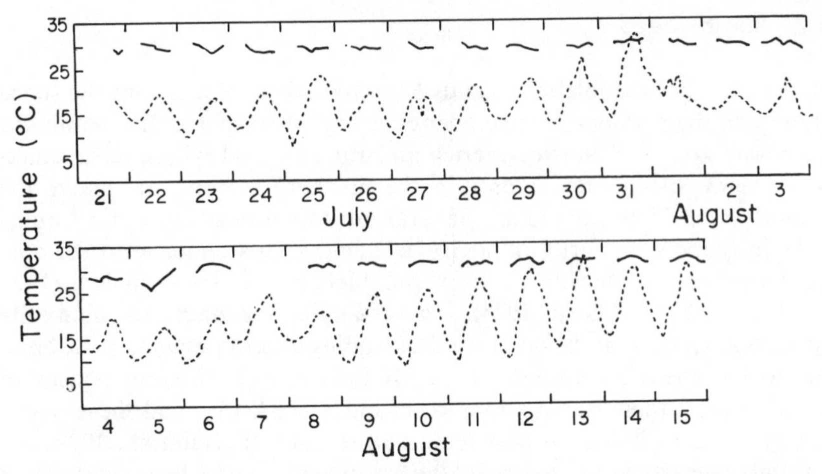

Fig. 12 Temperatures in a nest (solid line) of B. agrorum from July 21 to August 16 versus air temperature (dashed line). (From Himmer, 1933.)

temperature of the brood stabilized at between 29 and 33°C with T_A ranging from 5 to 11°C and averaging about 10°C. But upon the queen's departure, the brood temperature dropped at about 0.22–0.30°C/min and dipped to as low as 11°C, though generally remaining above 20°C. Upon the queen's return the brood temperature rose at about 0.27–0.43°C/min. The greater the drop in brood temperature, the longer the queen remained in the nest. It was not determined whether queens took shorter foraging flights at low than at high T_A, thereby minimizing cooling of the brood.

During the period of equilibrium for Richards' (1973) study colonies (when they had each gained a labor force of worker bees) brood temperatures were continuously maintained between 27 and 33°C, even at near frost conditions. Similarly, when Himmer (1933) monitored the temperature close to the brood cells in a shallow (2–3 cm of soil cover) underground colony of *Bombus agrorum* in Erlangen, Bavaria, he observed that the nest temperature generally stayed at about 30°C, rarely going below 29°C or rising above 32.5°C. The temperature of the soil adjacent to the nest was generally between 15 and 20°C and never exceeded 22.5°C (Fig. 12).

The time of relatively high and stable nest temperatures is followed by a period of decline in the precision of temperature control. This period begins with large variations in nest temperature, which gradually come to track variations in the ambient temperature. For example, near the end of the life of his colonies, Richards (1973) found their brood temperatures ranging from 10 to 20°C, just 5–8°C above T_A. Also, Hasselrot (1960) monitored a colony of *Bombus terrestris* in Sweden and observed that, whereas from June 29 to July 25 the nest temperature ranged just from 29 to 35°C, averaging 32.5°C (period of equilibrium), during the period of decline from July 26 to August 14 the nest temperature ranged from 10 to 34°C, averaging just 24°C.

The pattern of change in nest temperature control, from instability to stability and finally decline, has been recorded with various levels of rigor for a wide variety of species in North America and Europe in addition to the species mentioned above: Nielsen (1938) for *Bombus hypnorum*; Cumber (1949) for "both underground and surface-nesting species" in England; Fye and Medler (1954) for *B. fervidus*; Hasselrot (1960) for *B. lapidarius*, *B. hypnorum*, *B. lucorum*, *B. soroeensis*, and *B. pratorum*; and Wójtowski (1963) for *B. agrorum*, *B. derhamellus*, *B. terrestris*, *B. silvarum*, and *B. lapidarius*.

5.2.2 Mechanisms of Temperature Control

How do bumblebees achieve their often impressive nest temperature control? One of the primary reasons may be that they are well adapted behaviorally and physiologically (see Kammer, this volume) for direct incubation of their brood (Fig. 13). In 1837 Newport had already observed that some bees "seem to be occupied almost solely in increasing the heat of the nest and communicating warmth to the nymphs in the cells by crowding upon them and clinging to them closely." Hoffer stated in 1882 that, "while incubating, the queen frequently lies on the egg-cell in such a way that it warms it with its abdomen, as a hen does her eggs, the abdomen being closely pressed against the cell. Moreover, she also resorts to brooding in the case of older clumps of eggs and larvae, and the cocoons." Although incubation is a primary thermoregulatory technique of bumblebees, there are also many other factors of bumblebee social organization and ecology that influence temperature control in colonies.

The majority of bumblebees nest in vacated nests of small mammals—such as those of mice, squirrels, and shrews. However, dense clumps of leaves, grass, and moss are sometimes also used to form nests in some species. To build her nest, a queen bumblebee pulls together locally available fibrous material and forms a tightly knit mass, about the size of a tennis ball, with a small cavity in the center. Later in the colony cycle, as the nest cavity enlarges, the bees cover it with a waxen canopy. Ventilation holes are opened or closed in the nest ceiling, apparently in accordance with nest overheating or cooling (see Heinrich, 1979a). Although no

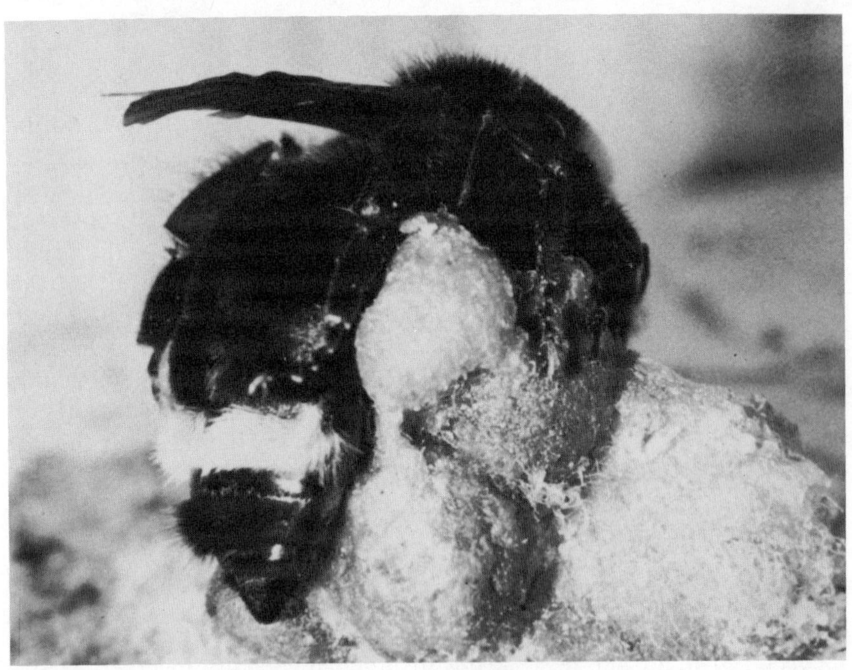

Fig. 13 Postures of resting bumblebee, **Bombus edwardsii** (top), showing elevation of the body above the substrate, in contrast to an incubating bumblebee, **B. vosnesenskii**, (bottom) pressing her extended abdomen against the brood. The queen is perched on a brood clump containing pupae and a packet of eggs (near right hind tarsus). (Photographs by B. Heinrich.)

one has investigated the matter, it seems highly likely that (1) the nest provides significant insulation and (2) nest insulation is varied depending on the needs of the colony.

Nest site selection may also contribute to creating a nest microenvironment favorable for incubation. Bumblebees avoid nesting in damp ground and often inhabit sunny places (Alford, 1975). And in the arctic, *B. polaris* generally uses surface nest sites instead of rodent burrows (the usual nest site of temperate zone *Bombus*), perhaps because the former are often 5°C warmer and noticeably drier than the latter. Furthermore, the majority of the nests of this species had their entrance direction in the 180–270° quadrant, thus facing the sun when the soil surface temperature was at its daily maximum (1300–1600 hr; sun bearing 195–240°) (Richards, 1973).

Inside the fuzzy ball of nest material of the initial nest the queen builds a brood cell of wax containing about 8–14 eggs provisioned with pollen. She constructs another wax cell, a honey pot, which she fills with nectar. From the time the eggs are laid until the time the workers eclose, the queen steadily incubates the brood clump, only occasionally pausing to forage for food or sip nectar from the honey pot (Fig. 14). The honey pot provides a fuel store during the night or inclement weather and is refilled when foraging is possible. Orientation to the brood clump appears to be facilitated by a pheromone which the queen herself deposits (Heinrich, 1974a), and she is able to locate brood clumps in total darkness (Fig. 15).

Direct temperature measurements by means of a thermocouple implanted in a brood clump of a *B. vosnesenskii* queen (laboratory conditions: uninsulated nest, diluted honey provided *ad libitum*) revealed that, when T_A was reduced to 5°C, the brood was maintained near 24°C; and when T_A was raised to 25°C, the brood was held near 32°C (Fig. 14). These data indicate that the brood clump is indeed incubated and that its temperature is to some degree independent of ambient temperature.

Possibly there are no specific temperature set points for the brood that the queen attempts to maintain. An incubating bumblebee queen regulates her abdominal temperature (Heinrich, 1974b) and, by replacing heat into the abdomen as soon as it is lost to the brood and elsewhere, the stabilization of a high brood temperature results automatically, without the bee's being aware of brood temperature as such.

An important part of the brood temperature regulation by incubating queens is their adjustment of their rate of heat production. The greater the temperature excess of the brood a bee maintains, the greater her metabolic rate (Fig. 16). The metabolic rate during incubation may approach that observed in flight (Heinrich, 1974b). Further details of the physiology

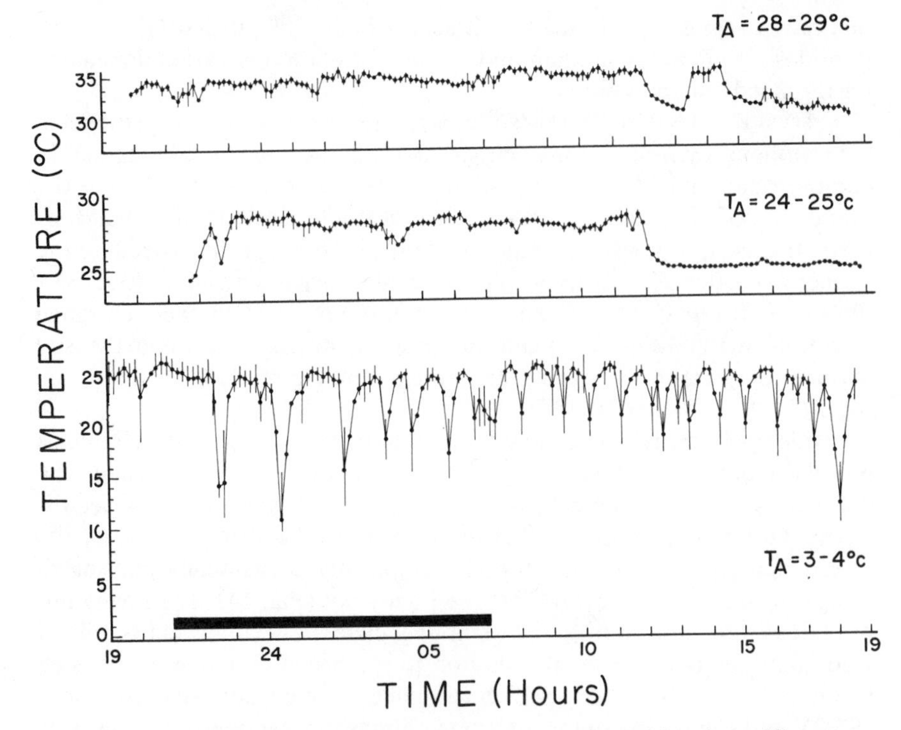

Fig. 14 Temperatures of an uninsulated brood clump maintained at three different ambient temperatures by one bumblebee queen (*B. vosnesenskii*) over three full days in the laboratory. Black bar = night. (From Heinrich, 1974b.)

of incubation by bumblebee queens are examined elsewhere (see Heinrich, 1979a; Kammer, this volume).

The periods of brood incubation by queens in the laboratory (Heinrich, 1974b) are often many hours in length and continue through both day and night. One bee incubated for at least 24 hr, pausing only for about a minute every $\frac{1}{2}$–3 hr to feed from the diluted honey provided (Fig. 14). It was calculated that, given a full honey crop and a full honey pot, a queen could maintain the metabolic rate of incubation (T_A = 5°C) for about 4 hr. In the field the queen's metabolic rate may not have to be so high because the nest is insulated and, if so, her maximum incubation duration could be extended. When bees have exhausted their food they enter torpor and cease to incubate. In 1912 Sladen (1912) observed: "Occasional periods of semistarvation, lasting for a day or two do not harm a colony of bumblebees: the bees simply become drowsy, remaining in a state of suspended animation."

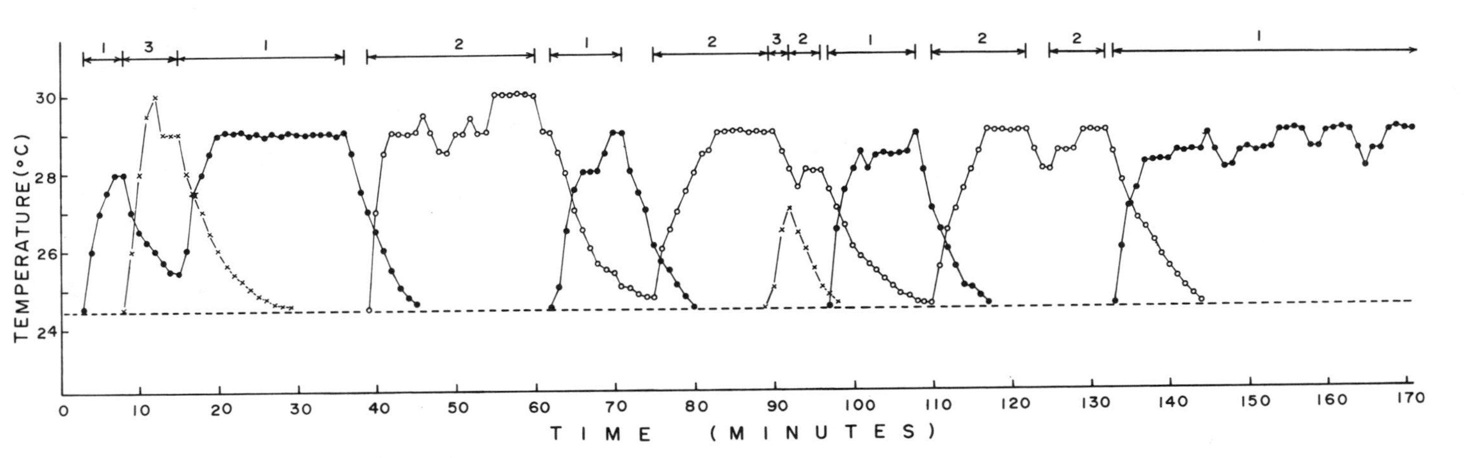

Fig. 15 Temperature increases and decreases in three **B. vosnesenskii** brood clumps (designated 1, 2, and 3) from three different queens, alternately incubated by one queen. In this experiment one **B. vosnesenskii** queen was placed in a closed 8 × 9 cm wooden box in a darkened photographic dark room. She incubated not only her own brood (2), but also that of the two other queens (1 and 3). (From Heinrich, 1974a.)

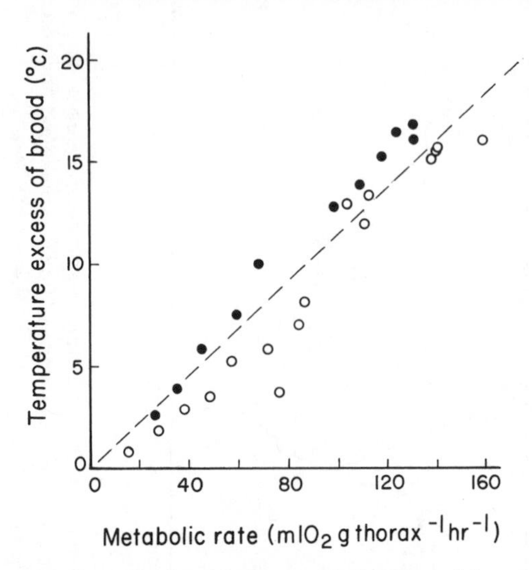

Fig. 16 Temperature excess of a brood clump as a function of the metabolic rate of the incubating B. vosnesenskii queens. Open and solid circles refer to two different individuals tested on the same brood clump. (From Heinrich, 1974b.)

Worker bumblebees, drones, and new queens also incubate brood. They all show the same extended abdomen posture while sprawling on brood comb. Direct temperature measurements on a section of comb containing cocoons and honey enclosed by 15 *Bombus edwardsii* workers found the brood temperatures elevated 7–10°C above T_A (22–23°C). This lasted until the honey was depleted, whereupon the brood comb temperature declined to that of the surrounding air (Heinrich, 1972). It may be that workers and drones also possess many of the physiological adaptations for brood incubation found in queens (see Bartholomew, this volume), though this has not been studied.

That the eggs, larvae, and pupae are essentially poikilothermic in the temperate zone species *B. vosnesenskii* and *B. edwardsii* is indicated by immediate cooling and approach to T_A by the brood clump when either the queen departs (Heinrich, 1974b) or, in a colony with workers, the colony's food becomes exhausted in the night (Heinrich, 1972). However, for *B. polaris,* whose range extends north of the Arctic Circle, there are data suggesting that larvae produce significant heat (Richards, 1973). Nests with larvae, relative to those with pupae, cooled down more slowly (0.22 versus 0.30°C/min, respectively) when the queen was away foraging and warmed more rapidly (0.43 versus 0.27°C/min) upon her return. However, the measurements were made in the field where T_A, nest mass, and other factors could not be controlled, and the matter requires further investigation.

As previously stated, when the queen is joined by daughter workers, the nest temperature becomes more stable than when she is working alone. There are several reasons behind this improved thermoregulation. The larger number of adult bees provides a greater source of heat but, probably more importantly, helps ensure that there will always be one or more individuals attending to control of the nest temperature. Oster and Wilson (1978) have formally described such improvements in system reliability in terms of the difference between "serial" and "parallel" operations. The solitary founding queen must perform her various duties—brood incubation, nest repair, food collection, nest defense—sequentially, one at a time, but a queen together with workers can address two or more of these tasks concurrently.

A further basis for the relatively high temperature stability in established colonies is their ability to gather and store significant amounts of honey. This food reserve of course helps ensure that colonies can continue heat production through times when food collection is impossible or at least cannot keep up with the energy needs for nest heating. Vacated cocoons and special waxen honey pots on the brood comb periphery provide storage vessels. *Bombus perplexus* and *B. fervidus* add the enzyme glucose oxidase to their stored honey, which protects it from fermentation by generating hydrogen peroxide (Burgett, 1974).

Both the quality and quantity of the stored food may improve as colonies grow in population size. In his study on *B. polaris,* Richards (1973) observed that two nests in the queen stage each contained one honey pot with capacities of 1.12 and 1.73 ml. Following worker emergence he found up to 14 honey pots per nest, each usually one-half to three-fourths full (volumes not stated). Furthermore, once the workers appeared, pots of thick honey (73–81% sugar) were found, whereas in the earlier queen nests the honey was noticeably thinner (59–72% sugar).

Thermoregulation may be a major line of labor allocation for worker bumblebees. Brian (1952) observed two small nests during two summers (26 and 40 workers total, respectively) and concluded that slightly less than two-thirds, on average, of these colonies' workers stayed inside the nest all day. The bees rarely fed the larvae; often Brian observed less than one larval feeding per hour. However, colony defense against ants (Hasselrot, 1960), birds, rodents, or other enemies might also require a large labor investment by bumblebee colonies, and this, instead of nest warming, might underlie the heavy allocation of workers to spend time in the nest.

Although individual queens must invest considerable time and energy specifically for the regulation of nest temperature at low ambient temperatures, these costs are considerably reduced in large established colonies. In a colony with a large population of workers enough heat may be generated as a by-product of the activities of nest inhabitants that no bees need take time out to incubate, and the energetic cost of active thermoregulation may be small. The time and energy costs of thermoregulation are reduced by varying nest insulation to trap heat. Plowright (1977) reported that, for a wide range of *Bombus* species, cold stress stimulated the production of wax used to construct an insulating canopy over the nest. At T_A of 12°C, nests with internal temperatures of 30°C lost heat 20% less rapidly with than without a wax canopy.

Under conditions of high environmental temperatures, bumblebees foster heat loss by fanning out their overheated nests with their wings, crawling outside the nest, and forming ventilation holes in the insulation surrounding the brood. One or more of these responses has been reported for *B. terrestris* (Hasselrot, 1960; Lindhard, 1912), *B. muscorum* (Jordan, 1936), *B. lapidarius* (Alford, 1975; Hasselrot, 1960; Postner, 1951), and *B. hypnorum* (Hasselrot, 1960), and thus they appear to be a general characteristic of bumblebees. Fanning, the first response to nest overheating, appears to start when the nest interior temperature exceeds about 33°C.

Toward the end of a colony's life large variations in the nest temperature begin to occur, and the average nest temperature drops and eventually tracks the ambient temperature. The causes of this decline in thermoregulation are not known with any precision. Hasselrot (1960) noted less stored honey in older colonies showing wide temperature variations relative to young colonies with a stable nest temperature and guessed that the newly reared reproductives were using up the food supply. A shortage of bees available for incubation may also be involved. Worker production generally ceases with the onset of the rearing of a colony's crop of queens and males, and mortality among foraging workers is high, about 30% in 5 days (Brian, 1952; Heinrich, 1976). Thus once worker production stops, their number may quickly diminish. However, new queens help incubate in nests (Hasselrot, 1960; Allen et al., 1978), thus in part compensating for the loss of workers.

5.3 Stingless Bees

The stingless honeybees (tribe Meliponini) are a strictly tropical and subtropical group of bees with advanced sociality. Their colonies range in size from small to moderately large, being comprised of about 500–4000 adults in species of the genus *Melipona* and 300–100,000 in the case of *Trigona* and related genera (Michener, 1974). The complexity of their social behavior—measured in terms of colony size, nest architecture, magnitude of queen-worker difference, complexity of communication codes, and altruistic behavior among colony members—appears to be at approximately the same level as that of the more familiar, true honeybee *(Apis mellifera)* (Sakagami, 1971).

Studies on nest temperature among the stingless bees are almost nonexistent, but several basic features of their biology suggest that temperature control at respectable levels of precision will be found in future studies. First, they reside in well-insulated nests. Most Meliponini nest in cavities which they enter through a small aperture. The majority of species nest in tree cavities. Other species use a variety of enclosed nest sites, including hollows among tree roots; abandoned nests of ants, termites, and subterranean rodents; and even walled-off chambers in occupied ant and termite nests. A very small minority of species build exposed nests in the branches of trees and bushes. These nests show a strong convergence with the aerial nests of vespine wasps in possessing a multilayered outer shell which provides protection against physical disturbance as well as thermal insulation. However, instead of paper, stingless bees use a mixture of their own wax, plant resins, and sometimes mud or vegetable matter as their nest-building material (Michener, 1974).

Other features of meliponine biology that apparently predispose them to colony thermoregulation are their flight muscles capable of considerable heat production, their often large colonies, their practice of storing honey as a food reserve, and of course their wings which can be used for nest ventilation and possibly for flights to collect nest-cooling water.

One of the few meliponine species whose nest temperatures have been studied is *Trigona spinipes,* a species with colonies sometimes containing over 100,000 members (Lindauer and Kerr, 1958). Nests of this species are built in tree branches, where they are exposed to the wind and sun, but they are protected by numerous hard layers of nest material surrounding the central brood nest. Mature nests are ovoid, up to 85 cm in length and 55 cm in diameter. Zucchi and Sakagami (1972) measured every $\frac{1}{2}$ hr for 24 hr the ambient and brood nest temperatures of a mature *T. spinipes* colony. Whereas T_A varied from 15.5 to 28°C, the colony's brood area was maintained at 34.1 to 36°C, thus at a considerably higher level and with less fluctuation than in the environment. Observations over a period of several months found T_A ranging over a fairly broad span, 8.2–30.3°C, while the nest's brood chamber varied only between 33.3 and 36.2°C. Clearly, *T. spinipes* is capable of warming its nests and has precision in temperature regulation comparable to that of the honeybee.

Not all stingless bees practice precise nest temperature control. Zucchi and Sakagami (1972) measured brood nest temperatures in nests of six other species of stingless bees—*Melipona rufiventris*, *M. quadrifasciata*, *Trigona depilis, T. droryana, T. varia,* and *T. mülleri*—during the 24-hr period mentioned above. In one species, *T. mülleri,* nest temperature varied nearly as much as T_A, from 16.5 to 29.2°C. The other species showed stabilities and elevations in nest temperature intermediate between those of *T. spinipes* and *T. mülleri.*

Nest cooling is probably accomplished among the Meliponini, as in other social bees, through a two-tiered process of wing fanning for ventilation and water collection for evaporative cooling. In some cases there may also be adjustments of the nest structure to facilitate nest ventilation. *Dactylurina staudingeri,* an African meliponine that constructs aerial nests, has adjustable pores in the nest covering (Darchen, 1973). The pores are approximately 1-mm-diameter openings lined with sticky resin. The number of open pores varies with the air temperature, increasing during the day. For example, at 0900 hr on 3 days, when T_A was about 23°C, the density of open pores was 5.29, 4.57, and 4.33 holes/9 cm^2. Later in the same 3 days, at 1500 hr when the air had warmed to 27–28°C, the density rose to 8.29, 10.00, and 11.44 holes/9 cm^2. The pores are all closed at night. In addition to the pores, the nest has, along with its main entrance, a second large opening near the base of the nest which can be adjusted open or closed depending on temperature and/or humidity.

5.4 The Honeybee (*Apis mellifera*)

Of all the systems of nest temperature control found among the social insects, that of the honeybee has been studied in the greatest detail. A practical significance of these studies is their contribution to applied beekeeping. In nature (Seeley, 1978), and probably under apicultural conditions as well, the winter season is the period of heavy mortality among honeybee colonies, and a thorough understanding of nest temperature control during the winter may help beekeepers develop improved techniques for overwintering their hives.

We will here begin our analysis of thermoregulation in the honeybee colony by reviewing its development across evolutionary time. Before humans carried *A. mellifera* around the world, its distribution was apparently the whole of Africa, all but northernmost Europe, and western Asia. Its ultimate place of origin was somewhere in the Asian or African tropics or subtropics. This is indicated by the perennial nature of its colonies and its feature of colony reproduction by swarming, properties that distinguish it from most social bees endemic to cold temperate areas but which are common among tropical social bees. Strictly tropical or subtropical species of *Apis (A. florea* and *A. dorsata)* (Lindauer, 1956; Morse and Laigo, 1969; Akratanakul, 1975) are probably quite limited in withstanding long-term cold. Also, at least one tropical race of *A. mellifera*, *A. m. adansonii*, may be unable to overwinter in temperate regions (Taylor, 1977).

Apis mellifera mellifera and the other European races of the honeybee have penetrated colder climates through the evolution of advanced techniques of colony thermoregulation. Genetic components of the thermoregulatory response have been examined (Brückner, 1975, 1976). In this chapter we consider colony thermoregulation in just the European races of *A. mellifera*. In these bees nest temperature control is highly precise and consequently the brood is highly temperature-sensitive. Himmer (1927) reared capped brood (pupae and late-stage larvae) in an incubator at various temperatures and found that few bees emerged at 26°C. Those that emerged at 28–30°C had a high incidence of shriveled wings and malformed mouthparts. Those reared at 32–36°C were normal. At 37°C some of the brood perished, and at 38–39°C it all died.

5.4.1 Nest Choice and Construction

A honeybee colony makes an important first step toward thermoregulation by selecting a protective nest site, such as a tree cavity, which will tightly enclose the colony's combs and workers. (Swarms sometimes build their nests from bare branches when no nest cavities are available, but such unprotected nests do not survive the winter in temperate regions.) Avoidance of undersized nest holes is ensured by the bees measuring the volume of potential nest cavities and, for European races of the honeybee, rejecting those below approximately 20 liters in volume. Measurement of a cavity's volume involves a bee's walking about the cavity's inner surfaces and then integrating information on the distances and directions of these walking movements (Seeley, 1977). The importance to colony thermoregulation of inhabiting an enclosed, protective nest site is indicated by comparing the nesting habits and geographical distributions of species in the genus *Apis*. Two species that build open-air nests on the undersides of tree branches, *A. florea* and *A. dorsata*, occur only within the tropics and warm subtropics. The two cavity-nesting species, *A. mellifera* and *A. cerana*, both have ranges that have expanded beyond the tropics into cold temperature areas (Ruttner, 1968a). A further indication of the thermal protection provided by a cavity's walls is Corkin's (1930) report of differences as great as 25°C between the outside air and the air inside a hive but outside the cluster of bees.

Besides cavity volume, several properties of a nest site's entrance opening are also evaluated during nest site selection. Bees prefer south-facing entrances smaller than about 60 cm^2 and which lie at the base of the nest cavity (Seeley and Morse, 1978). All three of these preferences apparently facilitate nest thermoregulation. A southward entrance in the Northern Hemisphere is probably sunnier and warmer than one facing north, and a small, low entrance may help keep a nest warm by inhibiting heat loss by convection currents. However, careful meteorological measurements on experimental hives permitting variation in these properties are needed to test these hypotheses about the adaptive significance of nest site properties. Anderson (1948) provides a valuable preliminary observation in reporting that an uninhabited, 42-liter hive heated inside by a 15-W light bulb was 3–8°C warmer on a winter afternoon when its entrance area was 9.3 cm^2 instead of 81.4 cm^2. Hazelhoff (1954) demonstrated that convection was a potentially important mechanism of heat loss. He measured an airflow of 0.17 liters/sec by convection alone out the top of an inhabited hive with both top and bottom entrance openings.

Bees can improve the protection provided by a nest cavity by sealing off unnecessary openings with sticky plant gums and resins, collectively termed "propolis." Because of this ability, apparently, they place little importance during site selection on the presence of numerous small holes in the walls of a potential nest (Seeley and Morse, 1978). One race of honeybees, *Apis mellifera caucasica*, where winters are very cold, even reduces the nest entrance in the autumn with curtains of propolis, apparently as an adaptation to climate conditions (Ruttner, 1968b).

Inside the nest cavity the bees suspend a series of parallel, waxen combs from the ceiling and walls. The upper and peripheral regions of the nest are devoted to honey and pollen storage, while the nest's core is the brood nest (Seeley and Morse, 1976). This arrangement (Fig. 17) may simply serve to minimize stress within the wax combs by placing the heavy honey near the points of comb attachment. But it also has the effect of surrounding the brood nest with a jacket of temperature-buffering material (Hess, 1926). Anderson (1948) observed this buffering effect when he compared the temperature dynamics on a winter day inside two hives, each of which was heated by a 15-W bulb at its center; one contained just empty honeycombs while the other's combs contained 11.8 kg of honey. Observed from 0700 to 2300 hr, the first hive showed a temperature rise and fall of 24.4 and 15.0°C, respectively, whereas the second hive's respective temperature changes were just 12.8 and 2.2°C.

One source of difficulty in understanding thermoregulation by honeybee colonies is the changing pattern of control that appears over a year. In general, the thermoregulatory challenge faced by a colony reflects the state of two variables: the presence or absence of brood, and ambient temperature. If a colony is rearing brood, then it must regulate its brood nest between 32 and 36°C, and if the environmental temperature drops below about 9°C, then a colony must keep its members above the chill coma temperature near 6°C (Free and Spencer-Booth, 1960). The first setting involves temperature control in the nest's central brood nest, whereas the second condition requires temperature maintenance above a

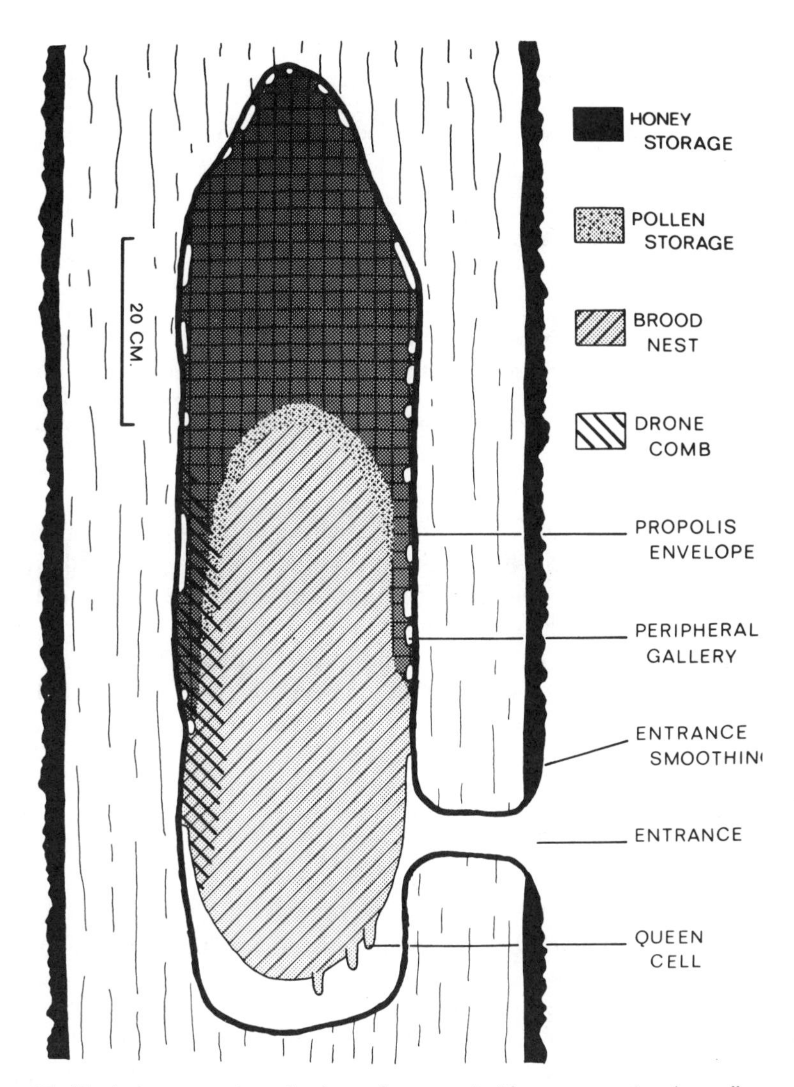

Fig. 17 Vertical cross section of a honeybee nest inside a tree cavity. A small opening through the nest's thick walls provides the nest entrance. A large portion of the nest is devoted to storage of honey, the food reserve of colonies that enables them to maintain elevated nest temperatures throughout the winter. (From Seeley and Morse, 1976.)

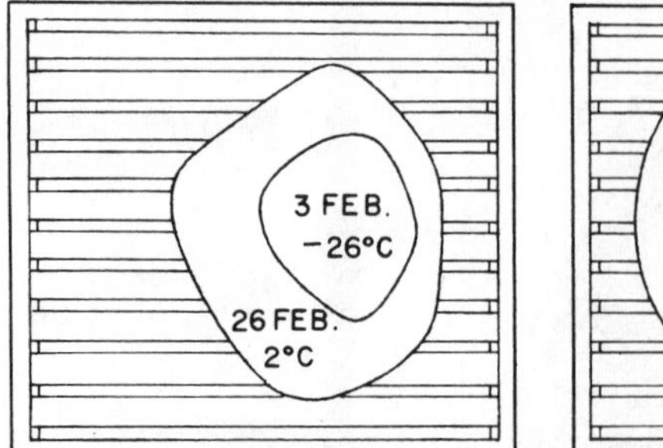

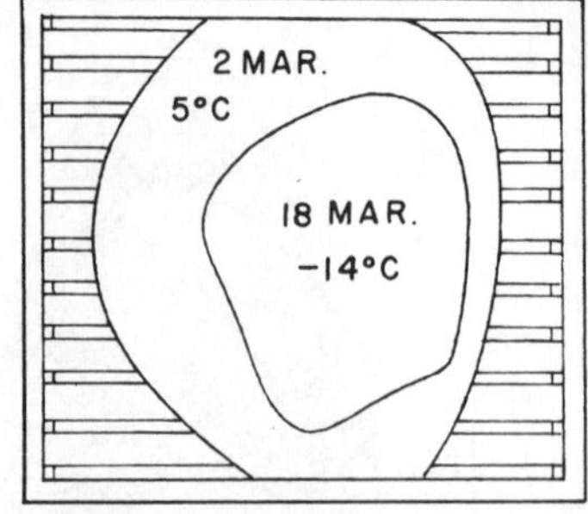

Fig. 18 As the ambient temperature falls during the winter, the honeybees in a winter cluster group together more tightly and the cluster shrinks. Shown here are the outlines of a cluster drawn against the background of the hive interior on various winter days. (Based on Wilson and Milum, 1927.)

vital minimum at the nest periphery. Usually just three combinations of these two variables occur for significant periods in European and North American climates: (1) colonies with brood in a warm environment, (2) colonies without brood in a cool environment, and (3) colonies with brood in a cool environment. In northern and middle latitude North America, where brood rearing runs from about January to October (Nolan, 1925; Avitabile, 1978), these three colony conditions correspond approximately with the summer, fall, and winter-spring seasons, respectively. What follows is a more detailed description of this annual pattern of colony thermoregulation.

5.4.2 The Broodless Winter Cluster

At T_A of 18°C or less the bees begin to cluster, and the tightness of the cluster increases with decreasing T_A (Fig. 18). Below 14°C the cluster has a compact outer shell of quiet bees (Gates, 1914; Phillips and Demuth, 1914; Wilson and Milum, 1927). Cluster contraction undoubtedly conserves heat by diminishing the area of cooling surface and probably also by diminishing internal convection currents (Corkins, 1930). Undisturbed colonies killed in the winter and dissected have revealed that a cluster consists of two parts, an outer shell of quiet, tightly packed bees and an inner core where the bees have room to move about. The shell can be several bee layers thick, and the bees arrange themselves with their heads inward (Farrar, 1943). Based on readings from a large number of thermocouples spaced regularly in a hive, Owens (1971) estimated that the boundary between shell and core was roughly at the 25°C isotherm.

Unlike in the brood nest, the temperature of broodless winter clusters fluctuates widely, although the bees engage in a variety of thermoregulatory behaviors. Temperature fluctuations (of up to 30°C) inside winter clusters are due to a complex of interactions involving feeding movements and changes in environmental temperature. The bees generally cluster on an empty comb. This allows them to crawl into the empty cells and to form a tighter cluster; it also reduces conductive heat loss to the honey stores. However, the bees must periodically move to the periphery of the honey-filled comb in order to feed. They are dependent on occasional warming of the nest interior (to above about 18°C) to break cluster temporarily and feed on nearby honeycombs. Colonies may die of "cold starvation"; hives containing plenty of honey may die out because protracted cold prevents movement of the bees from the cluster (Haydak, 1958). At temperatures above 18°C the bees periodically break cluster to feed, when they generate heat and produce transient fluctuations in hive temperature.

Changes in air temperature also affect the temperature of the broodless cluster independent of feeding movements. In general, as the temperature is lowered below 14°C, the bees generate transient heatings of the hive. Cluster temperature is maintained above a minimum of 13°C (Wilson and Milum, 1927; Corkins, 1930). The lower critical temperature of 13°C within the cluster may be a consequence of maintaining the bees on the outside of the cluster at least at 7–8°C (Hess, 1926; Owens, 1971) or about 1°C above the minimum body temperature required to cling to the cluster. At ambient temperatures above 14°C (outside the low-temperature danger zone), cluster temperature generally fluctuates *with* environmental temperature (Gates, 1914; Phillips and Demuth, 1914; Himmer, 1926, 1927). The temperature lability of the broodless winter cluster is probably not due to a lack of ability to keep warm. Rather, it is a mechanism of reducing the rate of fuel utilization.

Although it is clearly seen that a bee cluster contracts when ambient temperature decreases and expands when the surrounding air warms (Fig. 18), the relationship is not a rigid one. Cluster diameter reaches a lower limit at about 0 to −5°C, and below this temperature bees maintain cluster temperature by increasing heat production. This was demonstrated by monitoring the carbon dioxide production of a colony in a low-temperature room at a range of temperatures from 20 to −39°C. Carbon dioxide production was at a minimum at 10°C, about 0.1 mg carbon dioxide per bee per hour and rose in either direction, to about 1.0 and 2.5 mg carbon dioxide per bee per hour at −39° and 20°C, respectively (Free, 1977). Corkins and Gilbert (1932) employed a similar technique for the temperature range of −5 to 20°C but found no change in metabolic rate, perhaps because they used a different size of bee cluster or because the insulation was different.

Another indication of increased heat production at low temperatures is the relationship between cluster temperature (T_{Cl}) and T_A. In colonies in hives out-of-doors, and when T_A is less than about 0°C, there is an inverse relation between T_A and T_{Cl} (Gates, 1914; Hess, 1926; Phillips and Demuth, 1914; Himmer, 1926; Southwick and Mugaas, 1971). Because cluster contraction has already reached its limit at these low temperatures, and thus a colony has probably reached its limit in reducing heat loss, it seems reasonable to conclude that the increases in temperature represent increases in heat production. The goal of the cluster at these times is maintenance of its surface temperature above the vital minimum of about 9°C (and maybe also the cluster core above 18°C). This inverse relationship between T_A and T_{Cl} changes to a positive one at about T_A of 8°C. At this point the cluster apparently no longer has to fight the cold at its periphery and moreover can regulate its temperature by cluster contraction and expansion (Himmer, 1932).

The inverse relationship between T_A and T_{Cl} apparently holds only under short-term fluctuations in T_A, such as occur within the course of a winter day. Corkins (1930, 1932) found no such inverse correlation when he made long-term comparisons between T_A and T_{Cl}. For example, he made daily measurements of T_{Cl} for three to six colonies through four winters when there were very low ambient temperatures (−14° to −40°C). Over the whole period there was a slight but not significant positive correlation (0.053 ± 0.026) between temperatures in the cluster center and the environment. A synthesis of all these observations is that a decrease in outside temperature below about 0°C at first produces a rise in the temperature in a cluster's center, apparently by eliciting increased heat production, but that this temperature elevation is evidently not maintained. Some more slowly acting mechanism for compensating for low temperatures must operate when a low outside temperature persists. Simpson (1961) suggests that improvement in a cluster's self-insulation occurs as the bees in the cluster's outer shell become packed together more tightly, but further study is needed to solve this puzzle.

Not all features of cluster operation are designed to minimize heat loss. Some fanning in winter clusters can be heard (Simpson, 1950) or detected as drafts of humid air at the entrance (Bruman, 1928). Apparently the carbon dioxide that accumulates to levels of 3–4% inside the dense clusters is the stimulus for this ventilation (Es'kov, 1974), just as it can be during the summer (Seeley, 1974).

5.4.3 Temperature of the Brood Nest

A dramatic change occurs after the queen lays eggs in the spring. Rather than regulating the temperature of the *periphery* of the cluster (near 9°C), the bees now regulate cluster temperature itself. Hive temperatures in the brood nest are maintained between 30 and 35°C (see Fig. 9), despite large changes in external air temperature. Brood rearing can occur in the coldest weather, as well as the hottest, at air temperatures from −40 to 40°C or more (Corkins, 1930; Farrar, 1943; Ribbands, 1953; Simpson, 1950; Wohlgemuth, 1957). Although the brood is maintained above 30°C, temperatures in the broodless areas of the nest may fluctuate widely.

Broodright colonies in late winter or early spring consume much larger amounts of their honey stores than when broodless earlier in the winter. They also may now demonstrate the power of their thermoregulatory ability. On one remarkably cold winter evening Gates (1914) recorded the temperature in a colony's brood nest at 31°C while the outside air temperature was −28°C, a difference of 59°C.

On warm summer days, temperature control in honeybee nests reaches its peak in precision. Himmer (1927) monitored the temperature in a colony's brood nest continuously from August 8 to September 15 in Erlangen, Germany, and found it ranged only from 33.2 to 36.0°C, averaging 34.8°C and showing a mean daily range of just 0.6°C (Fig. 9). This nest's temperature was on average 16.4°C above T_A, ranging from 2.0 to 32.5°C above the environmental temperature. Similar observations have been reported by many other investigators, notably Gates (1914), Hess (1926), Dunham (1929), and Büdel (1960, 1968). The brood nest's center is generally the hottest part of the colony, and toward its periphery the temperature is lower and less constant. On one day with a mean outdoor temperature of 28°C, Dunham (1933) observed mean temperatures of 34.1°C (±0.4°C) in the brood nest's center, 31.8°C (±1.8°C) in the outer area of the brood nest, and 30.8°C (±2.8°C) in the broodless area of the combs covered by bees. These levels of temperature stability can be maintained in the face of severe environmental conditions.

5.4.4 The Individual Bee's Role in Nest Heating

For honeybees, as for bumblebees, the principal source of a colony's heat is the flight muscles of adults. The earliest evidence of this was the observation in 1895 by Ciesielski (see Himmer, 1932), and later by Pirsch (1923), Himmer (1925), and others, that the thoracic temperature of flying bees could be 10°C or more above their abdominal temperature, and 20°C

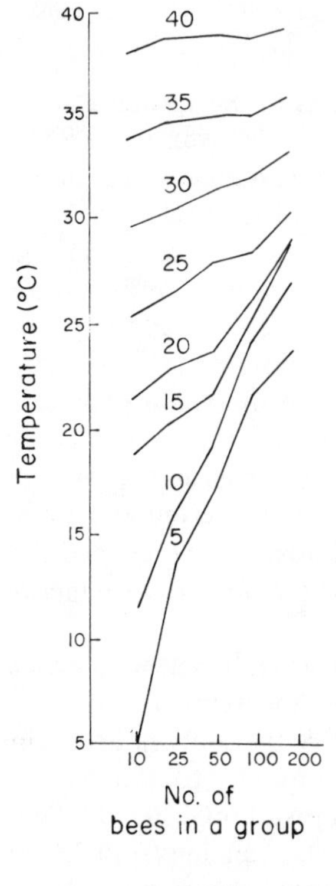

Fig. 19 Temperatures of groups of 10–200 honeybees at environmental temperatures of 5°C (bottom) to 40°C (top). (From Free and Spencer-Booth, 1958.)

or more above ambient temperature (Heinrich, 1979b). More recently the muscle physiology and neurobiology involved in this heat production have been the object of intensive research by Sotavalta (1954), Esch (1960, 1964), Roth (1965), Esch and Bastian (1968), and Bastian and Esch (1970). Simultaneous recordings of T_{Th}, flight muscle action potentials, and wing movements show that heat production is always accompanied by electrical activity, but that flight muscle activity need not always involve wing movements.

The rate of heat production by individuals is highly variable, as could be expected when they are examined out of the context of their social environment. The temperatures maintained by groups of bees may be quite different from that of individuals. The larger the group, the more readily the bees maintain elevated body temperatures at low T_A (Fig. 19). The absence or presence of brood, however, undoubtedly also influences whether bees will be at rest or attempt to thermoregulate. When a bee is at rest, its thoracic temperature approaches T_A (Himmer, 1925; Esch, 1960).

Attempts to compare the metabolic rates of resting and active individual bees (Kosmin et al., 1932; Jongbloed and Wiersma, 1934) have produced quite different results, apparently because of the high variability of an individual's metabolic rate and the difficulty of keeping it at a certain level for measurements with a Warburg apparatus. However, the broad trends are clear. For example, Jongbloed and Wiersma (1934) found that individual bees at 18°C consumed 2.87 and 146 μl oxygen per bee per minute when resting and flying, respectively, a 51-fold difference. A bee that is apparently resting but is producing heat through flight muscle activation is probably respiring at a level close to that of a flying bee (Bastian and Esch, 1970). Another important measurement for understanding heat production in colonies is that of the metabolic rate of a bee that is performing labor in the nest but is not specifically generating heat. Kosmin and collaborators (1932) measured the respiration rates of bees at several levels of activity besides resting and flying. Their values for a bee moving about grooming itself and moving about occasionally vibrating its wings were 25.0 and 68.0 μl oxygen per bee per minute and thus were 5.4 and 14.8%, respectively, of their value for a flying bee, 460 μl oxygen per bee per minute. These findings suggest that the heat generated by bees working in the nest with inactive flight muscles (nursing the brood, shaping combs, removing dead bees) may be quite small relative to that of bees producing heat for nest temperature regulation. By continuous recording for up to 3 weeks of the thoracic temperature of a few workers and their ambient temperatures in an observation hive, Esch (1960) was able to gain some insight into the pattern of colony heating at the level of individuals. When his experimental bees were "unemployed" and inactive on the comb, their T_{Th} matched T_A and ranged from 20 to 36°C; when they actively crawled about, T_{Th} still rose only 1–2°C above T_A. In contrast, "employed" thermoregulating bees, though they were usually indistinguishable behaviorally from unemployed bees, showed numerous, 1- to 4-min-long jumps in T_{Th}, often up to 38°C and thus 10°C above T_A, followed by a return to T_A as the heat radiated away.

As in resting adults, the metabolic rate of brood is but a small fraction of that of an adult with active flight muscles. Himmer (1927, 1932) measured the temperature of larvae thermoelectrically and found it to be the same as T_A for T_A 32–37°C, and approximately 0.5°C above T_A for T_A 15–32°C. Estimates of an immature's metabolic rate (Melampy and Willis, 1939; Allen, 1959; Stussi, 1972) are in fairly good agreement at about 1 μl oxygen per bee per minute, when averaged over all stages of development, and thus are of the same order of magnitude as for a resting adult.

The adult bees are a colony's principal heat source, but the ability of individuals to contribute toward heating the nest varies with age. The respiration rate increases progressively from newly emerged adults to foragers (Himmer, 1925; Allen, 1959). Body temperature excess and met-

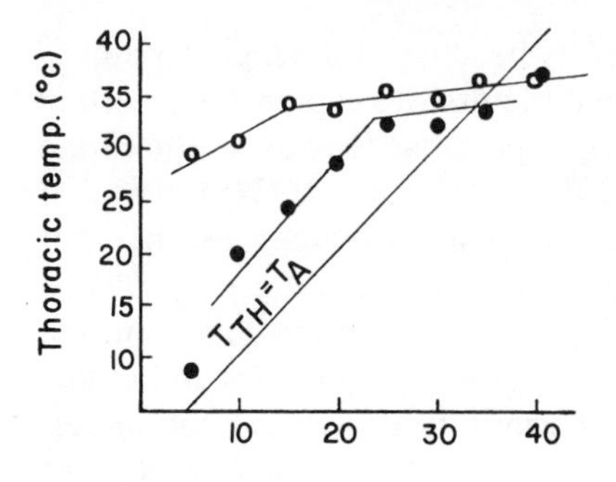

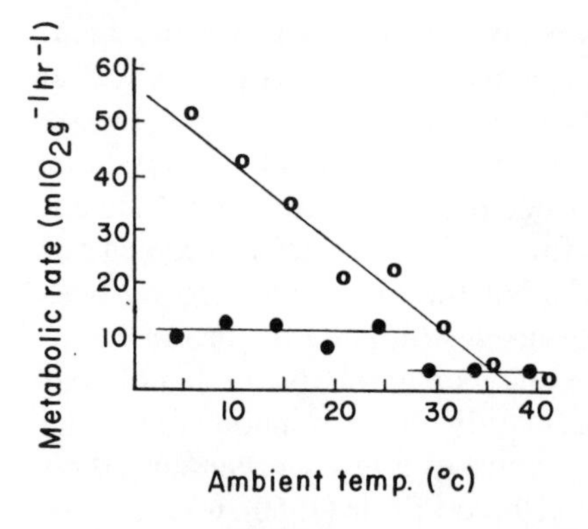

Fig. 20 Mean thoracic temperatures (top) and metabolic rates (bottom) of small groups (7–10) of honeybee workers (open circles) and drones (solid circles) exposed to various ambient temperatures (Cayhill and Lustic, 1976).

abolic rate of mature adult workers increase as T_A drops from 33 to 20°C, but the response is little developed in bees 3–4 days old (Himmer, 1925) or in drones (Fig. 20).

The maximum metabolic rate is observed somewhere between 15 and 20°C and below this level drops off abruptly, probably because when chilled below about 18°C honeybees have great difficulty in activating their flight muscles. Fortunately, these observations on individual bees in respirometers have been confirmed with bees under far more natural conditions. For example, bees in groups of up to 200 individuals were kept at various temperatures from 5 to 40°C, and their rates of sugar consumption were measured. At 10 to 35°C they consumed 33.9 and 9.5 mg sucrose per bee per day respectively (Free and Spencer-Booth, 1958). Also, when Kronenberg (1979) followed the carbon dioxide production of a small (one-frame) colony as T_A dropped from 40 to 10°C over 5 hr, she detected an increase in metabolic rate from 28 to 108 W/kg when the temperature inside the experimental hive dropped from 28 to 15°C.

Another pattern of variation in heat production is a daily rise and fall in a colony's total metabolism (Stussi, 1972; Kronenberg, 1979; Heussner and Roth, 1963) when under moderate cold stress. Kronenberg, following the metabolic rate for 48 hr of a one-frame observation colony kept at 19°C, observed that metabolic rate peaked during the day at about 80 W/kg and decreased to about half that rate during the night. Most importantly, the temperature of the brood (measured by a thermocouple inside a cell of capped brood) remained stable at 34.0 ± 1.5°C. The same circadian rhythm is detectable in the magnitude of the thermogenic response by individual bees in respirometers (Heussner and Roth, 1963; Heussner and Stussi, 1964). The adaptive significance of this daily variation in rate of heat production per bee may serve as an economy measure. Colonies can reduce their per-bee heat production at night when all the bees are gathered in the nest and thus can provide better insulation and a more massive heat source than during the day.

Details of an individual bee's behavior in the winter cluster have been analyzed by Esch (1960). He followed the course of individuals with chronically implanted thermocouples inside an observation hive for 14–21 days. When in the outer shell, a bee's T_{Th}, T_{Ab} (abdominal temperature), and T_A (measured 0.5 cm above the thorax) were close to one another with T_{Th} only about 1–2°C above the other two. Such a bee rested almost motionlessly, just lightly pumping its abdomen. Individuals in the shell would from time to time pass into the center, where, for example, T_{Th} rose from 24.5 to 33–37°C, thus raising it 3–6°C above T_{Ab} and 4–7°C above T_A. When inside the core of the cluster, numerous jumps in T_{Th} were observed, just as in incubating bees during the summer. Time spent in the cluster core was of variable duration, up to 12 hr, and upon leaving the center the jumps in T_{Th} disappeared. From these observations it seems clear that significant heat production occurs only after bees reach the loose cluster core. However, it is not clear what stimulates them to leave the shell, enter the center, and there actively generate heat. During the entire period of observation (December 1958–March 1959), at no time did Esch record a bee with $T_{Th} < 18$°C, even though at times T_A at the nest's entrance was less than −5°C. This is consistent with the observation of Esch and Bastian (1968) that the interval between muscle action potentials increases greatly as the flight muscle temperature drops to 20°C or less. The lowest flight muscle temperature at which bees observed "at various times of the year" could generate action potentials was 18°C, and thus it appears that this is near the lower limit of a cluster's core temperature.

One aspect of nest warming that is little studied but is essential to a full understanding of honeybee thermoregulation involves the stimuli used by bees to orient their warming activities. Some preliminary studies have been made. Koeniger (1975) observed that both chemical and mechanical stimuli were necessary to elicit worker incubation of the special, peanut-shaped cells in which queens are reared. Empty cells were ignored, but replacing a pupa shortly after its removal with a small stone restored the attractiveness of a cell for incubation, and extracts of queen and worker pupae also helped restore the attractiveness of long-empty queen cells. Kronenberg (1979) has observed that, in small colonies with brood, the workers tend to cluster on capped brood and have a higher metabolic rate there than on honey. Whether chemical and/or tactile signals are involved in recognizing capped brood is not clear. Ritter and Koeniger (1977) have shown that it can be the brood itself, and not just the time of year, that stimulates warming the nest to brood nest temperatures. In the summer, colonies with brood regulated their entire brood nest at 33 ± 2°C, whereas at the same time control colonies without brood had a temperature above 30°C in only a small, central region of the nest.

Whether or not bees produce heat (incubate) is presumably not only a function of their own temperature but also of that of the brood. The workers have "external" temperature receptors on the last five antennal segments, the sensilla ampullacea and sensilla coeloconia. These pit-peg receptors are "cold receptors"; their impulse frequency increases with decreasing temperature (Lacher, 1964). Possibly they can function in monitoring changes in brood temperature or in orienting in temperature gradients. Behavioral experiments in a temperature "organ"—a long chamber whose floor is heated at one end and cooled at the other to produce a temperature gradient—have demonstrated that bees can resolve temperature differences as fine as 0.25°C (Heran, 1952).

5.4.5 Cooling the Nest

Although the rate of heat production by brood and resting or laboring adults is low relative to that of an actively incubating adult, at times of high ambient temperatures honeybee colonies face the problem of their brood nests overheating. As described, the upper limit for brood nest temperature is about 36°C, with long-term temperature excesses of just 1–2°C severely disrupting brood metamorphosis. When Lindauer (1954) placed a hive in full sunlight on a lava field near Salerno, southern Italy, the hive's maximum internal temperature never exceeded 36°C, even though the outside temperature rose to 60°C. To counter overheating, colonies employ several mechanisms of nest cooling in a graded response, starting with simple dispersal of the adults in the nest and proceeding through fanning, water evaporation (Fig. 21), and finally partial evacuation of the nest.

Adult bee dispersal within the nest at increasing T_A is probably simply an extension of the cluster expansion that starts when T_A rises above about 0°C. The precise environmental temperature at which this cluster

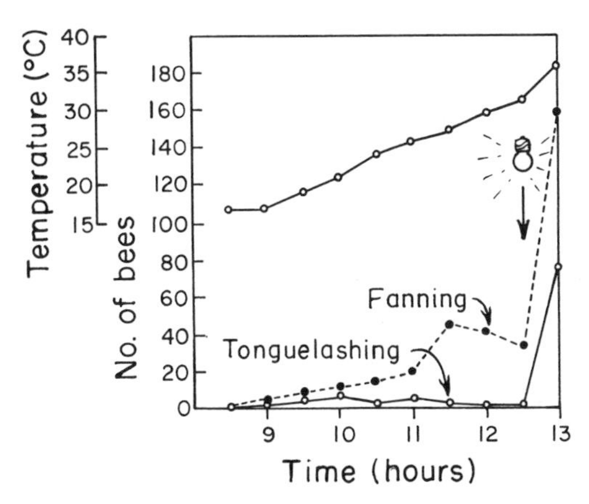

Fig. 21 The number of honeybees fanning and tongue-lashing with extruded fluid droplet as a function of air temperature, and after the colony was artificially heated. The bees were fed dilute (1 molal) sucrose solution throughout the experiment. (From Lindauer, 1954.)

expansion begins to be supplemented by fanning is probably quite variable and depends upon such factors as the nest's sun exposure, its insulation, and colony strength. Hess (1926) and Wohlgemuth (1957) report fanning when the nest temperature reached 36°C if not before. The fanning bees deploy themselves throughout the brood nest, aligning themselves in chains to drive air along existing (unidirectional) currents. Gangs of fanners face inward at the nest entrance and expel currents of warm air from the nest. Without entrance fanners there would probably be little exchange between the nest atmosphere and the outside air through the small entrance opening.

High rates of airflow are created by these fanners. Hazelhoff (1954) constructed a hive with two openings, one at the top connected to an anemometer, and one at the bottom for the hive's entrance. Using this hive he estimated the flow rates produced by fanning bees. Once, when there were 12 strongly fanning bees spaced evenly across the 25-cm-wide entrance, the rate of air flow inward through the top entrance was 0.8–1.0 liters/sec. Fanning bees use a markedly different pattern of wing movements than flying bees. The plane of wing movements, wingbeat frequency, and angle of attack all differ in the two behaviors (Neuhaus and Wohlgemuth, 1960).

The entrance fanners are separated from the central sites of nest overheating. What then are the stimuli evoking fanning at the entrance? Neuhaus and Wohlgemuth found that air streams with a velocity of about 2 m/sec and with a vibration frequency of 140–160 Hz (as presumably would be created by fanning bees inside the nest) were sufficient to elicit fanning at the entrance. Quite curiously, the temperature of the air stream is of little importance; currents of 22–40°C could stimulate fanning.

When cluster expansion and ventilation cannot cool the nest sufficiently, the powerful cooling technique of water evaporation is also brought into play. The potential for nest cooling afforded by water evaporation is demonstrated by the observations of Chadwick (1931) in California. One day in June when the midday air temperature rose to 48°C, the bees brought large amounts of water into their nests and there was little melting of combs. By 2100 hr the temperature had dropped to 29.5°C, but at midnight a hot breeze from the desert raised the air temperature to 38°C. The supplies of water in colonies became exhausted. No more water could be collected until daylight, and many wax combs softened and collapsed. Pure beeswax melts at 61.7°C.

The precise handling of water and the regulation of its collection formed the subject of intensive studies by Lindauer (1954) and Kiechle (1961). Water for nest cooling is distributed about the nest as small puddles in depressions on capped cells, smeared as a thin layer over the roofs of open cells, or placed as hanging droplets in these cells. It may also be rapidly evaporated via "tongue-lashing" behavior by bees which hang over the brood cells and steadily extend their probosces back and forth. Each time they do this they press a drop of water from their mouths and spread it with the proboscis into a film which has a large evaporating surface.

A remarkable communication system is employed to regulate a colony's water collection. The fundamental complication in this operation is that the bees that sense the overheating and become "water sprinklers" are the young nurse bees in the central brood nest and not the older forager bees which can gather water. This division of labor was demonstrated by heating a small, central area of comb in an observation hive with a narrow beam of intense light and observing that this stimulated water collection by the colony's foragers without their contacting this region of the nest.

How are foragers informed of the need for cooling water? First, Lindauer (1954) found that colonies collected some water each day from spring to fall for use in diluting honey to make brood food if not for cooling of the nest. Thus there are almost always a few individuals foraging for water. To transmit the information whether or not the colony needs more water, the hive bees make use of the short moment when they contact the water collectors during water delivery near the entrance hole. When a water shortage exists, a homecoming water forager is met at the entrance by hive bees which rush up to her and quickly suck up her extruded water droplet. Such a stormy reception informs the water forager that there is a pressing need for more water. On the other hand, when the overheating begins to subside, the hive bees show less interest in the water foragers. The latter now have to run around the hive themselves, trying to find a hive bee that will accept their loads of water, and the delivery time increases. The delivery time appears to be an accurate gauge of a colony's water needs, though the behavior of the hive bees toward foragers could also transmit this information. With delivery times of up to 60 sec, water collection continues, but if foragers cannot unload their water in 60 sec their eagerness for collecting decreases rapidly. Water collecting almost disappears when delivery takes longer than 180 sec. A second point concerning recruitment is that, when delivery times are very short (up to about 40 sec), the water collectors even perform recruitment dances after each collecting flight, and this stimulates other foragers to collect from the water source.

Finally, when under conditions of extremely high temperature and high humidity, the bees may partially evacuate their nest and form a "beard" of hanging, clustered bees at the nest entrance. Dunham (1931) observed the start of this when the coolest region of a colony's brood nest reached more than 34°C, and presumably ventilation and evaporative cooling were already being applied in full force within the nest.

6 SUMMARY

Regulation of nest temperature is a general capacity of the highly social wasps, ants, bees, and termites. It is little developed in species forming small colonies, but conspicuous levels of temperature control are universal among species with large and complex societies.

Several properties of the solitary ancestors of social insects provided important stepping stones in the evolution of colony thermoregulation. These include nest site selection and nest construction behaviors, heat production as a by-product of metabolism, wings capable of fanning, massive flight muscles for heat production, and mechanisms of individual thermoregulation (see Kammer, this volume). These were elaborated into a diverse array of techniques for nest temperature control.

The array of techniques employed for nest temperature control fall into two general categories: (1) automatic, long-term control over a wide range of environmental conditions through nest location and design, and (2) short-term control through behavioral and physiological responses of individual colony members to minute-by-minute environmental changes.

The second generalization is that the particular ensemble of thermoregulatory techniques used by each of the major groups of social insects reflects certain constraints and capabilities. For example, because termite and ant workers lack wings, they are unable to ventilate their nests by fanning and they cannot readily commute to sources of open water to collect water for evaporative nest cooling. On the other hand, both of these cooling techniques are commonplace among bee and wasp societies whose workers bear wings. Even more important, the lack of large, functional flight muscles in ant and termite workers severely limits them in warming their nests through vigorous generation of body heat in the manner of workers in yellow jacket wasp, bumblebee, and honeybee societies. Instead, termites and ants rely heavily upon the location and design of their nests for automatic temperature regulation. In addition, they can transport their brood within their nests to reach spots best suited for development. The brood in bee and wasp societies, in contrast, is characteristically reared in fixed cells, so temperature regulation of the nursery areas in these societies cannot involve shifting the nursery as local conditions change. More specifically, there are also differences within each of the major groups of social insects.

Termite colonies rely primarily upon the location and design of their nests for temperature regulation. The primitive, wood-dwelling termites of the families Kalotermitidae and Hodotermitidae depend upon the insulating properties of their nest substrate and upon their ability to move from chamber to chamber in their nests as local conditions demand. The more advanced termites—some genera of the Hodotermitidae, the Rhinotermitidae, and the Termitidae—exercise a wider and more precise choice of nest sites in wood or soil and often construct elaborate and well-insulated nests. In species such as *N. exitiosus* and *C. acinaciformis*, whose colony populations can number in the millions, nest temperatures can be elevated by pooling the metabolic heat from colony members and of decaying vegetation. Two striking examples of nest design adapted to temperature control are the anvil-shaped, north-south-aligned mounds of *A. meridionalis* in Australia and the meter or more tall, castle-shaped mounds of *M. bellicosus* in Africa, which enclose a labyrinth of airways surrounding a central nest nursery. Nests of the former design have minimal mound exposure to strong midday sun. The latter nest plan promotes nest ventilation by channeling convection currents inside the nest.

Ants, like termites, are principally wood- and soil-dwelling insects and regulate their nests' temperatures through nest site selection, nest architecture, and movement of brood to nest chambers with the best available temperatures. Some ants are fugitive species. They inhabit short-lived sites and readily move on when the temperature or some other condition of a nest site deteriorates. Most species occupy more stable sites. They typically nest in rotting wood in the wet tropics, in deep underground galleries in deserts, just under the leaf litter in warm temperate regions, and under sunny stones or mounds in cooler areas. Sunlit stones and mounds (which act as solar heat collectors) warm up more rapidly than their surrounding soil and so provide unusually warm microenvironments. The very large mounds of species such as *F. rufa* and *F. polyctena* heat up relatively rapidly under the sun's rays but cool relatively slowly because of their large bulk. Thus they act as storage devices for solar energy. Army ants (genus *Eciton*), whose colonies can contain hundreds of thousands of workers, form open-air nests consisting entirely of the ants' intertwined bodies suspended as a mass from a tree or log. Inside this mass of ants, temperatures are higher because of the ants' metabolism, and they are more stable than in the surrounding air.

Temperature regulation in the colonies of wasps and bees differs radically from that in termite and ant colonies. Bees and wasps rear their brood in fixed cells, thus they cannot move it about at will to favorable microenvironments. Instead, they often build well-insulated nests and heat or cool them as needed. Warmth is produced by activating massive flight muscles, but without wing movements. Cooling techniques include wing fanning to ventilate nests, and evaporation of water droplets placed inside nests.

Paper wasps (genus *Polistes*) construct open-air, nonenclosed nests which they cannot themselves heat. They rely instead upon building nests in sunny locations and cooling them with water evaporation when overheating threatens. Yellow jackets and hornets (subfamily Vespinae) build nests enclosed either by the walls of tree or soil cavities or by a multilayered, paper jacket. These nests can retain metabolically produced heat, and heat production involves both larvae and adult wasps. Adults incubate pupae. The larvae can raise their own temperature by muscular activity and provide, via sugary secretions regurgitated to adults, a colony food reserve.

The large, hairy bumblebees (genus *Bombus*) are notable among the social insects for their incubation powers. Unlike vespine wasps, their larvae generate little or no heat, and adults incubate eggs, larvae, and pupae. The brood cluster is warmed by heat conducted from the bodies of adults sprawled over the brood rather than secondarily from warming the nest atmosphere. Honey pots scattered around the brood clump help ensure a supply of fuel for nearly continuous incubation through times when foraging is impossible.

The honeybee (*A. mellifera*) possesses the best studied, and possibly the most advanced, system for nest temperature control. Two different strategies in temperature regulation are employed depending on whether or not the colony is rearing brood. Colonies containing eggs, larvae, and pupae maintain the brood nest at 32–36°C, even at ambient temperatures of −30°C or less to 48°C or above by heating (flight muscle activity) or cooling (fanning and water evaporation) as needed. Overwintering bees without brood assemble in a tight cluster on their combs. The clustering bees do not regulate the temperature of the cluster interior, but that of its perimeter. This temperature is not allowed to descend below about 9°C, the vital minimum for honeybees. Cluster temperature is controlled by regulating heat loss through cluster contraction or expansion and by adjusting heat production inside the cluster.

REFERENCES

Abraham, M. and Pasteels, J. M. (1977). Nest-moving behaviour in the ant *Myrmica rubra*. *Proc. Eighth Int. Cong. Int. Union Stud. Soc. Insects (Wageningen),* p. 286.

Akratanakul, P. (1975). The natural history of the dwarf honey bee, *Apis florea* F. in Thailand. Ph.D. Thesis, Cornell University, Ithaca, N.Y.

Alford, D. V. (1975). *Bumblebees*. Davis-Poynter: London.

Allen, M. D. (1959). Respiration rates of larvae of drone and worker honey bees, *Apis mellifera*. *J. Econ. Entomol.* **52,** 399–402.

Allen, M. D. (1959). Respiration rates of worker honeybees at different ages and temperatures. *J. Exp. Biol.* **36,** 92–101.

Allen, T., Cameron, S., McGinley, R. M., and Heinrich, B. (1978). The role of workers and new queens in the ergonomics of a bumblebee colony (Hymenoptera: Apidae). *J. Kans. Entomol. Soc.* **51,** 329–342.

Anderson, E. J. (1948). Hive humidity and its effect upon wintering of bees. *J. Econ. Entomol.* **41,** 608–615.

Andrews, E. A. (1927). Ant mounds as to temperature and sunshine. *J. Morphol. Physiol.* **44,** 1–20.

Avitabile, A. (1978). Brood rearing in honeybee colonies from late autumn to early spring. *J. Apic. Res.* **17,** 69–73.

Bastian, J. and Esch, H. (1970). The nervous control of the indirect flight muscles of the honey bee. *Z. Vergl. Physiol.* **67,** 307–324.

Benois, A. (1972). Étude écologique de *Camponotus vagus* Scop. (=*pubescens* Fab.) (Hymenoptera, Formicidae) dans la région d'Antibes: Nidification et architecture des nids. *Insectes Soc.* **19,** 111–129.

Bigley, W. S. and Vinson, S. B. (1975). Characterization of a brood pheromone isolated from the sexual brood of the imported fire ant, *Solenopsis invicta*. *Ann. Entomol. Soc. Am.* **68,** 301–304.

Bouillon, A. (1970). Termites of the Ethiopian region. In *Biology of Termites*, vol. 2, K. Krishna and F. M. Weesner, Eds. Academic: New York.

Brian, A. D. (1952). Division of labour and foraging in *Bombus agrorum* Fabricius. *J. Anim. Ecol.* **21,** 223–240.

Brian, M. V. (1952). The structure of a dense natural ant population. *J. Anim. Ecol.* **21,** 12–24.

Brian, M. V. (1956). The natural density of *Myrmica rubra* and associated ants in West Scotland. *Insectes Soc.* **3,** 474–487.

Brian, M. V. (1973). Temperature choice and its relevance to brood survival and caste determination in the ant *Myrmica rubra* L. *Physiol. Zool.* **46,** 245–252.

Brian, M. V. (1975). Larval recognition by workers of the ant *Myrmica. Anim. Behav.* **23,** 745–756.

Brian, M. V. and Brian, A. D. (1951). Insolation and ant populations in the west of Scotland. *Trans. R. Entomol. Soc. Lond.* **102,** 303–330.

Brückner, D. (1975). Die Abhängigkeit der Temperaturregulierung von der genetischen Variabilität der Honigbiene (*Apis mellifera* L.). *Apidologie* **6,** 361–380.

Brückner, D. (1976). Vergleichende Untersuchungen zur Temperaturpraeferenz von ingezüchteten und nichtingezüchteten Arbeiterinnen der Honigbiene (*Apis mellifera*). *Apidologie* **7,** 139–149.

Bruman, F. (1928). Die Luftzirkulation im Bienenstock. *Z. Vergl. Physiol.* **8,** 366–370.

Büdel, A. (1960). Bienenphysik. In *Biene und Bienenzucht*, A. Büdel and E. Herold, Eds. Ehrenwirth: Munich.

Büdel, A. (1968). Le Microclimat de la ruche. In *Traité de Biologie de l'Abeille*, vol. 4, R. Chauvin, Ed. Masson: Paris.

Burgett, D. M. (1974). Glucose oxidase: A food protective mechanism in social Hymenoptera. *Ann. Entomol. Soc. Am.* **67,** 545.

Cayhill, K. and Lustic, S. (1976). Oxygen consumption and thermoregulation in *Apis mellifica* workers and drones. *Comp. Biochem. Physiol.* **55A,** 355–357.

Ceusters, R. (1977). Social homeostasis in colonies of *Formica polyctena* Foerst. (Hymenoptera, Formicidae): Nestform and temperature preferences. *Proc. Eighth Int. Congr. Int. Union Stud. Soc. Insects (Wageningen)*, pp. 111–112.

Chadwick, P. C. (1931). Ventilation of the hive. *Glean. Bee Cult.* **59,** 356–358.

Cheema, P. S., Das, S. R., Dayal, H. M., Koshi, T., Maheshwari, K. L., Nigam, S. S., and Ranganathan, S. K. (1960). Temperature and humidity in the fungus garden of the mound-building termite *Odontotermes obesus* (Rambur). In *Termites in the Humid Tropics*. UNESCO: Paris.

Coaton, W. G. H. (1948). *Trinervitermes* species. The snouted harvester termites. *U.S. Afr. Dep. Agric. Bull.* **290,** 1–24.

Corkins, C. L. (1930). The metabolism of the honey bee colony during winter. *Bull. Wyo. Agr. Exp. Sta.* **175,** 1–54.

Corkins, C. L. (1932). The temperature relationship of the honeybee cluster under controlled temperature conditions. *J. Econ. Entomol.* **25,** 820–825.

Corkins, C. L. and Gilbert, C. S. (1932). The metabolism of honey-bees in winter. *Bull. Wyo. Agric. Exp. Sta.* **187,** 1–30.

Cowles, R. B. (1930). The life history of *Varanus niloticus* (Lin.) as observed in Natal, South Africa. *J. Entomol. Zool.* **22,** 1–31.

Cumber, R. A. (1949). The biology of bumble-bees with special reference to the production of the worker caste. *Trans. R. Entomol. Soc. Lond.* **100,** 1–45.

Darchen, R. (1973). La thermorégulation et l'écologie de quelques espèces d'abeilles sociales d'Afrique (Apidae, Trigonini et *Apis mellifica* var. *adansonii*), *Apidologie* **4,** 341–370.

Délye, G. (1967). Physiologie et comportement de quelques fourmis (Hym. Formicidae) du Sahara en rapport avec les principaux facteurs du climat. *Insectes Soc.* **14,** 323–338.

Doflein, F. (1906). Termites truffles. *Spolia Zeylan.* **3,** 203–209.

Dreyer, W. A. (1942). Further observations on the occurrence and size of ant mounds with reference to their age. *Ecology* **23,** 486–490.

Dunham, W. E. (1929). The influence of external temperature on the hive temperatures during the summer. *J. Econ. Entomol.* **22,** 798–801.

Dunham, W. E. (1931). A colony of bees exposed to high external temperatures. *J. Econ. Entomol.* **24,** 606–611.

Dunham, W. E. (1933). Hive temperatures during the summer. *Glean. Bee Cult.* **61,** 527–529.

Ebner, R. (1926). Einige Beobachtungen an Termitenbauten. *Denkschr. Akad. Wiss. Wien, Math.-Nat. Kl.* **100,** 75–76.

Elton, C. (1932). Orientation of the nests of *Formica truncorum* in north Norway. *J. Anim. Ecol.* **1,** 192–193.

Esch, H. (1960). Über die Körpertemperaturen und den Wärmehaushalt von *Apis mellifica*. *Z. Vergl. Physiol.* **43,** 305–335.

Esch, H. (1964). Über den Zusammenhang zwischen Temperatur, Aktionspotentialen und Thoraxbewegungen bei der Honigbiene (*Apis mellifica* L.). *Z. Vergl. Physiol.* **48,** 547–551.

Esch, H. and Bastian, J. (1968). Mechanical and electrical activity in the indirect flight muscles of the honey bee. *Z. Vergl. Physiol.* **58,** 429–440.

Escherich, K. (1911). *Termitenleben auf Ceylan.* Fischer: Jena.

Es'kov, E. R. (1974). The microclimate of the hive and the biological conditions of the colony (in Russian). *Pchelovod. Mosk.* **94,** 19–21.

Fabritius, M. (1976). Experimentelle Untersuchung des Wärmeverhaltens der Hornissen (*Vespa crabro*). Dissertation, Frankfurt.

Farrar, C. L. (1943). An interpretation of the problems of wintering the honeybee colony. *Glean. Bee Cult.* **71,** 513–518.

Fielde, A. M. (1904). Observations on ants in their relation to temperature and to submergence. *Biol. Bull.* **7,** 170–174.

Forel, A. (1874). *Les Fourmis de la Suisse.* Société Helvétique des Sciences Naturelles: Zurich.

Free, J. B. (1977). *The Social Organization of Honeybees.* Arnold: London.

Free, J. B. and Butler, C. G. (1959). *Bumblebees.* Collins: London.

Free, J. B. and Spencer-Booth, H. Y. (1958). Observations on the temperature regulation and food consumption of honey-bee (*Apis mellifica*). *J. Exp. Biol.* **35,** 930–937.

Free, J. B. and Spencer-Booth, H. Y. (1960). Chill-coma and cold death temperatures of *Apis mellifica*. *Entomol. Exp. Appl.* **3,** 222–230.

Frisch, K. von. (1974). *Animal Architecture.* Harcourt Brace Jovanovich: New York.

Fye, R. E. and Medler, J. T. (1954). Temperature studies in bumblee (*sic*) domiciles. *J. Econ. Entomol.*, **47,** 847–852.

Gates, B. N. (1914). The temperature of the bee colony. *Bull. U.S. Dep. Agric.* **96,** 1–29.

Gaul, A. T. (1952). Additions to vespine biology IX. Temperature regulation in the colony. *Bull. Brooklyn Entomol. Soc.* **47,** 79–82.

Gaul, A. T. (1952). Metabolic cycles and the flight of vespine wasps. *J. N.Y. Entomol. Soc.* **60,** 21–24.

Gay, F. J. and Calaby, J. H. (1970). Termites from the Australian region. In *Biology of Termites*, vol. 2, K. Krishna and F. M. Weesner, Eds. Academic: New York.

Gay, F. J. and Wetherly, A. H. (1970). The population of a large mound of *Nasutitermes exitiosus* (Hill) (Isoptera: Termitidae). *J. Aust. Entomol. Soc.* **9,** 27–30.

Geyer, J. W. (1951). A comparison between the temperatures in a termite supplementary fungus garden and in the soil at equal depth. *J. Entomol. Soc. S. Africa* **14,** 36–43.

Gibo, D. L., Yarascavitch, R. M., and Dew, H. E. (1974). Thermoregulation in colonies of *Vespula arenaria* and *Vespula maculata* (Hymenoptera: Vespidae) under normal conditions and under cold stress. *Can. Entomol.* **106,** 503–507.

Gibo, D. L., Dew, H. E., and Hajduk, A. S. (1974). Thermoregulation in colonies of *Vespula arenaria* and *Vespula maculata* (Hymenoptera: Vespidae). II. The relation between colony biomass and calorie production. *Can. Entomol.* **106,** 873–879.

Gibo, D. L., Temporale, A., Lamarre, T. P., Soutar, B. M., and Dew, H. E. (1977). Thermoregulation in colonies of *Vespula arenaria* and *Vespula maculata* (Hymenoptera: Vespidae). III. Heat production in queen nests, *Can. Entomol.* **109,** 615–620.

Grassé, P. P. (1944). Recherches sur la biologie des termites champignonnistes (Macrotermitinae). *Ann. Sci. Nat. Zool.* **6,** 97–171; **7,** 115–146 (1945).

Grassé, P. P. (1949). Ordre des Isoptères ou Termites. In *Traité de Zoologie*, vol. 9, P. P. Grassé, Ed. Masson: Paris.

Grassé, P. P. and Noirot, C. (1958). Le comportement des termites à l'égard de l'air libre. L'atmosphère des termitières et son renouvellement. *Ann. Sci. Nat. Zool.* **20,** 1–28.

Greaves, T. (1964). Temperature studies of termite colonies in living trees. *Aust. J. Zool.* **12,** 250–262.

Greaves, T. (1967). Experiments to determine the populations of tree-dwelling colonies of termites *Coptotermes acinaciformis* (Froggat) and *C. frenchi* (Hill), Division of Entomology Technical Paper No. 7. Commonwealth Scientific and Industrial Research Organization: Australia.

Greaves, T. and Florence, R. G. (1966). Incidence of termites in blackbutt regrowth, *Aust. For.* **30,** 153–161.

Grigg, G. C. (1973). Some consequences of the shape and orientation of "magnetic" termite mounds. *Aust. J. Zool.* **21,** 231–237.

Grigg, G. C. and Underwood, A. J. (1977). An analysis of the orientation of "magnetic" termite mounds. *Aust. J. Zool.* **25,** 87–94.

Hasselrot, T. B. (1960). Studies on Swedish bumblebees (genus *Bombus* Latr.). *Opuscula Entomol. Suppl.* **17,** 1–200.

Haydak, M. H. (1958). Wintering of bees in Minnesota. *J. Econ. Entomol.* **51,** 332–334.

Hazelhoff, E. H. (1954). Ventilation in a bee-hive during summer. *Physiol. Comp. Oecol.* **3,** 343–364.

Heinrich, B. (1972). Patterns of endothermy in bumblebee queens, drones and workers. *J. Comp. Physiol.* **77,** 65–79.

Heinrich, B. (1974a). Pheromone-induced brooding behavior in *Bombus vosnesenskii* and *B. edwardsii* (Hymenoptera: Bombidae). *J. Kans. Entomol. Soc.* **47,** 396–404.

Heinrich, B. (1974b). Thermoregulation in bumblebees. I. Brood incubation by *Bombus vosnesenskii* queens. *J. Comp. Physiol.* **88,** 129–140.

Heinrich, B. (1974c). Thermoregulation in endothermic insects. *Science* **185,** 747–756.

Heinrich, B. (1976). The foraging specializations of individual bumblebees, *Ecol. Monogr.* **46,** 105–128.

Heinrich, B. (1979a). *Bumblebee Economics.* Harvard University Press: Cambridge, Mass.

Heinrich, B. (1979b). Thermoregulation of African and European honeybees during foraging, attack, and hive exits and returns. *J. Exp. Biol.* **80,** 217–229.

Heinrich, B. and Kammer, A. E. (1973). Activation of the fibrillar muscles in the bumblebee during warm-up, stabilization of thoracic temperature and flight. *J. Exp. Biol.* **58,** 677–689.

Heran, H. (1952). Untersuchungen über den Temperatursinn der Honigbiene unter besonderer Berücksichtigung der Wahrnehmung strahlender Wärme. *Z. Vergl. Physiol.* **34,** 179–206.

Hess, W. R. (1926). Die Temperaturregulierung im Bienenvolk. *Z. vergl. Physiol.* **4,** 465–487.

Heussner, A. and Roth, M. (1963). Consommation d'oxygène de l'abeille à différentes températures. *C. R. Hebd. Séance Acad. Sci., Paris* **256,** 284–285.

Heussner, A. and Stussi, T. (1964). Métabolisme énergétique de l'abeille isolée. Son rôle dans la thermoregulation de la ruche. *Insectes Soc.* **11,** 239–266.

Hill, G. F. (1942). *Termites (Isoptera) from the Australian Region.* Commonwealth Scientific and Industrial Research Organization: Melbourne.

Himmer, A. (1925). Körpertemperaturmessungen an Bienen und anderen Insekten. *Erlanger. Jahrb. Bienenk.* **3,** 44–115.

Himmer, A. (1926). Der soziale Wärmehaushalt der Honigbiene. I. Die Wärme im nichtbrütenden Wintervolk. *Erlanger. Jahrb. Bienenk.* **4,** 1–51.

Himmer, A. (1927). Ein Beitrag zur Kenntnis des Wärmehaushalts in Nestbau sozialer Hautflügler. *Z. Vergl. Physiol.* **5,** 375–389.

Himmer, A. (1931). Über die Wärme im Hornissennest. *Z. Vergl. Physiol.* **13,** 748–761.

Himmer, A. (1932). Die Temperaturverhältnisse bei den sozialen Hymenopteren. *Biol. Rev.* **7,** 224–253.

Himmer, A. (1933). Die Nestwärme bei *Bombus agrorum* F. *Biol. Zentralbl.* **53,** 270–276.

Hoffer, E. (1882). Die Hummelbauten. *Kosmos.* **12,** 412–421.

Holdaway, F. G. and Gay, F. J. (1948). Temperature studies of the habitat of *Eutermes exitiosus* with special reference to the temperature within the mound. *Aust. J. Sci. Res.* **B1,** 464–493.

Holdaway, F. G., Gay, F. J., and Greaves, T. (1935). The termite population of a mount colony of *Eutermes exitiosus* Hill. *J. Counc. Sci. Ind. Res.* **8,** 42–46.

Hubbard, M. D. and Cunningham, W. G. (1977). Orientation of mounds in the ant *Solenopsis invicta* (Hymenoptera, Formicidae, Myrmicinae). *Insectes Soc.* **24,** 3–8.

Huber, P. (1810). *Recherches sur les Moeurs des Fourmis Indigènes.* Paschoud: Paris.

Ishay, J. (1972). Thermoregulatory pheromones in wasps. *Experientia* **28,** 1185–1187.

Ishay, J. (1973). Thermoregulation by social wasps: Behavior and pheromones. *Trans. N.Y. Acad. Sci.* **35,** 447–462.

Ishay, J. and Ikan, R. (1968). Food exchange between adults and larvae in *Vespa orientalis* F. *Anim. Behav.* **16,** 298–303.

Ishay, J. and Ikan, R. (1968). Gluconeogenesis in the oriental hornet, *Vespa orientalis* F. *Ecology* **49,** 169–171.

Ishay, J. and Ruttner, F. (1971). Thermoregulation in Hornissennest. *Z. Vergl. Physiol.* **72,** 423–434.

Ishay, J., Bytinski-Salz, H., and Shulov, A. (1967). Contributions to the bionomics of the oriental hornet (*Vespa orientalis* Fab.). *Israel J. Entomol.* **2,** 45–106.

Jackson, W. B. (1957). Microclimate patterns in the army ant bivouac. *Ecology* **38,** 276–285.

Janet, C. (1895). Études sur les Fourmis, les Guêpes et les Abeilles. Neuvième note. Sur *Vespa crabro* L. *Mém. Soc. Zool. Fr.* **8,** 1–140.

Jongbloed, J. and Wiersma, C. A. G. (1934). Der Stoffwechsel der Honigbiene während des Fliegens. *Z. Vergl. Physiol.* **21,** 519–533.

Jordan, R. (1936). Beobachtung der Arbeitsteilung im Hummelstaate (*Bombus muscorum*). *Arch. Bienenk.* **17,** 81–91.

Josens, G. (1971). Variations thermiques dan les nids de *Trinervitermes geminatus* Wasmann, en relation avec le milieu extérieur, dans la savane de Lamto (Côte d'Ivoire). *Insectes Soc.* **18,** 1–14.

Katô, M. (1939). The diurnal rhythm of temperature in the mound of an ant, *Formica truncorum truncorum* var. *Yessenni* Forel, widely distributed at Mt. Hakkôda. *Sci. Rep. Tôhoku Univ. Sendai*, **14,** 53–64.

Kemp, P. B. (1955). The termites of north-eastern Tanganyika: Their distribution and biology. *Bull. Entomol. Res.* **46,** 113–135.

Kiechle, H. (1961). Die soziale Regulation der Wassersammeltätigkeit im Bienenstaat und deren physiologische Grundlage. *Z. Vergl. Physiol.* **45,** 154–192.

Kneitz, G. (1964). Untersuchungen zum Aufbau und zur Erhaltung des Nestwärmehaushaltes bei *Formica polyctena* Foerst. (Hym. Formicidae). Dissertation, Würzburg.

Koeniger, N. (1975). Experimentelle Untersuchungen über das Wärmen der Brut bei *Vespa crabro* und *Apis mellifica*. *Verh. Dtsch. Zool. Ges.*, 148.

Kosmin, N. P., Alpatov, W. W., and Resnitschenko, M. S. (1932). Zur Kenntnis des Gaswechsels und des Energieverbrauchs der Biene in Beziehung zu deren Aktivität. *Z. Vergl. Physiol.* **17,** 408–422.

Kronenberg, F. (1979). Characteristics of colonial thermoregulation in honey bees. Ph.D. Thesis, Stanford University, Stanford, Calif.

Lacher, V. (1964). Elektrophysiologische Untersuchungen an einzelnen Rezeptoren für Geruch, Luftfeuchtigkeit und Temperatur auf den Antennen der Arbeitsbiene und der Drohne. *Z. Vergl. Physiol.* **48,** 587–623.

Lange, R. (1959). Experimentelle Untersuchungen über den Nestbau der Waldameisen. Nesthügel und Volkstärke. *Entomophaga* **4,** 47–55.

Lee, K. E. and Wood, T. G. (1971) *Termites and Soils*. Academic: New York.

Levieux, J. (1972). Le microclimat des nids et des zones de chasse de *Camponotus acvapimensis* Mayr. *Insectes Soc.* **2,** 63–79.

Lindauer, M. (1954). Temperaturregulierung und Wasserhaushalt im Bienenstaat. *Z. Vergl. Physiol.* **36,** 391–432.

Lindauer, M. (1956). Über die Verständigung bei indischen Bienen. *Z. Vergl. Physiol.* **38,** 521–557.

Lindauer, M. and Kerr, W. E. (1958). Die gegenseitige Verständigung bei den stachellosen Bienen. *Z. Vergl. Physiol.* **41,** 405–434.

Linder, C. (1908). Observations sur les Fourmilières-Boussoles. *Bull. Soc. Vaud. Sci. Nat., 5 Ser.* **44,** 303–310.

Lindhard, E. (1912). Humlebien som Husdyr. Spredte Traek af nogle danske Humlebiarters Biologi. *Tidsskr. Plavl.* **19,** 335–352.

Loos, R. (1964). A sensitive anemometer and its use for the measurement of air currents in the nests of *Macrotermes natalensis* (Haviland). In *Études sur les Termites Africains,* A. Bouillon, Ed. Masson: Paris.

Lüscher, M. (1951). Significance of "fungus gardens" in termite nests. *Nature* **167,** 34–35.

Lüscher, M. (1955). Der Sauerstoffverbrauch bei Termiten und die Ventilation des Neste bei *Macrotermes natalensis* (Haviland). *Acta Trop.* **12,** 289–307.

Lüscher, M. (1961). Air-conditioned termite nests. *Sci. Am.* **205,** 138–145.

Martin, M. A. and Martin, J. S. (1978). Cellulose digestion in the midgut of the fungus-growing termite *Macrotermes natalensis*: The role of acquired digestive enzymes. *Science* **199,** 1453–1455.

Maschwitz, U. (1966). Das Speichelsekret der Wespenlarven und seine biologische Bedeutung. *Z. Vergl. Physiol.* **53,** 228–252.

Maschwitz, U., Koob, K., and Schildknecht, H. (1970). Ein Beitrag zur Funktion der Metathoracaldrüse der Ameisen. *J. Insect. Physiol.* **16,** 387–404.

Melampy, R. M. and Willis, E. R. (1939). Respiratory metabolism during larval and pupal development of the female honey-bee. *Physiol. Zool.* **12,** 302–311.

Meudec, M. (1977). Le comportement de transport du couvain lors d'une perturbation du nid chez *Tapinoma erraticum* (Dolichoderinae). Rôle de l'individu. *Insectes Soc.* **24,** 345–353.

Michener, C. D. (1974). *The Social Behavior of the Bees. A Comparative Study.* Harvard University, Cambridge, Mass.

Michener, C. D. and Wille, A. (1961). The bionomics of a primitively social bee, *Lasioglossum inconspicuum. Univ. Kans. Sci. Bull.* **42,** 1123–1202.

Michener, C. D., Lange, R. B., Bigarella, J. J., and Salamuni, R. (1958). Factors influencing the distribution of bees' nests in earth banks. *Ecology* **39,** 207–217.

Milum, V. G. (1930). Variation in time of development of the honey bee. *J. Econ. Entomol.* **23,** 441–447.

Möglich, M. (1978). Social organization of nest emigration in *Leptothorax* (Hym., Form.). *Insectes Soc.* **25,** 205–225.

Montagner, H. (1964). Étude du comportement alimentaire et des relations trophallactiques des mâles au sein de la société de guêpes, au moyen d'un radio-isotope. *Insectes Soc.* **11,** 301–316.

Montagner, H. (1966). Le mécanisme et les conséquences des comportements trophallactiques chez les guêpes du genre *Vespa.* Thesis, Nancy.

Morimoto, R. (1960). Experimental study on the trophallactic behavior in *Polistes* (Hymenoptera, Vespidae). *Acta Hymenoptera* **1,** 99–103.

Morse, R. A. and Laigo, F. M. (1969). *Apis dorsata* in the Philippines. *Monogr. Philipp. Assoc. Entomol.* **1,** 1–96.

Neuhaus, W. and Wohlgemuth, R. (1960). Über das Fächeln der Bienen und dessen Verhältnis zum Fliegen. *Z. Vergl. Physiol.* **43,** 615–641.

Newport, G. (1837). On the temperature of insects and its connexion with the functions of respiration and circulation in this class of invertebrate animals. *Phil. Trans.* **B127,** 259–338.

Nielsen, E. T. (1938). Temperatures in a nest of *Bombus hypnorum* L. *Vidensk. Medd. Naturhist. Foren. Kbh.* **102,** 1–6.

Noirot, C. (1958–1959). Remarques sur l'écologie des termites. *Ann. Soc. R. Zool. Belg.* **89,** 151–169.

Noirot, C. (1970). "The nests of termites. In *Biology of Termites*, vol. 2, K. Krishna and F. M. Weesner, Eds. Academic: New York.

Nolan, W. J. (1925). The brood-rearing cycle of the honeybee. *Bull. U.S. Dep. Agric.* **1349,** 1–56.

Nye, P. H. (1955). Some soil-forming processes in the humid tropics. IV. The action of the soil fauna. *J. Soil Sci.* **6,** 73–83.

Ofer, J. (1970). *Polyrachis simplex*, the weaver ant of Israel. *Insectes Soc.* **17,** 49–82.

Oster, G. F. and Wilson, E. O. (1978). *Caste and Ecology in the Social Insects*, Princeton University Press: Princeton, N.J.

Otto, D. (1962). *Die Roten Waldameisen.* Ziemsen: Wittenberg.

Owens, C. D. (1971). The thermology of wintering honey bee colonies, *Tech. Bull. U.S. Dep. Agric.* **1429,** 1–34.

Park, O. W. (1925). The storing and ripening of nectar by honey-bees. *J. Econ. Entomol.* **18,** 405–410.

Phillips, E. F. and Demuth, G. S. (1914). The temperature of the honeybee cluster in winter. *Bull. U.S. Dep. Agric.* **93,** 1–16.

Pirsch, G. B. (1923). Studies on the temperature of individual insects, with special reference to the honey bee. *J. Agric. Res.* **24,** 275–287.

Plowright, R. C. (1977). Nest architecture and the biosystematics of bumble bees. *Proc. Eighth Int. Cong. Int. Union Stud. Soc. Insects (Wageningen)*, pp. 183–185.

Pontin, A. J. (1960). Field experiments on colony foundation by *Lasius niger* (L.) and *L. flavus* (F.) (Hym., Formicidae). *Insectes Soc.* **7,** 227–230.

Pontin, A. J. (1963). Further considerations of competition and the ecology of the ants *Lasius flavus* (F.) and *L. niger* (L.). *J. Anim. Ecol.* **32,** 565–574.

Postner, M. (1951). Biologisch-Ökologische Untersuchungen an Hummeln und ihren Nestern. *Veroff. Mus. Bremen* **1,** 46–86.

Quinlan, R. J. and Cherrett, J. M. (1978). Aspects of the symbiosis of the leaf-cutting ant *Acromyrmex octospinosus* (Reich) and its food fungus. *Ecol. Entomol.* **3,** 221–230.

Raignier, A. (1948). L'économie thermique d'une colonie polycalique de la fourmi des Bois (*Formica rufa polyctena* Foerst). *Cellule* **51,** 281–368.

Réaumur, R. A. F. de. (1742). *Mémoires pour Servir à l'Histoire des Insectes*, vol. 6. Royale: Paris.

Ribbands, C. R. (1953). *The Behaviour and Social Life of Honeybees.* Bee Research Association: London.

Richards, K. W. (1973). Biology of *Bombus polaris* Curtis and *B. hyperhoreus* Schönherr at Lake Hazen, Northwest Territories (Hymenoptera: Bombini). *Quest. Entomol.* **9,** 115–157.

Ritter, W. and Koeniger, N. (1977). Influence of the brood on the thermoregulation of honeybee colonies. *Proc. Eighth Int. Cong. Int. Union Stud. Soc. Insects (Wageningen)*, pp. 283–284.

Roland, C. (1969). Rôle de l'involucre et du nourissement au sucre dans la régulation thermique à l'intérieur d'un nid de Vespides. *C. R. Hebd. Séance Acad. Sci. Paris* **269,** 914–916.

Roth, M. (1965). La production de chaleur chez *Apis mellifica* L. *Ann. Abeille.* **8,** 5–77.

Ruelle, J. E. (1964). L'architecture du nid de *Macrotermes natalensis* et son sens fonctionel. In *Études sur les Termites Africains,* A. Bouillon, Ed. Masson: Paris.

Ruttner, F. (1968a). Systématique du genre *Apis.* In *Traité de Biologie de l'Abeille*, vol. 1, R. Chauvin, Ed. Masson: Paris.

Ruttner, F. (1968b). Les Races d'Abeilles. In *Traité de Biologie de l'Abeille*, vol. 1, Chauvin, Ed. Masson: Paris.

Sakagami, S. F. (1971). Ethosoziologischer Vergleich zwischen Honigbienen und Stachellosen Bienen. *Z. Tierpsychol.* **28,** 337–350.

Sakagami, S. F. and Hayashida, K. (1960). Biology of the primitive social bee *Halictus duplex* Dalle Torre. II. Nest structure and immature stages. *Insectes Soc.* **7,** 57–98.

Sakagami, S. F. and Hayashida, K. (1961). Biology of the primitive social bee *Halictus duplex* Dalle Torre. III. Activities in spring solitary phase. *J. Fac. Sci. Hokkaido Univ. Ser. 6, Zool.* **14,** 639–682.

Sakagami, S. F. and Michener, C. D. (1962). *The Nest Architecture of the Sweat Bees.* University of Kansas: Lawrence.

Sands, W. A. (1969). The association of termites and fungi. In *Biology of Termites,* vol. 1, K. Krishna and F. M. Weesner, Eds., Academic: New York.

Scherba, G. (1958). Reproduction, nest orientation and population structure of an aggregation of mound nests of *Formica ulkei* Emery. *Insectes Soc.* **5,** 201–213.

Scherba, G. (1962). Mound temperatures of the ant *Formica ulkei* Emery. *Am. Midl. Nat.* **67,** 373–385.

Schneirla, T. C., Brown, R. Z., and Brown, F. C. (1954). The bivouac or temporary nest as an adaptive factor in certain terrestrial species of army ants. *Ecol. Monogr.* **24,** 269–296.

Seeley, T. D. (1974). Atmospheric carbon dioxide regulation in honey-bee (*Apis mellifera*) colonies. *J. Insect Physiol.* **20,** 2301–2305.

Seeley, T. D. (1977). Measurement of nest cavity volume by the honey bee (*Apis mellifera*). *Behav. Ecol. Sociobiol.* **2,** 201–227.

Seeley, T. D. (1978). Life history strategy of the honey bee, *Apis mellifera. Oecologia* **32,** 109–118.

Seeley, T. D. and Morse, R. A. (1976). The nest of the honey bee (*Apis mellifera* L.). *Insectes Soc.* **23,** 495–512.

Seeley, T. D. and Morse, R. A. (1978). Nest site selection by the honey bee. *Apis mellifera. Insectes Soc.* **25,** 323–337.

Simpson, J. (1950). Humidity in the winter cluster of a colony of honey-bees. *Bee World* **31,** 41–44.

Simpson, J. (1961). Nest climate regulation in honey bee colonies. *Science* **133,** 1327–1333.

Skaife, S. H. (1955). *Dwellers in Darkness.* Longmans Green: London.

Sladen, F. W. L. (1912). *The Humble-bee, Its Life-History and How to Domesticate It, with Descriptions of All the British Species of Bombus and Psithyrus.* Macmillan: London.

Snyder, T. E. (1926). Preventing damage by termites or white ants. *U.S. Dep. Agric. Farm. Bull.* **1472,** 1–21.

Sotavalta, O. (1954). On the thoracic temperature of insects in flight. *Ann. Zool. Soc. Vanamo* **16,** 1–22.

Southwick, E. E. and Mugaas, J. N. (1971). A hypothetical homeotherm: The honeybee hive. *Comp. Biochem. Physiol.* **40A,** 935–944.

Steiner, A. (1924). Über den sozialen Wärmehaushalt der Waldameise (*Formica rufa* var. *rufo-pratensis* For.). *Z. Vergl. Physiol.* **2,** 23–56.

Steiner, A. (1926). Temperaturmessungen in den Nestern der Waldameise (*Formica rufa* var. *rufo-pratenis*) und der Wegameise (*Lasius niger*) wahrend des Winters. *Mitt. Naturforsch. Ges. Bern* 1–19.

Steiner, A. (1929). Temperaturuntersuchungen in Ameisennestern mit Erdkuppeln, im Nest von *Formica exsecta* Nyl. und in Nestern unter Steinen. *Z. Vergl. Physiol.* **9,** 1–66.

Steiner, A. (1930). Die Temperaturregulation im Nest der Feldwespe (*Polistes gallica* var. *biglumis* L.). *Z. Vergl. Physiol.* **11,** 461–502.

Steiner, A. (1932). Die Arbeitsteilung der Feldwespe *Polistes dubia* K. Z. *Vergl. Physiol.* **17,** 101–152.

Stuart, A. M. (1977). A polyethic and homeostatic response to a simple stimulus in a tropical termite. *Proc. Eighth Int. Cong. Int. Union Stud. Soc. Insects (Wageningen)*, pp. 149–151.

Stussi, T. (1972). Réaction de thermogenèse au froid chez la guêpe ouvrière et autres Hyménoptéres sociaux. *C. R. Hebd. Séance Acad. Sci., Paris*, **274,** 2687–2689.

Stussi, T. (1972). L'heterothermie de l'abeille. *Arch. Sci. Physiol.* **26,** 131–159.

Sudd, J. H., Douglas, J. M., Gaynard, T., Murray, D. M., and Stockdale, J. M. (1977). The distribution of wood-ants (*Formica lugubris* Zetterstedt) in a northern English forest. *Ecol. Entomol.* **2,** 301–313.

Taylor, O. R. (1977). The past and possible future spread of Africanized honeybees in the Americas. *Bee World* **58,** 19–30.

Vanderplank, F. L. (1960). The bionomics and ecology of the red tree ant *Oecophylla* sp., and its relationship to the coconut bug *Pseudotheraptus wayi* (Brown) (Coreidae). *J. Anim. Ecol.* **29,** 15–33.

Veith, H. J. and Koeniger, N. (1978). Identifizierung von *cis*-9-Pentacosen als Auslöser für das Wärmen der Brut bei der Hornisse. *Naturwissenschaften* **65,** 263.

Waloff, N. and Blacklith, R. E. (1962). The growth and distribution of the mounds of *Lasius flavus* (Fabricius) (Hym., Formicidae) in Silwood Park, Berkshire. *J. Anim. Ecol.* **31,** 421–437.

Walsh, J. P. and Tschinkel, W. R. (1974). Brood recognition by contact pheromone in the red imported fire ant, *Solenopsis invicta. Anim. Behav.* **22,** 695–704.

Wasmann, E. (1915). *Das Gesellschaftsleben der Ameisen.* Aschendorfsche: Münster.

Weir, J. S. (1973). Air flow, evaporation and mineral accumulation in mounds of *Macrotermes subhyalinus* (Rambur). *J. Anim. Ecol.* **42,** 509–520.

Wellenstein, G. (1928). Beiträge zur Biologie der roten Waldameise (*Formica rufa* L.) mit besonderer Berücksichtigung klimatischer und förstlicher Verhältnisse. *Z. Angew. Entomol.* **14,** 1–68.

Wellenstein, G. (1967). Zur Frage der Standortansprüche hügelbauender Waldameisen (*F. rufa*-Gruppe). *Z. Angew. Zool.* **54,** 139–166.

Weyrauch, W. (1928). Beitrag zur Biologie von *Polistes. Biol. Zentralbl.* **48,** 407–427.

Weyrauch, W. (1936). Das Verhalten sozialer Wespen bei Nestüberhitzung. *Z. Vergl. Physiol.* **23,** 51–63.

Wilson, E. O. (1959). Some ecological characteristics of ants in New Guinea rain forest. *Ecology* **40,** 437–447.

Wilson, E. O. (1971). *The Insect Societies.* Harvard University, Cambridge, Mass.

Wilson, H. F. and Milum, V. G. (1927). Winter protection for the honeybee colony. *Res. Bull. Wis. Agric. Exp. Sta.* **75,** 1–47.

Wohlgemuth, R. (1957). Die Temperaturregulation des Bienenvolkes unter regeltheoretischen Gesichtpunkten. *Z. Vergl. Physiol.* **40,** 119–161.

Wójtowski, F. (1963). Studies on heat and water economy in bumble-bee nests. *Zool. Poloniae* **13,** 19–36.

Woodworth, C. E. (1936). Effect of reduced temperature and pressure on honeybee respiration. *J. Econ. Entomol.* **29,** 1128–1138.

Yung, E. (1900). Combien y a-t-il de fourmis dans une fourmiliere (*Formica rufa*)? *Rev. Sci.* **14,** 269–272.

Zahn, M. (1958). Temperatursinn, Wärmehaushalt und Bauweise der Roten Waldameise. *Zool. Beiträge N.F.* **3,** 127–194.

Zucchi, R. and Sakagami, S. F. (1972). Capacidade termo-reguladora em *Trigona spinipes* e em algumas outras espécies de abelhas sem ferrão. In *Homenagem à Warwick E. Kerr.* Rio Claro: Brazil.

7
Ecological and Evolutionary Perspectives

BERND HEINRICH

1 INTRODUCTION

Insects stand premier among land animals in inhabiting the earth's extreme temperature environments. Their ability to live from hot deserts to the High Arctic depends, in large part, on various levels of response to temperature. These responses range from immediate behavioral and physiological adjustments to long-term seasonal synchronizations of life cycles. Mechanisms of the short-term responses used to regulate body temperature (the main focus of this volume) have been discussed in detail in the other chapters. I will here delve into the other, largely long-term, responses that ultimately affect body temperature and activity. In addition, I will review some aspects of insect physiology to lay a foundation for speculation on the evolution of insect temperature regulation and on aspects of its functional significance in this group or organisms. Some of the topics, presented for the sake of providing a broad context, must necessarily be here covered in a less than comprehensive fashion.

2 RESISTANCE TO FREEZING

Most insects are unable to survive the low temperatures of winter in temperate and arctic regions without making physiological adjustments. These adjustments have so far been shown to involve two phenomena: first, the avoidance of internal freezing (frost resistance) and, second, the ability to tolerate internal freezing (frost tolerance). Rather than attempting to review the rich literature on the physiology of overwintering in insects (see recent reviews by Downes, 1965; Salt, 1961, 1969; Asahina, 1969) and other organisms (Crowe and Clegg, 1978), I will here merely outline the main features to indicate the range of responses.

2.1 Supercooling

In most northerly regions inhabited by insects, the lowest winter temperatures usually range from −20 to −50°C, although temperatures of −60°C are known (Asahina, 1969). Nevertheless, a majority of hibernating insects avoid internal freezing by supercooling. Supercooling (the ability to remain in the unfrozen state *below* the melting point) is not necessarily a result of cold acclimation or cold-hardening, since insects may supercool many degrees whether or not they are cold-adapted. The supercooling point (the temperature at which "instant" freezing of a slowly cooled insect that is not seeded with ice crystals or other nucleating agents occurs) may be −10 to −20°C or as low as −40°C in the presence of antifreeze compounds.

Supercooling is due to the unavailability of nucleation sites for ice crystal formation, and supercooled insects freeze instantly ("flash") when seeded with ice crystals. Since the gut contents normally freeze first, they provide a source of seeding crystals which reduce the amount of supercooling possible. Gut evacuation prior to hibernation results in an immediate lowering of the supercooling point (Salt, 1966). In the carpenter ant, *Camponotes obscuripes,* however, two supercooling points have been demonstrated (Ohyama and Asahina, 1972). When the ants are cooled slowly, the first supercooling point is reached at −8.5°C, when the foregut contents freeze, and the other is reached at −20°C, when the rest of the body fluids freeze and the ant becomes hard and brittle. In general, insects not protected by substantial amounts of antifreeze compounds are not able to supercool to temperatures near −30°C.

2.2 Antifreeze

The freezing point depression is normally determined by the *number* of dissolved particles. A 1 molal solution typically lowers the freezing point 1.86°C. However, 1 mole of sodium chloride added to 1 liter of water lowers the freezing point only 3.38°C, rather than the 3.72°C predicted if all the sodium chloride dissociates to Na^+ and Cl^-.

It might be predicted that organisms would, for energy economy, package many small molecules in preference to a few large ones. However, there are also biochemical considerations; many of the antifreeze compounds of insects and other animals lower the freezing point more than predicted only on the basis of their colligative properties. For example, the parasitic wasp *Bracon cephi* accumulates glycerol in the blood before winter to a concentration of 5 molal (Salt, 1959). This amount of solute depresses the freezing point of the blood to −17.5°C, rather than to −9.3°C.

In some instances relatively low concentrations of antifreeze compounds can cause remarkable freezing point depressions while not lowering the melting point: They promote supercooling. Antifreeze substances of this type in the blood of polar marine teleost fishes are 200–500 times as effective as sodium chloride in preventing the formation of ice crystals in the blood. In the antarctic ice fish *Trematomus borchgrevinki* (DeVries, 1970), as well as in some other arctic and antarctic fish (see Patterson and Duman, 1979, for references), these antifreeze compounds are large glycoprotein molecules currently thought to bind to embryonic ice crystals, thus eliminating nucleation sites so that supercooling can occur (DeVries, 1970). They are not effective in freezing protection against large ice crystals. They keep the animal in the supercooled state instead of acting primarily to lower the freezing point. Recently Duman (1979) and Patterson and Duman (1979) have indicated that antifreeze agents like those in the polar fish also accumulate in the blood of various species of insects, particularly overwintering beetles. Larvae of the beetle *Tenebrio mollitor* contain six different antifreeze proteins (Patterson and Duman, 1979).

Most of the antifreeze compounds so far found in insects lower not only the freezing point in a noncolligative manner but also the supercooling point. However, it is not clear whether they also all enhance supercooling by masking nucleation sites such as embryonic ice crystals.

To add more complexity to an already complex subject, there is also evidence that some of the insects that *tolerate* intracellular ice formation have proteinaceous solutes that *promote* ice crystal formation by *preventing* supercooling. These proteins apparently provide a template for the growth of embryonic ice crystals (see discussion in Duman, 1979). The adaptive significance of these compounds, which also aid in freezing survival, is not clear. However, when supercooled insects eventually do freeze, they freeze nearly "instantly," and this presumably results in intracellular damage (see discussion in next section). Freezing slowly can be ensured under natural conditions only if supercooling is avoided.

The relatively low-molecular-weight antifreeze compounds so far documented in insects include two polyhydric alcohols (polyols), glycerol and sorbitol, and the disaccharide trehalose. The different proteinaceous antifreeze macromolecules have been only partially characterized (Duman, 1979).

Polyols have been found to accumulate in diapausing eggs, larvae, pupae, and adults. Glycerol was first discovered in the hemolymph of diapausing *Hyalophora cecropia* pupae by Wyatt and Kalf (1958). Chino (1957) then found it in diapausing eggs of *Bombyx mori*, and it is present in overwintering larvae of the slug caterpillar, *Monema flavescens* (Asahina, 1969), and in the overwintering adult female ichneumon wasp, *Pterocormas molitorius* (Asahina and Tanno, 1968). Glycerol is now known to accumulate in the hemolymph of a large variety of overwintering insects (Asahina, 1969). Concentrations of up to 25% fresh weight of glycerol have been observed (Salt, 1959; Asahina, 1969). In most insects lowering of melting and supercooling points, and increases in cold hardiness, are almost directly correlated with glycerol concentration (Sømme, 1964; 1965). However, it is likely that there are additional complexities and mechanisms of frost resistance that are still little understood (Crowe and Clegg, 1978). For example, a seasonal increase in glycerol content without a lowering of the supercooling point has also been observed (Ohyama and Asahina, 1972).

Gycerol generally accumulates in diapausing insects as a result of various stimuli, often regardless of temperature. For example, *H. cecropia,* which enter an obligatory pupal diapause, accumulate glycerol when reared at 25°C and survive temperatures to −70°C (Asahina and Tanno, 1966). Nevertheless, glycerol content normally increases and decreases in parallel with seasonal changes. In some insects that may spend more than 1 year in diapause the glycerol disappears in the summer and is resynthesized in the fall (Sømme, 1965). It generally disappears shortly after the breaking of diapause, even at low temperatures. In larvae of the darkling beetle, *Meracantha contracta,* antifreeze is produced in response to short photoperiod and low relative humidity, and it is lost in response to high temperature and long photoperiod (Duman, 1977). In the third-instar larvae of the gall fly, *Eurasta solidagensis,* glycerol accumulates while temperatures are still relatively high in summer and fall. In addition, the larvae also synthesize other antifreeze compounds, sorbitol and trehalose, in response to low temperatures in the fall and midwinter, respectively (Morrissey and Baust, 1976).

3 FROST TOLERANCE

Some insects hibernate in exposed places, such as those under the light snow cover in the High Arctic that are without the benefit of insulation. They may there be subjected to nearly a full range of air temperatures that can reach −50°C (Downes, 1965). Freezing of the tissues under these extreme conditions cannot be avoided, even with substantial supercooling and with antifreeze compounds in the blood.

The mechanisms that allow some insects to survive freezing and thawing are not well understood. It appears, however, that causes of death include protein denaturation as well as mechanical disruption of cells or cell organelles by ice crystals. Mechanical damage could be reduced by high solute concentrations or by dehydration, but dehydration in turn could disturb the microstructure of the tissues. Glycerol is one of the solutes that has been implicated in survival. This compound, which as already mentioned protects tissues from freezing by lowering the freezing and supercooling points, may in addition also be beneficial in protecting sulfhydryl groups involved in preventing low-temperature protein denaturation (Levitt, 1978). Glycerol has long been used to protect frozen human red blood cells and frozen bull spermatozoa, and this alcohol is found in most freezing-tolerant insects when they are overwintering.

Nevertheless, it is also found in some insects that are killed by freezing (Asahina, 1969), suggesting that there are a number of mechanisms involved and/or that there are conditions that must be met for it to be effective.

One of the conditions affecting survival of freezing is the rate of cooling. For example, the overwintering prepupae of the Japanese poplar sawfly, *Trichiocampus populari,* and many other insects (Asahina, 1969) can survive freezing in liquid nitrogen as long as initial cooling is very slow. Prepupae cooled instantly (at 327°C/min) were all dead upon thawing. Those cooled at 4°C/min had nearly all their visceral cells frozen intracellularly and, although they lived many days after thawing, did not transform to pupae. Freezing in liquid nitrogen preceded by cooling at 0.8°C/min resulted in intracellular freezing of only one-sixth of the visceral cells, and half of the animals developed into adults. When the animals were slowly prefrozen before being dropped into liquid nitrogen, there was no intracellular freezing and they all completed development into adults after thawing. The extracellular fluids were frozen in all treatments (Tanno, 1967, 1968). Recently the cooling rate has been shown to be even more critical than previously thought. Adult beetles, *Upis ceramboides,* all survived freezing to −50°C, after they had been cooled at 0.28°C/min, but mortality was 100% for 0.35°C/min cooling to the same temperature (Miller, 1978).

An explanation for the above results appears to be related to changes in solute concentrations in the intra- and extracellular compartments. The blood is more dilute than the cell contents and has an approximately 10°C higher freezing point; with slow cooling, ice crystals first form in the blood surrounding the cells. As water is "removed" from the extracellular environment by ice crystal formation, the osmotic concentration of the remaining fluid increases and creates an osmotic gradient that withdraws water from the intracellular compartment, lowering the cells' melting and supercooling points. Extracellular freezing occurs in every case of tissue freezing, but intracellular freezing only occurs after extracellular freezing, if at all. When cooling is rapid, there is not sufficient time for the osmotic exchange of water, and thus a large proportion of cells freeze intracellularly and the insect immediately dies on thawing. When intracellularly frozen cells are thawed, there is cytological evidence of destruction of cell structures, and the more rapid the cooling, the greater such damage (Asahina, 1969).

Given the above mechanism, it is tempting to speculate that high concentrations of glycerol, or other solutes in the extracellular component such as the blood, also afford frost tolerance by dehydrating the cells.

Possibly when dehydrated or glycerinated cells finally do freeze, the ice crystals are small and less damaging than the large, jagged crystal lattice resulting during the freezing of solutions of low solute concentrations. Nevertheless, injection of glycerol into freezing-susceptible insects does not necessarily make them tolerant of freezing (Takehara and Asahina, 1960).

Freezing tolerance may thus ultimately be limited by the ability to withstand dehydration. This is supported by the observation that the ability to withstand desiccation automatically confers cold hardiness, even in tropical insects. For example, the larvae of the midge, *Polypedilum vanderplanki,* that inhabit temporary pools in equatorial Africa, can withstand almost "complete" drying (body water content to 3%), and while the larvae are desiccated they can be cooled, with no apparent damage, to −270°C in liquid helium (Hinton, 1960a,b). When rehydrated and warmed, these larvae immediately resume feeding and regain their sensitivity to high and low temperatures.

The capacity to become "immediately" active when conditions are suitable is typical also of other insects. The Alaskan carabid beetle, *Pterostichus brevicornis,* for example, freezes brittle when overwintering in the wild. When thawed (though saturated with a 25% body weight content of glycerol), it immediately walks about (Miller, 1969).

Maintenance in the frozen (or dried) state can afford the insect extraordinarily long survival times. Metabolism is extremely low (Scholander et al., 1953), and food reserves are not rapidly exhausted. However, death following long freezing is not necessarily always due to exhaustion of energy reserves. The slug caterpillar, *M. flavescens,* survives freezing down to −70°C, but the length of time frozen affects survival. When frozen to −20°C, all were alive after 100 days, but none were alive after 200 days. This difference can hardly be due to exhaustion of food reserves, since these caterpillars can survive for at least 275 days at 0°C (Asahina, 1955, 1969). If kept at −190°C for 100 days, all caterpillars survived and fed for several months on thawing, but they failed to complete metamorphosis.

4 SEASONAL SYNCHRONIZATION

The physiological adjustments needed to survive temperature extremes generally appear to be profound enough to preclude normal growth and activity. It is therefore not surprising that only one stage in the insect's life cycle has evolved the specialized capacity to make these adjustments.

Most insects inhabiting areas of seasonal temperature extremes have evolved the capacity to arrest development in either the egg, larval, pupal, or adult stage. Each species has one of these stages physiologically specialized for the cold hardiness described in the preceding pages. What determines the timing of the cold hardiness within a given developmental stage?

At one end of the spectrum of response are species from warm temperate and tropical areas that continue development without interruption in any stage, while at the other end are species from boreal areas that obligatorily enter a deep, long-term dormancy regardless of ambient conditions. Between these two extremes there are various degrees of facultative developmental arrest (Mansingh, 1971; Jungreis, 1978). Temporary arrest, or quiescence, may be proximally induced by the direct effect of environmentally stressful conditions. In addition, however, many insects anticipate the oncoming severe conditions by decoding environmental stimuli and making the appropriate physiological responses in advance. Long-term physiological dormancy, or diapause, ensures that the developmental stage specialized for cold hardiness is present in the winter, and the active stage is present during optimal ecological conditions. The timing of dormancy induction is an important physiological feature of an insect's long-term adaptation to the thermal environment. How are life cycles synchronized with the dominant temperature and other ecological changes to form the first level of defense against the temperature extremes?

4.1 Response to Environmental Cues

In the Far North, the sun is above the horizon throughout the growing season and an insect's life cycle cannot be synchronized by photoperiod. In addition, prevailing temperatures even in the growing season are often so low that many insects cannot complete even a single life cycle in 1 year. The moth caterpillars of *Gynaephora* spp., for example, may take 5 years or more to complete development but, if conditions are good, they can pass through several instars in one season (Downes, 1965). *Gynaephora,* as well as many other insects of the Far North, survive by being able to hibernate in almost any larval instar. They are opportunistic, taking advantage of brief warm periods and then often freezing solid to await another warm, favorable period. Numerous special adaptations are required to live and reproduce in the High Arctic environment, and Downes (1965) suggests that these insects live close to their physiological limits, in part because there are relatively few species in the insect fauna of these regions. Nevertheless, the limited productivity of the vegetation and low number of niches are undoubtedly also contributing factors to low species number.

In temperate regions the seasons can be reliably predicted by day length, and the primary cue for dormancy induction is photoperiod, temperature, or a combination of both. (In some species nutritional factors and moisture are additional cues.) Termination of dormancy is not necessarily triggered by the same cues as induction, though generally the primary cues for both are photoperiod and temperature, as in vertebrate hibernators. Except for quiescence due to direct and immediate environmental effects, most dormancies are long-term responses mediated by the endocrine system. Such endocrine-mediated dormancy, or diapause, which allows the insects to escape deleterious winter conditions (hibernation) often begins in late summer (estivation).

There is considerable variation in response by individual insects of any one species to diapause-inducing and diapause-breaking stimuli (Hoy, 1978). This variation may be pronounced within a given population or vary among populations. For example, in the univoltine populations of the spruce sawfly, *Gilpinia polytoma,* about 50% of the prepupae terminate diapause after 1 year, 18% terminate after 4 years, and 0.04% terminate after 6 years (Prebble, 1941). The introduced European corn borer, *Ostrinia nubilalis*, now has numerous physiological races over North America, each adapted to local temperature conditions by means of adjustments of its photoperiodic response that "predict" the season (Beck et al., 1962). Selection may be rapid; univoltine strains of *Antheraea pernyi* hibernate in response to short-day photoperiods, but it has been possible to produce a race by selection that hibernates in response to long-day conditions (Danileviskii, 1965).

The linking of a sensory stimulus, such as photoperiod, to hormonal output and diapause response is highly complicated in that it most likely involves a circadian clock. For example, in the cabbage butterfly, *Pieris brassicae,* the larvae apparently "sample" the environment for photic cues at about 14.5 hr after lights-on. Lights-on is a *Zeitgeber,* and under continuous dark the photosensitive "window" continues on a circadian basis. If there is darkness during the photosensitive window at 14.5 hr after lights-on, then the larvae develop into diapausing pupae. A single flash of light at this critical time, however, will prevent diapause induction (Bünning and Joerrens, 1960). Manipulation of photoperiod to either prevent diapause or break it prematurely may be a promising method of insect control; diapause was prevented in 70–90% of the larvae of the corn borer, *O. nubilalis*, in maize plants in experimental plots illuminated in the autumn with fluorescent lamps or mercury vapor lights (Schechter et al., 1971).

4.2 Pupal Diapause

The hormones involved in diapause are for the most part the same ones found in normal developmental processes. In the giant silk moth, *H. cecropia,* for example, the pupae enter an obligatory diapause. Diapause involves a shutdown of brain hormone release, hence a lack of ecdysone production, so that further development is arrested (Williams, 1969). The pupae must be chilled for several months before the neuroendocrine system can be reactivated. In another silkmoth, *A. pernyi,* pupal diapause is facultative, being prevented if the larvae are subjected to long (>14 hr) days. In this moth, as presumably in other lepidopterans with pupal diapause, the key physiological step for diapause also involves a shutdown of production or release of brain hormone.

4.3 Larval Diapause

In the southwestern corn borer, *Diatrae grandiosella* (a recent United States immigrant from the tropics), diapause occurs in the larval stage. At 30°C, larvae progress through six molts and pupate in 16–18 days. But at 23°C larvae may not pupate for 20 weeks, undergoing a series of stationary (larvae→larva) molts. Consequently the larval diapause is not simply due to a shutdown of the brain–thoracic gland endocrine axis, although diapausing larvae may have low levels of some key enzymes (see Odesser et al. 1972). Rather, diapause is maintained by the continued activity of the corpora allata, which results in relatively high titers of juvenile hormone (JH) which prevent the larva-to-pupa transformation (Chippendale, 1977).

4.4 Adult Diapause

Adult, or reproductive, diapause is also controlled by JH. Indeed, in terms of physiological mechanisms, adult diapause involves relatively minor variations of the normal reproductive processes controlled by JH (Engelmann, 1970). Furthermore, in some species such as the milkweed bug, *Oncopeltus fasciatus,* similar endocrine processes may trigger migratory flight (another means of escaping deleterious conditions) as well as reproductive diapause (Rankin, 1978).

In the Colorado potato beetle, *Leptinotarsa decemlineata,* the sensory stimuli that produce the reproductive quiescence of the overwintering adults are short daily photoperiod, low temperature, and senescent foliage for food. Aside from shutting down the ovaries, however, the low titers or lack of JH is correlated with degeneration of the flight muscles and with the behavioral response of burrowing into the soil for overwintering (deWilde, 1969). In contrast, in the milkweed bug, *Oncopeltus fasciatus,* where short photoperiod and low temperature also trigger adult diapause via the JH axis, low JH titers cause the insects to fly *upward,* where they are carried by the prevailing winds and "migrate" to areas of suitable temperature and food (Rankin, 1978). High JH titers, which are produced when the insects encounter longer days, higher temperatures, and food, cause the bugs to stop flight and to reproduce.

4.5 Egg Diapause

Many insects have an egg diapause, but the hormonal basis of this diapause is known only for the bivoltine race of the commercial silkworm, *B. mori.* In this species diapause is determined in the eggs before they are laid. The female parent produces a "diapause hormone" from her subesophageal ganglion when she is subjected to long-day conditions of summer, and this hormone enhances glycogen storage and diapause induction of the eggs in the ovary (Fukuda, 1952; Kohayashi and Ishitoya, 1964).

5 FOOD RESERVES

Survival in the unfrozen state over several months potentially exhausts metabolic reserves, and most insects accumulate glycogen and fat prior to hibernation (Downer and Mathews, 1976), which may, however, be interconvertible with glycerol (see Asahina, 1969). Temperature and photoperiod also stimulate lipid deposition prior to hibernation in mosquitoes (Harwood and Takaka, 1965), and 85% of these fat reserves are exhausted during hibernation in *Culex pipiens* (Buxton, 1935). Since these reserves are exhausted considerably faster at high than at low temperatures because of the Q_{10} effect (Tekle, 1960; Lange, 1963), there is a potential disadvantage to a mild winter unless the insect is supplied with very large food reserves accumulated before hibernation.

In addition to storing glycogen and fats prior to hibernation, some insects such as adult solitary bees, *Ceratina flavipes* and *C. japonica,* have extraordinarily high blood sugar in the winter (Tanno, 1964). This sugar could serve simultaneously as an energy source and as an antifreeze compound.

Another way of prolonging food supplies through the hibernation period, besides remaining in a cool place, is to reduce the metabolic drain. This is accomplished by quiescence and partial degeneration of the reproductive tract, as well as possibly other tissue degeneration. For example, in the Colorado potato beetle, *L. decemlineata,* which use up the major portion of their lipid and glycogen reserves during hibernation (deWilde, 1969), the flight muscles with their mitochondria degenerate during diapause. The flight muscles account for 80% of the resting metabolism (Stegwee, 1964), and in diapausing beetles metabolic rates are over 20-fold reduced primarily because the flight muscles degenerate and are converted to fat. The beetles thus convert a potential energy drain into an energy store. In the spring, the beetles resynthesize their flight muscles even before emerging from hibernation underground.

6 THERMAL ADAPTATION AND ACCLIMATION

Aside from having to make long-term cyclic adjustments to temperature extremes, insects also confront geographic and locally unpredictable thermal heterogeneity. Prevailing environmental temperatures when and where insects are active may differ widely, and species may adapt to them through evolution and proximally through acclimation. Both processes can proceed with surprising rapidity. Buffington (1969) reports that seasonal selection is in part responsible for the acclimation observed in the mosquito *C. pipiens,* and Meats (1973) maintains that when the fruit fly, *Dacus tryoni,* is transferred to lower temperature they acclimate "immediately" (within several minutes) so that the insects undergo daily cycles of acclimation and deacclimation. They apparently reset their thresholds for torpor and flight, as does the butterfly *Danaus plexippus* (Kammer, 1971), but the capacity for metabolic temperature compensation is lacking. With the exception of *D. plexippus*, all the insects so far investigated for acclimation are most likely poikilotherms whose body temperatures approximate that of the environment.

There is a wealth of information on thermal adaptation in poikilothermic, principally aquatic, invertebrate animals (Bullock, 1955; Hazel and Prosser, 1974; Wieser, 1973), but our understanding of temperature adaptation and acclimation in insects is relatively meager. Scholander et al. (1953) compared the metabolism of arctic and tropical insects (both larvae and adults) as well as other poikilotherms at 10 and 20°C. As expected, oxygen consumption was a function of mass and temperature, but it did not differ in arctic and tropical forms. On the other hand, they reported that *aquatic* poikilotherms (primarily fish and crustaceans) showed considerable temperature compensation, although the apparent cold adaptation in the fish may have been due to high levels of spontaneous activity (Holeton, 1974). Nevertheless it is probable that some cold-adapted insects, such as the grylloblattids inhabiting the edge of glaciers, have different thermal responses from, for example, the tenebrionid beetles inhabiting hot deserts. However, survival, as such, does not presuppose thermal adaptation in activity rate or metabolism. Cold-adapted insects may simply be living in slow motion. Indeed, this is in part the case in arctic insects. Many species that in temperate areas fly, only crawl in the Arctic (Downes, 1965).

The first evidence for temperature acclimation in several insect species was that of Mellanby (1940), who showed that chill coma temperatures could be lowered as much as 7.5°C in 24 hr. Subsequently there have been numerous reports of insect acclimation. For example, several species of arctic insects kept at higher than normal temperatures were subsequently immobilized by cold that had not harmed them previously. Larvae of the mealworm, *T. mollitor,* kept at 37 and 30°C had death points (after a 1-hr exposure) of 44 and 42°C, respectively (Mellanby, 1954). In the Queensland fruit fly, *D. tryoni*, the thresholds for cold torpor were changed by a maximum of 0.5°C/1°C change in acclimation temperature (Meats, 1976). German cockroaches, *Blatella germanica,* reared at 35, 25, 15, and 10°C, have thermal death points of 7.0, 5.5, 4.4, and 4.3°C, respectively (Calhoun, 1960). Roaches reared at 15°C remain active at lower temperatures than those reared at 25 and 30°C (Calhoun, 1954).

The physiological bases of both thermal adaptation and acclimation are not clear, although numerous possibilities have been discussed (Bullock, 1955; Wiesner, 1973). One possibility is that enzyme concentrations are increased at low temperature, compensating for the reduced Q_{10}-dependent rates of individual molecules. Singh and Das (1977) have shown that cold-acclimated *Periplaneta americana* have higher oxygen consumption rates than warm-acclimated animals at the same temperature. In this cockroach the number of mitochondria in the coxal muscles increases 25%, as does muscle apyrase activity (Mutchmor and Richards, 1961), as a result of cold acclimation (Thiessen and Mutchmore, 1967). Possibly the increased enzyme activity is due to the greater number of mitochondria. Chill coma does not appear to be due directly to failure of the central nervous system. In three species of cockroaches that acclimate, electrical activity was still demonstrable from isolated ganglia well below chill coma temperatures (Anderson and Mutchmore, 1968).

Knowledge of acclimation could have some value in insect control. For example, acclimation in laboratory-reared cultures has an important effect in the success of competitive matings using sterile males of the Queensland fruit fly, *D. tryoni* (Meats and Fay, 1976). Low temperature also significantly affects development rates and survival of weevils infesting stored food products (David et al., 1977).

Essentially nothing is known about possible acclimation in endothermic insects. Kammer (1971a) reported that monarch butterflies, *D. plexippus* (which are moderately endothermic in flight), kept at 4–5°C for a few days, shivered more readily when released at 15–16°C than those kept at 23–24°C.

7 HEAT TOLERANCE

In contrast to the large differences in the minimum temperatures insects can tolerate in the active as well as the inactive state, there is relatively little variability in the maximum temperatures their tissues can tolerate. The upper tolerable temperatures of protozoa and other animals, including insects, are near 40–50°C (Brock, 1967). (Notable exceptions include grylloblattids living on and under the edge of glaciers.)

There are few valid comparative data on heat tolerance, however, primarily because of lack of uniformity in measuring it. In insects that dry out rapidly at high air temperatures, it is often difficult to separate "heat torpor" from dehydration. In addition, the length of time an insect is subjected to a given temperature may have a large effect on whether or not it is temporarily and reversibly incapacitated, irreversibly incapacitated, or killed.

7.1 Environmental Differences

Insects whose body temperature depends in large measure on that of the environment appear to have higher lethal temperatures when they live in hot environments rather than in cool ones. Edney (1971) examined upper lethal temperatures of ground-dwelling tenebrionid beetles from the Namib Desert and from a mesic environment near Grahamstown, South Africa. The beetles were exposed for 30 min to a constant temperature at saturated vapor pressure. The criterion for survival was that the animals could walk away from a 50-W tungsten lamp into shade. The upper lethal temperatures of three diurnal desert forms (*Onymacris plana, O. rugatipennis,* and *O. laeviceps*) were 48–51°C, while those of two woodland species (*Trigonopus capicola* and *Trigonopus* sp.) were 42.5–45°C. Cloudsley-Thompson (1962), on the other hand, defined lethal temperatures of several desert arthropods as determined by 50% survival following 24-hr exposures at <10% relative humidity. He observed that the rate of water loss increased markedly near the lethal temperature and that survival was enhanced by preconditioning to high temperatures.

Vast microclimatological differences in any one environment may reduce the value of attempting to make large-scale geographic comparisons. A long-legged beetle, *Stenocarpa phalangium,* for example, lives in an entirely different thermal environment (about 4°C cooler) than those with short legs active at the same time and place in the Namib Desert (Henwood, 1975). Similarly, a perching butterfly opening its wings in the sun and a moth initiating flight both experience within seconds entirely different, and probably greater, thermal changes than they would likely encounter in a tropical versus a temperate habitat.

7.2 Endothermic Insects

Relatively few data are available for temperature acclimation in large flying insects that produce their own heat. Heath et al. (1971) have correlated decreasing heat torpor temperatures (from 45 to 42°C) in six Arizona cicadas with increasing elevation (from 400 to 2100 m). However, the minimum flight temperatures of the cicada are not significantly different ($p < .1$). It is possible that the relatively slight differences in the heat torpor temperatures are due to acclimation or some other factor. (The analogous case cited by Heath et al. (1971) for the *Manduca sexta* from the Mojave Desert and from the midwestern United States can be explained largely, if not fully, by differences in methods of measurement (see discussion by Heinrich, 1974a)). Watt (1968) finds that populations of *Colias* butterflies at low, medium, and high elevations in the Colorado Mountains and in British Columbia all regulate thoracic temperature near 36–38°C, like some butterflies from the tropical lowlands of New Guinea (Heinrich, 1972e).

Similarly, scarabeid dung beetles from Kenya near the equator weighing approximately 1 g fly with a thoracic temperature of 34–42°C at an ambient temperature of 25°C (Bartholomew and Heinrich, 1978), while other scarabeids, the *Plecoma* sp. rain beetles (0.6 g) from the coastal mountains of the western United States that fly at T_A about 20°C lower (near 0°C), after rain- and snowstorms, have nearly the same thoracic temperature (K. Morgan and G. A. Bartholomew, personal communication). Undoubtedly part of the rain beetles' ability to maintain such high body temperature is due to their thick layer of pile. (The African beetles lack insulating pile.) Church (1960) has shown that 65% of the temperature excess of bumblebees (at an airflow of 300 cm/sec) is due to their insulation.

Some insects that are not, or only occasionally, endothermic, yet live in hot environments, do not have temperature tolerances dramatically different from those of primarily endothermic species from cool environments. For example, the desert cicada, *Diceroprocta apache,* which is active at midday during the hottest part of the year in the Arizona low desert, loses motor control at 45.6°C (Heath and Wilkin, 1970). The cactus dodger, *Cacama vulvata,* another high-temperature cicada from the same environment, enters heat torpor at 44.5°C (Heath et al., 1972). Desert tenebrionid beetles, *Centrioptera muricata,* were invariably killed in 1 hr at 47°C and 10–20% relative humidity (Ahearn, 1970). These upper lethal temperatures are within 1–2°C of those routinely observed during flight in some sphinx moths, bumblebees, large beetles, and dragonflies flying in the temperate zone or at high elevations in the tropics (Bartholomew and Heinrich, 1973) at ambient temperatures down to 10°C or less. These endothermic insects fly with a T_{Th} close to their upper lethal tolerances. As a sweeping generalization, it may be safe to say that there is little evidence for ecological variation in the heat tolerance of *endothermic* insects.

8 ENDOTHERMY AND FLIGHT

In the preceding discussion I have pointed out ecological factors that stand out as selective pressures in the evolution of thermal responses of insects. Although ecological factors predominate in the thermal biology of poikilothermic animals, the overall focus of this volume is on flying insects that regulate their body temperature, and these species, particularly the large ones (Bartholomew, this volume), generate their own, sometimes enormous, heat loads in flight. This heat production from flight has resulted in a whole new set of evolutionary challenges.

Thermoregulation in endothermic insects largely involves the use, sometimes with only relatively slight modifications, of preexisting neuromuscular and circulatory systems adapted primarily for other functions. With the exception of the tymbal muscles of cicada (Josephson, this volume), all significant quantities of heat are produced by contractions of the major flight muscles. These muscles obligatorily produce heat in all insects during flight, though the smallest flyers, which necessarily rapidly lose heat by convection in flight, are not necessarily endothermic despite significant heat production. Endothermic heat production during preflight warm-up is a variation of the flight behavior itself. Furthermore, other activities associated with endothermy, such as singing in katydids (Heath and Josephson, 1970) and incubation in bees (Heinrich and Kammer, 1973), are similar to preflight warm-up, hence to flight.

Because of the close interrelationship among endothermy, temperature control, and flight, I will first review our understanding of the evolution of insect flight and then make inferences about the evolution of body temperature regulation that apparently developed incidental to flight evolution.

8.1 Evolution of Flight

The first wings are thought to have evolved either from gills or gill plates (Wigglesworth, 1976; Kukalova-Peck, 1978) or from lateral expansions (paranotal lobes) of the body wall that may have been first used for shielding (Crampton, 1916). Winged insects arose from thysanuran ancestors in Devonian or Lower Carboniferous times (Carpenter, 1971). In contrast to those of flying vertebrates, which did not evolve until Jurassic times, the wings of insects are not modified legs or homologous appendages. Rather, they developed from lateral outgrowths from the body wall. In the Upper Carboniferous, about 300 million years ago, the wings of insects were already highly specialized, and well-preserved fossils indicate that they resembled those of present-day insects. Three of the at least 11 orders—Ephemeroptera, Orthoptera, and Blattodea—from the Upper Carboniferous have survived until the present time.

According to the paranotal theory, the wings that arose from meso- and metathoracic lobes (like those found on the prothorax of some Paleozoic Paleodictyoptera, Protodonata, Empemeroptera, and Protorthoptera) could have also originally been used like parachutes, to delay descent, and then as aerofoils in gliding, after insects made their first aerial excursions (probably dropping from vegetation to avoid predators). But according to the gill theory, insects "learned" to fly in the water. All thoracic and abdominal segments of primitive Paleozoic insect nymphs had lateral structures resembling the gills and gill plates still found in present-day Ephemeroptera nymphs (Kukalova, 1968). These gill plates function not only in respiration. They are also used in rowing, for locomotion. To a limited extent, the wings of butterflies still function as respiratory organs (Portier, 1930). They are richly supplied with gas-filled tubes, and the air in them has a lower partial pressure of oxygen than atmospheric air.

Kukalova-Peck (1978) points out that the presumed prewings (which had been identified as paranotal lobes) in all primitive Paleozoic nymphs are *articulated* with the body wall, concluding that gliding did not come first in the evolution of flight. Rather, the gill plates, or gill covers, could have given rise to wings and flapping flight. The prewings from the abdomen were lost in the adult, while those in the thorax were enlarged and in part equipped with muscles derived from the leg coxae. Both

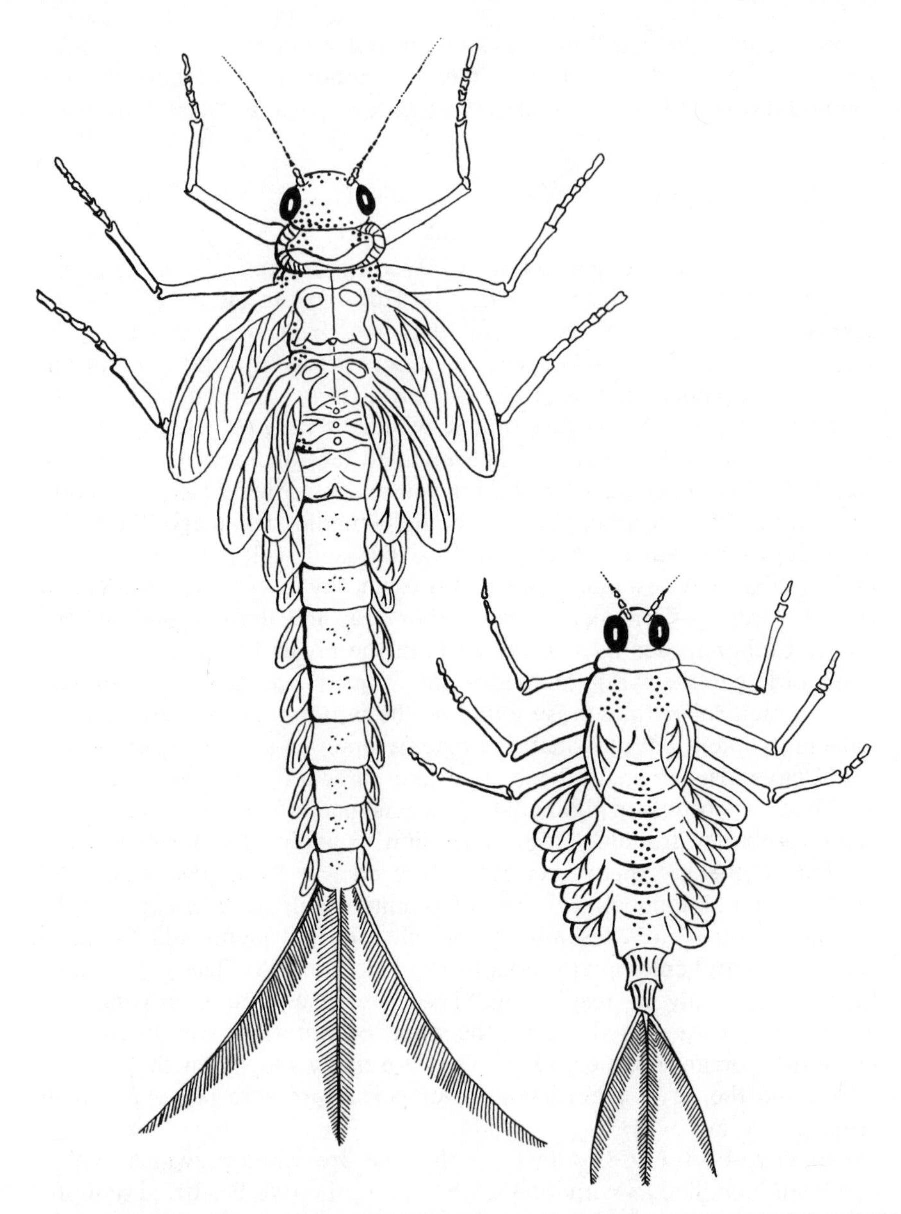

Fig. 1 Diagram of typical Paleozoic (Lower Permian) mayfly nymph (left) showing abdominal and thoracic prewings, presumably used as gills as well as for rowing under water. (After Kukalova, 1968.) Nymph at right is a contemporary species, **Siphlonorus** sp. with large abdominal gills.

Wigglesworth (1976) and Kukalova-Peck (1978) conclude that the apparent similarity between the pre- or prowings (and ultimately wings) and gill plates in Ephemeroptera is not just a superficial resemblance but a true homology. Present-day mayfly larvae, however, never develop gill plates on the thorax, presumably because the thorax is already committed to the development of wings from the gill progenitors.

Gill plates are movable and are supplied with muscles. They presumably developed from styli such as those found in thysanurans. The styli have muscles with attachments to the leg coxae, as do some of the direct flight muscles of primitive insects. In cockroaches, for example, depression of the wings during flight is accomplished largely by direct muscles (the basalars and the subalar), while the dorsal longitudinal muscles, the main wing depressors of modern insects with an indirect flight system, are weakly developed (Tiegs, 1955). In the cockroach the direct muscles used for wing depression are still also used for walking (Wilson, 1962).

8.2 Evolution of Endothermy

Along with large wing surfaces, insects needed to evolve the flight machinery to perfect strong flapping flight. The wings, and thoraces, of the insects from the Upper Carboniferous period, were distinctly larger than those in most modern insects. Members of the Protorthoptera and Protodonata had species with wing spreads up to 50 cm. Some of the present-day Odonata such as *Anax junius* (wing spread = 10 cm), which are small in comparison with the giant protodonate *Meganeura monyi* (wing spread 34 cm or more), generate a temperature excess of up to 17°C as a by-product of flight metabolism (May, 1976). It is probable that the large protodonates also heated up in flight, and it is therefore likely that the ancient coal forest dragonflies had solved the problem of excess endogenous heat generation during flight; they may have had to thermoregulate (get rid of excess heat, and warm up) in order to fly. However, we do not know whether they prevented overheating by gliding or by being intermittent flyers, like the present-day perching dragonflies, or whether they physiologically dissipated excess heat by means of the circulatory system so that they could remain on the wing without stopping like some members of the present-day Aeshnidae (Heinrich and Casey, 1978).

8.3 Design of the Flight Motor

Regardless of the precise origin of wings, it is probable that thermoregulation evolved gradually, as the evolution of aerofoil structures for flap-

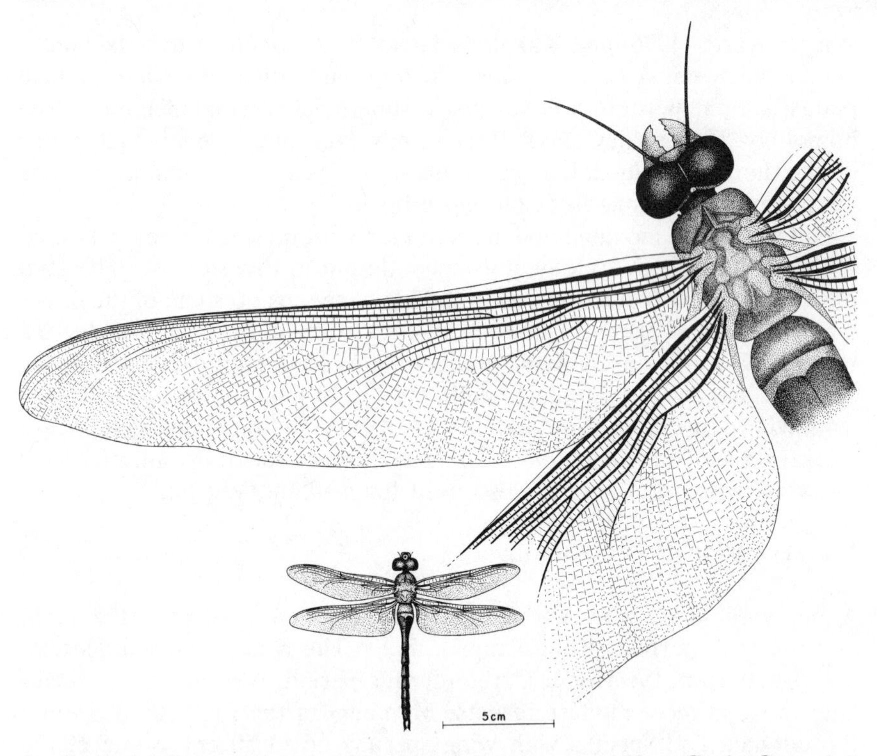

Fig. 2 Partial diagram of the protodonate **Meganeura monyi** (from Charles Brongniart, **Insectes Fossiles**, Saint-Etienne, France, 1893) in comparison with the contemporary cruising dragonfly, **Aeshna multicolor**, one of the largest North American species.

ping flight was accompanied by remarkable adaptations to power them. Flapping flight is without equal as an energy-demanding means of locomotion, and most insect flight muscle (although it is neither stronger nor faster than human muscle, Buchthal et al., 1957) can deliver at least 10 times more mechanical energy per unit time than maximum human muscular effort. The maximum power output from the muscles of members of an Olympic rowing crew, for example, is about 19 W/kg, while that of the locust flight muscle is 190 W/kg (Neville, 1965). What accounts for the insects' high power (and heat) output?

The work a muscle twitch accomplishes is a product of the load times the shortening distance, and in insects the shortening distance is very small. However, this is more than made up in the rapid contraction times made possible primarily by the rapid relaxation times; a low work output per twitch is compensated for by a high repetition rate of contraction. Numerous repetitive contractions are in part related to specializations in

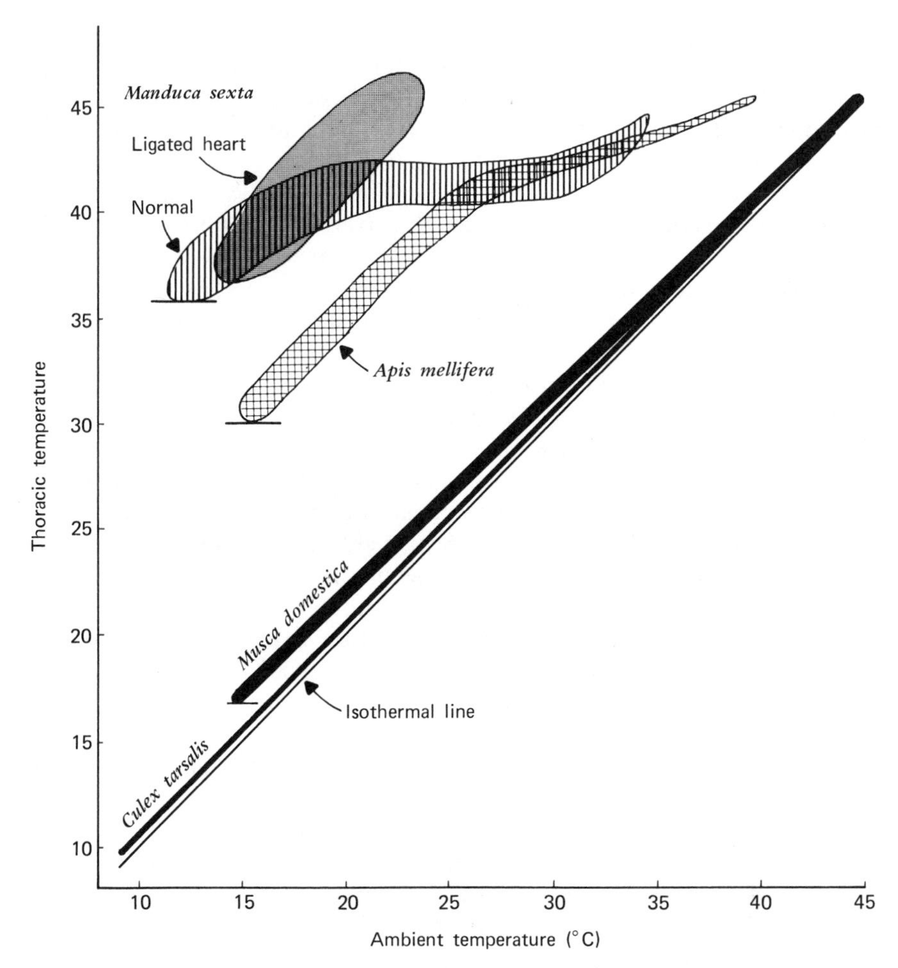

Fig. 3 Activity ranges for continuous flight, and thoracic temperatures during continuous flight. (Data for **M. sexta** from Heinrich, 1970a, for **Apis mellifera** from Heinrich, 1979a, and for **Musca domestica** and **Culex tarsalis** (Heinrich, unpublished.)

the endoplasmic reticular system (Elder, 1971; Josephson, this volume) that permit rapid relaxation. They are also related to the physical arrangement of the muscles in the thorax where the contracting wing depressor muscles stretch the wing elevators, and vice versa.

Second, an efficient ventilation system (Miller, 1974) ensures a dependable oxygen supply. Tracheoles penetrate between the myofibrils in many insects and indent the mitochondria in some (Smith, 1961). Ventilation is in part automated in some forms where deformations of the thorax during flight movement simultaneously cause airflow to and from the muscles. The oxygen supply is apparently always sufficient (at least at sea level), even for the most vigorous insect flight. Third, various hormonal mechanisms facilitate the mobilization of fuel to the flight muscles (Kammer and Heinrich, 1978).

Last, insect flight muscles are packed with an ample complement of biochemical machinery. The mitochondria (sarcosomes) are among the largest known, and they may occupy 40% of the muscle volume (Sacktor, 1974). The mitochondrial cristae incorporate supramolecular assemblies of respiratory enzymes for aerobic metabolism. These include cytochrome oxidase, other electron carrier molecules, and ATP synthase. Since the limited available intracellular space is used to stock enzymes used primarily for aerobic rather than anaerobic metabolism, it is not surprising that some of these enzymes are found in very high concentrations. The concentrations of cytochrome oxidase are more than 10 times higher than those of the mammalian heart (Sacktor, 1974). Similarly, hexokinase activity in some insect flight muscle is about 20 times that found in rat heart, a source of one of the highest activity of this enzyme in vertebrate animals (Crabtree and Newsholme, 1972). The condensing enzyme, which forms citrate from oxalacetate and acetyl fragments and is the gateway to the tricarboxylic acid (TCA) cycle, is 130 times more active in locust flight muscle than in rat gastrocnemius muscle (Zebe, 1960).

As a result of the many adaptations that enhance flight speed and maneuverability, insect flight has become one of the energetically most demanding activities known in multicellular organisms. Since at least 90% of the energy expended appears as heat within the working muscles of the thorax, insects have a large potential for endothermy. Endothermy, and body temperature control in large insects, thus probably evolved in parallel with flight.

8.4 Rate of Muscle Contraction

In part, the generation of a large endogenous temperature excess is related to the rate of muscle contraction, or wingbeat frequency. The neurogenic flight system (which persists in all the surviving "primitive" insects), where each muscle contraction is initiated by an impulse from the central nervous system, is necessarily limited in the wingbeat frequencies that are possible. However, this system is nevertheless adequate for achieving wingbeat frequencies near 100 Hz in some sphinx moths (Sotavalta, 1947), and the moths generate a large thoracic temperature excess. Later, a major step in the evolution of the flight system was the appearance of myogenic muscles (which contract upon stretching) allowing for more rapid wingbeats and some "automation" of the wingbeat cycles. The myogenic system, with which wingbeat rates in excess of 1000 Hz are possible in some small insects (Sotavalta, 1947), was probably the last most major evolutionary innovation of the flight system. Diptera and Hymenoptera, characterized by this flight mode, first made their appearance in the Triassic period, about 180 million years ago. But Coleoptera and Hemiptera, also having a myogenic system, are considerably more ancient. There is no evidence, however, that the myogenic system has had a great influence on endothermy and temperature regulation.

There has been a trend for small insects to be myogenic flyers. However, despite this trend, some of the largest present-day insects—beetles weighing up to 25 g (Bartholomew and Heinrich, 1978) and heteropteran bugs (belastomatids) weighing over 23 g (Barber and Pringle, 1966)—are myogenic flyers, and the weight-relative rates of energy expenditure, hence heat production, of myogenic flyers both large and small are comparable to those of neurogenic flyers (see Kammer and Heinrich, 1978).

9 ENZYME SPECIALIZATION AND ACTIVITY RATES

Regardless of the time intervals that were required for the evolution of flight, or the types of muscular systems driving the flight mechanisms that evolved, there are certain biochemical features that bear investigation because they link flight activity and the evolution of thermoregulation. As flight mechanisms were evolving for stronger and more vigorous flight, there was presumably selective pressure on the enzymes that drive the metabolism to improve their catalytic potential and efficiency. The absolute maximal catalytic rates any of the enzymes must evolve (as measured *in vitro*) must likely be set far higher than those they must actually perform *in vivo*. This is because substrate concentrations *in vivo* are seldom at saturating levels. Second, and more importantly, the enzymes must be subject to various compromises and controls *in vivo* to affect homeostasis with numerous external and internal variables (Somero, 1975). In other words, they must be able to operate at very high rates *despite* being usually curbed from operating at maximal rates.

Heat production as a by-product of flight metabolism would have increased with power output. The inevitably elevated thoracic temperature would in turn affect enzyme functioning and place selective pressure on biochemical readjustments and/or physiological mechanisms of heat dissipation.

Little is known about the biochemistry of temperature-related aspects of insect flight muscle, and what follows is based on inferences from biochemical systems in other animals. In all animals the temperature environment within which the biochemical machinery operates is an extremely important aspect affecting activity. For the most part, sustained aerobic activity, whether in birds, mammals, or insects, is confined within relatively narrow ranges of tissue temperature. Why?

From evolutionary and comparative perspectives, it is of interest, first of all, that the ranges of temperature within which different organisms may operate are wide. Some acellular organisms live and reproduce near −10°C, while others have evolved to be active at temperatures 90°C higher (Stokes, 1967; Brock, 1967). Some fish are active at −1.6°C, while others are active at temperatures 40°C higher (Hochachka and Somero, 1973). Some small dipterans, such as mosquitoes and midges, fly in the Arctic at T_A (and T_{Th}) less than 10°C (Downes, 1965), while other dipterans only fly with thoracic temperatures near 30°C (Heinrich and Pantle, 1975). Small scarab beetles fly with thoracic temperatures close to 25°C, while larger species in the same environment fly with thoracic temperatures near 45°C (Bartholomew and Heinrich, 1978).

9.1 Temperature and Rates of Muscle Contraction

Most insects that are active with relatively low rates of muscle contraction (as in walking) can function over a wide range of tissue temperatures, but the muscle temperature range for flight is commonly narrow, at least in large insects. Ground-dwelling tenebrionid beetles in deserts, for example, are active with preferred body temperatures near 45°C in the daytime (Hamilton, 1971; Edney, 1971; Henwood, 1975). However, in order to gain water from fog, the beetles also walk (slowly) at night at ambient (and body) temperatures down to 2°C (W. J. Hamilton, III, personal communication); they do not need to maintain a minimum walking rate. Ants also walk at any of a wide range of rates that are strictly temperature-dependent (Shapley, 1924). On the other hand, the minimum wingbeat frequency for flight in larger insects is achieved only above a minimum high muscle temperature.

In highly endothermic insects, as well as in those that generate little or no temperature excess during flight, the wingbeat frequency increases with increasing thoracic temperature, and wingbeat rate can be used to gauge T_{Th} (Sotavalta, 1947; Digby, 1955; Dorsett, 1962; Heinrich and Bartholomew, 1971). However, large endothermic flyers, such as sphinx moths, are unable to fly until their wing vibration rate during warm-up reaches the near maximum—that observed during flight (Heinrich and Bartholomew, 1971). In most sphinx moths, this maximum wingbeat frequency, the minimum for flight, is achieved only when thoracic tem-

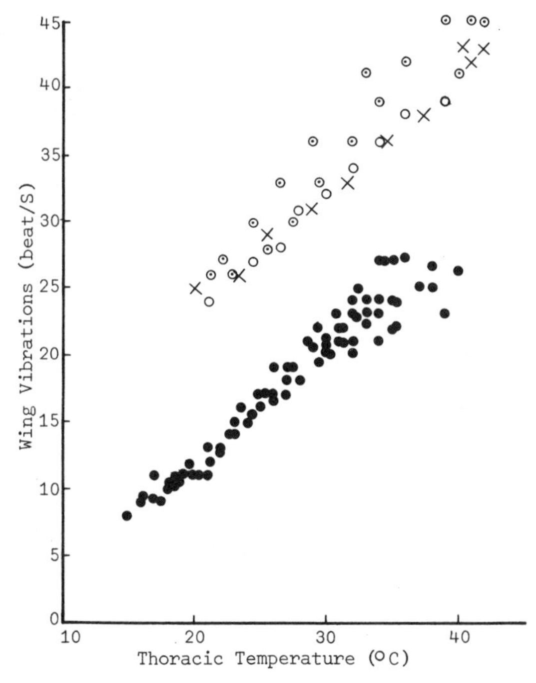

Fig. 4 Wing vibration rate as a function of thoracic temperature during preflight warm-up in sphinx moths. ●, **Manduca sexta.** (From Heinrich and Bartholomew, 1971.) ⊙, **Euchloron megaera** (2.5–2.6 g), x, **Pseudoclanis postica** (0.75 g), ○, **Deilephila nerii** (1.7 g). (From Dorsett, 1962.)

perature is at least 36°C. Small poikilothermic flyers, in contrast, are able to remain airborne at only a fraction of their maximum wingbeat frequency, and possibly also power output (Josephson, this volume). As a general trend small, light insects can remain airborne in flight over a wide range of muscle temperatures while large insects can fly only with muscle temperatures in a relatively narrow range of about 35–45°C. Although the optimum temperature for large insects is clearly defined, that for small poikilothermic insects is not. It remains to be seen to what extent their flight specializations are due to physical or physiological parameters of the flight motor.

9.2 Biochemical Options

Organisms that must rely on high-speed locomotion for survival generally require enzyme systems that are highly efficient (generate the most substrate turnover per unit time while still being subject to cellular control

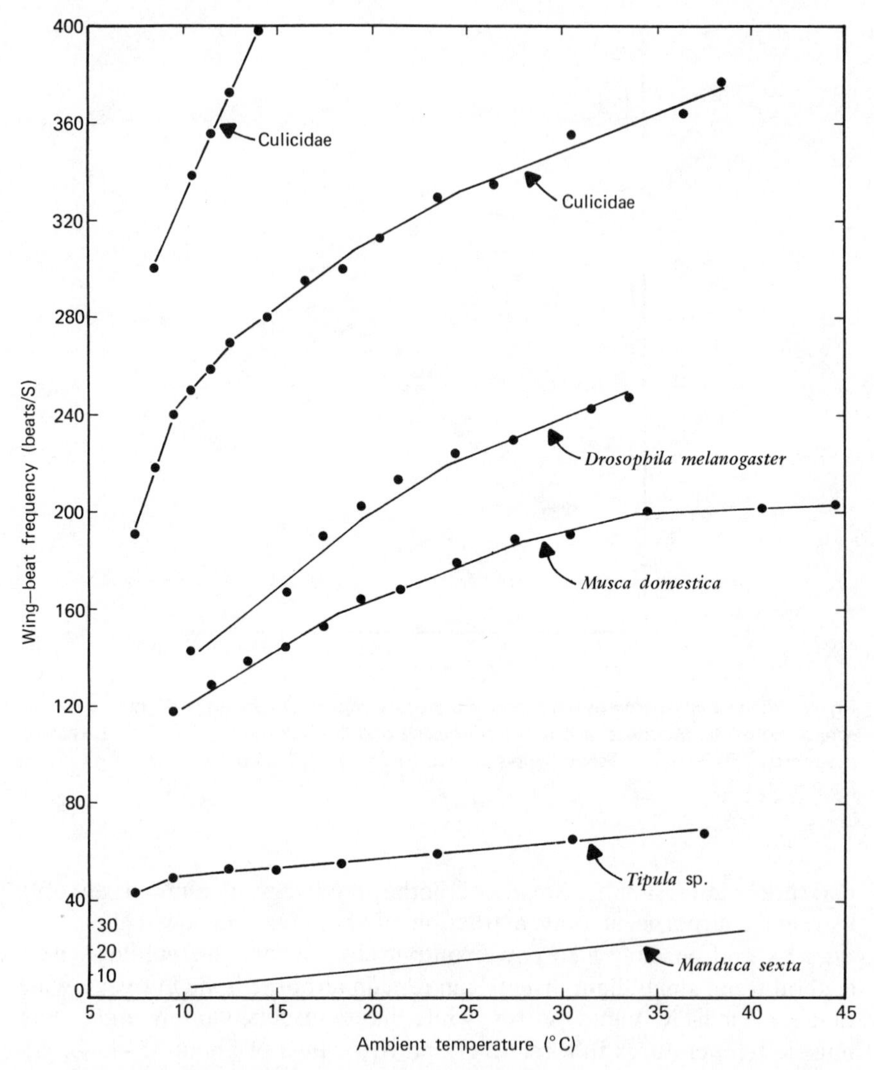

Fig. 5 Wingbeat frequencies as a function of ambient temperature in Diptera (adapted from Sotavalta, 1947), with M. sexta shown for comparison.

mechanisms). Enzyme activity to increase metabolic rate (hence work output) could conceivably be achieved by a variety of options, each with its specific cost and benefits.

9.2.1. Enzyme Duplication

One option for enhancing the activity rate is to leave the enzyme and its functioning unaltered but to increase the *amount* of enzyme present, as occurs commonly in invertebrate temperature acclimation (Hazel and Prosser, 1974). It must be noted, however, that this option soon reaches a ceiling—that determined by packaging. While this strategy can increase low (*basal*) metabolic rates, it has not been shown to affect upper, near maximal rates that are presumably limited by the amounts of enzyme present. This suggests that, when an animal has been under selective pressure to evolve the *maximal* possible rates of activity (as during flight), there are obvious limits to acclimation; when adapted to very high rates of activity, the cell is, as a *first* priority, already fully packed with a complement of enzymatic machinery, and no further packing (and acclimation) is possible.

Temperature-*independent* activity rates at any *one* time could conceivably be achieved by provisioning the cell with sets of duplicate enzymes (isozymes), each adapted to function at a different temperature (Hochachka and Somero, 1973). It is obvious, however, that while this strategy minimizes the Q_{10} effect, allowing activity over a wide range of temperatures, it also does so at the cost of reducing overall catalytic potential. This is because at any *given* temperature the cells would contain a quantity of functionless or "dead" molecules in direct proportion to the number of enzyme duplicates that increase the temperature range of activity. That is, the potential catalytic output per "average" individual molecule is necessarily low, at least in comparison to a system where every single molecule present is potentially maximally active at that temperature. One would predict, therefore, that this strategy would also have only limited feasibility in the evolution of the *maximum* possible rates of work output (flight) at a variety of temperatures.

9.2.2 Temperature Specialization

Given that selective pressure continues to produce the capability for very high rates of sustained work output, cells should not only be packaged with just one set of specialized enzymes, but these enzymes should be improved qualitatively. One way of increasing their catalytic potential is to specialize them further for operation over one particular temperature

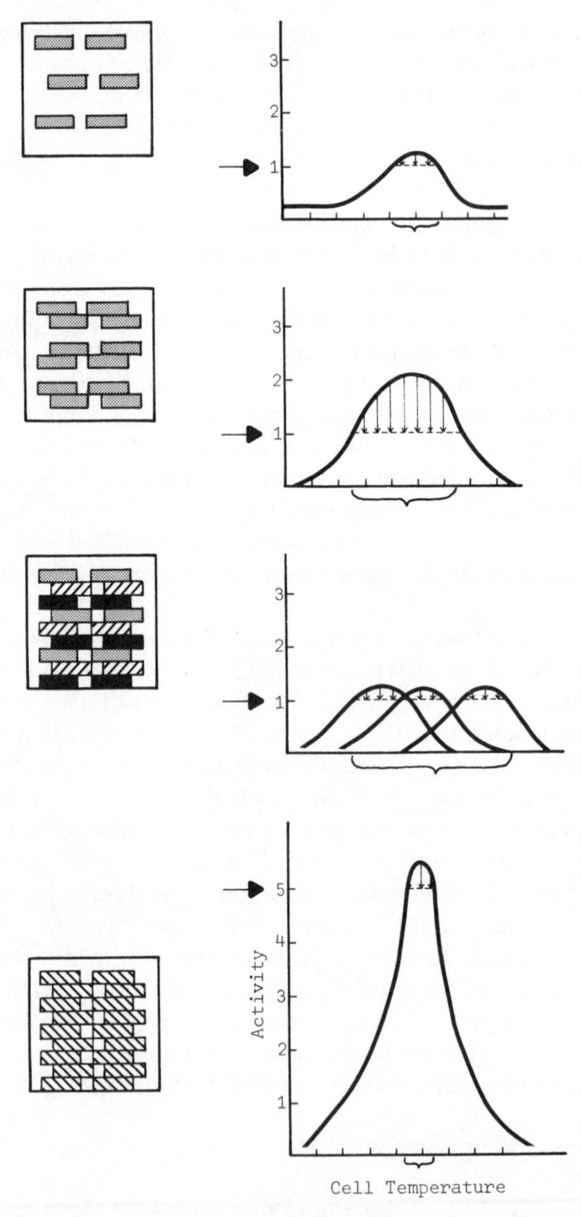

Fig. 6 Hypothetical packaging of different kinds and numbers of enzymes (or polymolecular enzyme complexes) at the left, and corresponding potential temperature–activity curves (right). First set: narrow temperature range for activity at "required" rate (arbitrarily set at 1). Second set: broader temperature range for same rate due to duplica-

range and then to regulate body temperature to regulate the cell's temperature environment.

The possibility for *maximal* enzyme catalysis rates over broad temperature ranges is severely limited because the specific tertiary and quaternary structure of enzyme molecules, which is essential for optimal functioning, is very sensitive to temperature changes (Somero, 1975). In order for enzymes (and presumably other macromolecules) to operate efficiently they must be able to undergo small conformational changes and maintain a balance between structural rigidity and flexibility. Rigidity, such as that maintaining tertiary and quaternary structure, is achieved by weak intramolecular bonds. These bonds are reversibly broken as the kinetic energy (temperature) of the molecule is increased. Molecules can be tailored to remain flexible and to function at relatively low temperatures, possibly by reducing the number of weak bonds (Somero, 1975; Hoffmann, 1976; Perutz, 1978). When the molecules are called upon for optimal functioning at *high* temperatures (as are those in the flight muscles of large insects), then they must of necessity contain sufficient intramolecular bonding to resist significant stereospecific distortion and ultimately denaturation. The more of these weak bonds (van der Waals forces, hydrogen bonds, ionic bonds, hydrophobic bonds) that are used to maintain the tertiary and quaternary structure of a molecule, the more *inactive,* though at the same time resistant to denaturation, it becomes (Low and Somero, 1976).

To get the most activity out of an enzyme molecule (or membrane-associated enzyme complex) involves a compromise between catalytic potential and structural stability (Somero, 1975). If it is to operate at low temperatures, it must be held together with a few weak bonds to retain flexibility for efficient functioning, rendering it susceptible to disruption at higher temperatures. But if it is to function at high temperatures, it must be held together by *numerous* bonds, making it relatively inflexible and thus functionally impaired at low temperatures.

The above considerations point to the conclusion that enzymes must be adapted to function optimally at the highest temperatures they experience during the animal's activity. High temperatures can be experienced either as a result of forced endothermy arising from the activity itself, from external heat input, or from a combination of both (Heinrich, 1977). If

---

tion of enzyme and suppression of full catalytic potential. Third set: broad temperature range for required activity achieved by packaging three different enzyme sets. Fourth set: Full packaging of cell, as in the third set, but of same enzyme to achieve a much higher required rate of activity (as in flight) at the cost of a narrow temperature range.

animals were adapted to function optimally at the low temperatures they experience, then their biochemical machinery would become inactivated or denatured at the high temperatures they experience. Insects adapted to high temperatures and cooled to temperatures below the optimum are simply consigned to a reduced rate of activity, unless they must maintain a minimum threshold of activity, as during flight, for example. As predicted by the above considerations, we observe that ground-dwelling beetles regulate near the upper temperatures encountered (Edney, 1971; Hamilton, 1971; Henwood, 1975). Many large flying insects that inevitably heat up 40°C or more as a by-product of their flight metabolism have their biochemical machinery adapted to operate at these high temperatures; they need them to maintain sustained high-intensity locomotor activity (Heinrich, 1974a), even though they must warm up before flight. Apparently it is more important to survive the high-temperature bottlenecks than it is to have biochemical machinery that can operate efficiently at low temperature; animals subjected to high tissue temperatures have little choice but to adapt to the high temperatures experienced. Indeed, Hoffmann and Marstatt (1977) have shown recently that the optimal catalytic properties of pyruvate kinase in crickets are set according to the highest temperatures experienced.

The second conclusion is that enzymes can be regulated to operate at temperature-independent rates, but only to achieve *submaximal* rates. Maximal catalytic rates are unregulated and temperature-dependent. Below its optimum temperature the enzyme is a victim of the Q_{10} effect—approximately halving its rate with every 10°C drop in temperature. Conceivably it could minimize the Q_{10} effect by becoming rigid enough to reduce the temperature effect on its configuration, but the loss of flexibility would also impair its catalytic rate. These ideas are indirectly supported by experiments on whole animals and tissue extracts. Newall (1969, 1973) provides a wealth of data on intertidal invertebrates amply demonstrating that metabolic rates of resting animals (which respond to acclimation) can be quite independent over a wide range of tissue temperatures. The elevated metabolic rates during *activity*, however, are sharply temperature-dependent; to achieve maximum output, the animal must choose a specific temperature.

10 BODY TEMPERATURE REGULATION

Ultimately, control of the enzyme's immediate temperature environment (body temperature regulation) becomes a means of circumventing the biochemical constraints and permits the enzyme to reach its catalytic potential (when it is not inhibited for control purposes) over a range of ambient temperatures. The cost of this strategy is the metabolic expense of keeping warm at low ambient temperatures. However, if the insect only heats up as a by-product of flight metabolism, then heat may be in abundant supply, although only at the particular times when near-maximal power output occurs.

The conservatism in flight temperatures is probably due to the fact that any change in temperature optima probably involves amino acid substitutions that alter the degree of weak bonding at other than the active sites of most of the body's proteins, hence requires many gene changes. But relatively small changes in behavior or physiology can effect large body temperature changes. In addition, insects in any one habitat are subjected to a large range of temperatures and, as discussed previously, they must adapt to the *highest* temperatures encountered during activity, and these highest temperatures may not be significantly different in tropical areas and in temperate areas in summer.

10.1 Set Points

If there has been selective pressure to increase the intensity and duration of physical performance—and flying insects exhibit some of the highest sustained metabolic rates known—then it is likely that the enzymes of aerobic metabolism are tailored to function in the relatively narrow range of body temperatures that can be maintained most easily, given the prevailing thermal environment and other constraints. Presumably the nervous system has established upper and lower temperature limits so that body temperature is maintained within this range by appropriate behavioral and physiological mechanisms.

One of the tasks in the elucidation of mechanisms of body temperature control is to identify and to locate the neurological centers that plot appropriate physiological and behavioral responses. So far neither task has been accomplished, and our present set point models with high- and low-temperature set points in the head and thorax may or may not be fully analogous to mechanical thermostat models. At the present time, however, we can examine presumed temperature set points, or lack of them, indirectly by observing the factors affecting body temperature.

There are, first of all, constraints of size (Bartholomew, this volume), insulation (Church, 1960), and metabolic rate (Kammer and Heinrich, 1978) that affect the maximum body temperatures that are possible. Insects such as mosquitoes, weighing 1–2 mg, heat up less than 1°C during flight, while houseflies *Musca domestica,* weighing 14 mg, heat up to 2°C above air temperature (Fig. 3). Large moths (0.8–2.5 g) and large (queen) bumblebees (0.4–0.6 g), on the other hand, heat up in flight as much as 35°C above ambient temperature. In general, the larger the insect, the higher its temperature during flight, at least until the high-temperature ceiling of 45–46°C is reached. In a large sample of flying moths at 15–17°C (arriving at a light), thoracic temperatures ranged from within 5°C of ambient temperature in those weighing 100 mg or less to 30–35°C in those weighing 0.5 g. Above 0.5 g and to 6.0 g thoracic temperatures approached 45°C and were independent of ambient temperatures (Bartholomew and Heinrich, 1973). Similarly, in a study of flight temperatures of beetles as a function of body size (at air temperatures of 23–27°C), animals weighing 0.5 or less were agile flyers having thoracic temperatures generally between 28 and 36°C (Bartholomew and Heinrich, 1978). Beetles weighing 10–13 g, however, generally had thoracic temperatures between 40 and 45°C.

There are considerable species differences in evolved flight temperatures independent of body size. For example, four species of moths weighing 0.5–0.9 g (the lasiocampid *Prorifrons* near *muelleri,* pericopid *Pericopis fortis,* sphingid *Amplypterus donysa,* and the arctiid *Amastus aconia*) flew with thoracic temperatures of 35–44°C, 30–36°C, 34–39°C, and 27–34°C, respectively (Bartholomew and Heinrich, 1973). Flight temperature differences are related to metabolic rate as well as insulation. Sphinx moths fly generally with a thoracic temperature 3–4°C higher than saturniid moths of the same size (Bartholomew and Heinrich, 1973), even though in flight saturniid moths are better insulated (Bartholomew and Epting, 1975). The higher thoracic temperatures of the flying sphinx moths, which are good hoverers, is due to their higher flight metabolism (Bartholomew and Casey, 1978).

In the honeybee, *Apis mellifera,* the African variety, *A. m. adansonii,* maintains nearly the same thoracic temperatures as the European, *A. m. mellifera,* even though it is only 66% of the mass of the European species (Heinrich, 1979a). The African bees appear to be more rapid flyers, suggesting that the comparable temperature excess they maintain during flight (15°C above air temperature) is due to a greater metabolic rate which counteracts their tendency for greater heat loss due to their smaller size.

In insect heterotherms, as well as in vertebrate homeotherms, body temperature is regulated between specific maxima and minima which are affected by activity state or work load. During heavy exercise in bumblebees (Heinrich, 1975a), as well as in laboratory rats (Gollnick and Ianuzzo, 1968), kangaroo rats (Wunder, 1974), and other animals, body temperature may stabilize, but at levels higher than those observed during rest or during light exercise. Regulated hyperthermia grades into unregulated hyperthermia, particularly at high ambient temperatures. For example, at ambient temperature near 40°C, the desert cicada can fly only for about 3 sec before it heats to 45°C and must stop flight (Heath and Wilkin, 1970). Similarly, cheetahs have a calculated sprint endurance of about 30 sec at 22°C, as their body temperature increases as a result of heat storage from 37 to 41°C (Taylor, 1974), while Thompson's gazelles, the cheetah's prey, can heat several degrees Celsius higher (Taylor and Lyman, 1972). Human subjects running on a treadmill at 70% maximum aerobic capacity at 23°C showed a continual rise in body temperature, until exhaustion occurred when they reached a rectal temperature of 39.5°C (MacDougall et al., 1974).

Unregulated hyperthermia, which may decisively limit endurance performance, is obviously related to body size and the metabolic rate of locomotion. Mosquitoes and other small Diptera cannot overheat, regardless of how vigorously they fly. However, the endurance of the 100 to about 1000 times heavier sphinx moths and bumblebees is limited to about 2 min of free flight at ambient temperatures above 35°C as a result of hyperthermia, despite their impressive ability to dissipate heat actively, which is used to regulate thoracic temperature at lower ambient temperature (Heinrich, 1971b).

Sphinx moths (Heinrich, 1971a) and bumblebees (Heinrich, 1975a) adjust the thoracic temperature they maintain in flight in part in response to the load carried. Raising thoracic temperature could have dual functional significance. On the one hand, it permits the generation of a higher rate of activity by capitalizing on the Q_{10} effect. On the other hand, increasing the temperature difference between body and air reduces the strain of the heat loss mechanisms and permits thermoregulation over a greater range of ambient temperatures. The latter reason for high temperature set points has repeatedly been suggested for vertebrate homeotherms that rely on their water supplies, as well as on metabolically demanding hyperventilation, to dissipate heat (see review by Taylor, 1974). In insects that generally do neither, however, there is little water or energy to be saved by regulating a higher body temperature. Work output, however, can clearly be increased, but only to a limited extent (Heinrich, 1975a).

It is not immediately obvious how insects "choose" specific muscle temperatures to regulate in flight. Hypothetically it is possible they use set points such as the hypothalamic centers in vertebrate homeotherms. However, in insects the answer appears to be at least in part related indirectly to visual input that affects whether or not the animal remains airborne. Changes in the visual field cause alterations in muscle activation to effect turning movements in moths (Kammer, 1971b) and other insects. Visual input also affects power output (hence heat production) in honeybees (Esch, 1976). Moths, beetles (Bartholomew, unpublished observations), honeybees (Esch, 1976), and bumblebees (Heinrich, 1975a) frequently try to lift off and fail but try again and again until they are airborne after thoracic temperature has presumably increased to provide greater power output. While in fixed flight, or flight on a roundabout flight mill (when thoracic temperature declines), the insects have no opportunity to be informed by vision or other sensors of their lowered flight effort; they reduce their flight effort as thoracic temperature declines, but they still remain "airborne."

Although flight effort as such could be an indirect monitor of thoracic temperature, it is unlikely that it is the only one. Incubating bumblebees maintain thoracic temperatures similar to those observed in flight (Heinrich, 1974b). At least in the saturniid silkmoth, *H. cecropia,* the transition from warm-up motor patterns to the flight motor pattern depends on the temperature of the thoracic ganglia (Hanegan and Heath, 1970b) which presumably closely track the temperature of the thoracic muscles.

The variables that *directly* affect the temperature of the flight motor pattern of individual insects may also have affected the evolution of set points among different species. Wing loading increases with the increasing weight (such as honey crop content) that an individual insect with a given wing area holds aloft. Wing loading also varies with species-specific wing area relative to given body weight. In moths, species with high wing loading generally have higher thoracic temperature during flight than those with low wing loading (Dorsett, 1962; Heinrich and Casey, 1973; Bartholomew and Heinrich, 1973). Sphinx moth species with 0.1 g body weight/cm^2 wing area fly with a thoracic temperature near 36°C, while those with a wing loading of 0.4 g/cm^2 fly with a thoracic temperature averaging 45°C (Heinrich and Casey, 1973). Since moths (Dorsett, 1962), beetles (Bartholomew and Casey, 1977; Bartholomew and Heinrich, 1978), bumblebees (Heinrich, 1975a), and presumably other insects warm up to temperatures near those maintained during flight, we can infer that the different temperatures during flight are also an evolutionary consequence of different temperature set points.

The specific temperature set points have meaning relative to specific ambient temperatures because they potentially play a large part in the kind and amount of regulation that is possible. The animal could evolve a low set point, one that is close to ambient temperature, thereby placing the major burden of temperature regulation on heat dissipation mechanisms. On the other hand, it could also evolve a high set point, one that is far above the commonly encountered ambient temperature, placing the major emphasis of temperature regulation on heat production. However, inasmuch as there is no evidence so far for the regulation of heat production during flight for temperature regulation (see Kammer, this volume), the primary cost of high temperature set points would be the time and energy needed for preflight warm-up. Maintaining thoracic temperature near ambient temperature during flight would necessitate the evaporation of large amounts of water (Heinrich, 1975b) which is generally precious. However, relatively great amounts of heat can be readily dissipated from the flight motor by the circulation of small amounts of blood through the cool abdomen, as long as the difference between thoracic and abdominal (and air) temperatures is great enough.

A compromise set point, one allowing operation (at the range of ambient temperatures encountered by the insect) of both heat dissipation and production (for warm-up) mechanisms should produce the least strain on either mechanism, thus allowing activity over a wider temperature range. Indeed, so far all large flying insects in which heat dissipation mechanisms for flight have been implicated—sphinx moths, bumblebees, some dragonflies, and large beetles—also have preflight warm-up. At least in dragonflies, species that have no endogenous warm-up also either have no or a very poorly developed physiological mechanism of heat dissipation (Heinrich and Casey, 1978; May, 1976). Not enough data are available from other taxa to make wide comparisons.

10.2 Shivering

10.2.1 Preflight Warm-up

During warm-up insects produce heat by contracitng the flight muscles largely against each other (shivering) rather than against the wings (see Kammer, this volume). Thoracic temperature may rise as much as 10°C/min, but abdominal temperature remains near ambient. Preflight warm-up involves cost of time and energy that are a necessary consequence of evolving the biochemical machinery to operate at higher than prevailing ambient temperatures. Without a warm-up capability, many large, strong-flying insects would have to remain grounded nearly continuously, or at least forego the capacity for nocturnal activity outside the tropics.

The proximal necessity of warm-up is to achieve sufficiently high thoracic temperature so that the minimum wingbeat frequency and power output per stroke for takeoff and flight can be generated (Josephson, this volume). At low thoracic temperature the muscles have long twitch durations and work against each other. The resulting inefficiency produces heat, as do other muscle contractions, but a large portion of the power is exerted against antagonistic muscles rather than on the wings.

The tendency of the antagonistic muscles, when they are cool, to contract against each other is accentuated during warm-up in neurogenic

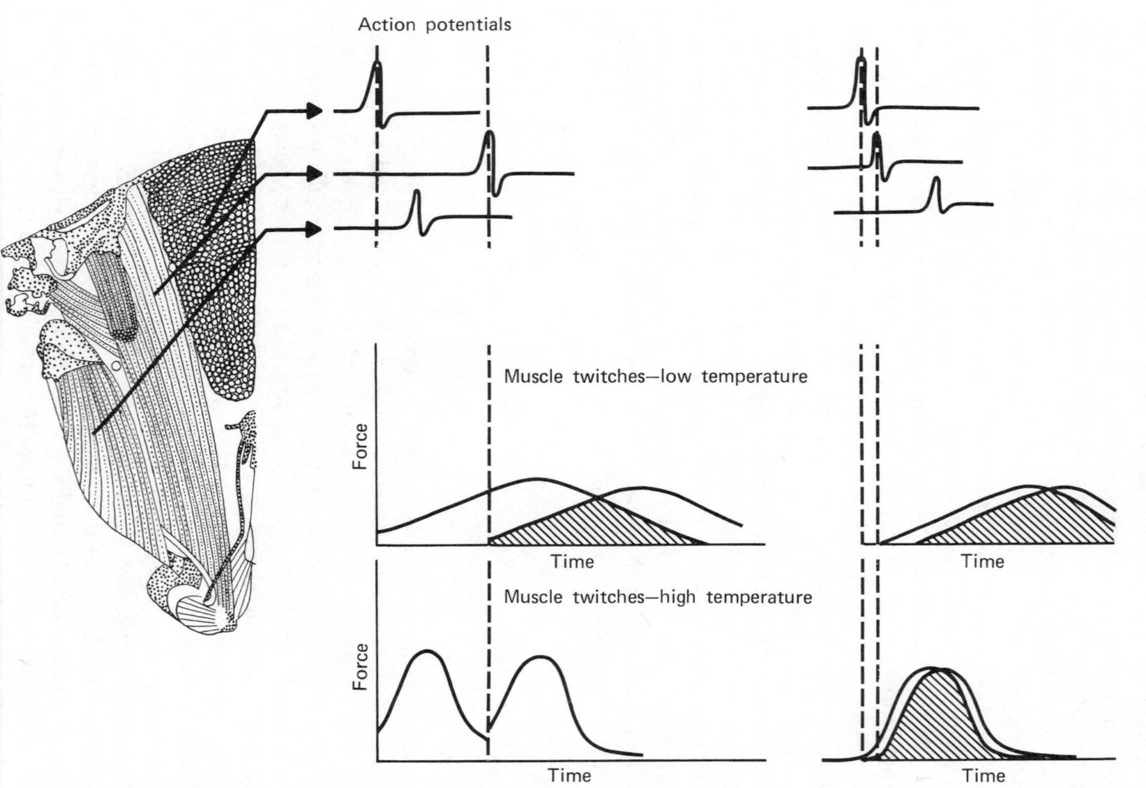

Fig. 7 Right side of cross section of a sphinx moth thorax. Lines lead from the dorsal longitudinal, dorsoventral, and basalar muscle, respectively, to action potentials recorded during flight (left) and warm-up (right). Below are shown the corresponding interrelationships of the muscle twitches of the main flight muscles, the wing depressors (dorsal longitudinal) and wing depressors (dorsoventrals). Cross-hatching indicates overlap of contraction where the muscles are working against each other rather than on the wing.

flyers by modifications in the timing of muscle activation from the central nervous system (see Kammer, this volume). It is doubtful that the inability of some insects to warm up by endothermy is due to evolutionary barriers of the central nervous system or muscles. The primary requirement for warm-up by endothermic shivering is the ability to activate the muscles at other than set phase relationships, and these phase relationships are relatively labile in flight in all insects studied. Indeed, in *Schistocerca*, which does not warm up by shivering, each of the five motor units of one muscle, the dorsal longitudinal, can be activated at independent frequencies. Thus the coupling of the motor units, within a given muscle and between muscles, is neither structurally nor functionally rigid (Neville, 1965). Wilson (1962), who found that the same motor units in thoracic muscles of grasshoppers could be used to move either the wings or the legs, also concluded that flight and related motor patterns were not due to a set of fixed connections between the motor neurons.

It is probable that warm-up evolved directly from flight. Flight itself can function as a means of warm-up. For example, if thrown into the air, many insects will initiate weak, clumsy flight which becomes increasingly stronger and coordinated with flight duration; thoracic temperature increases from the flight metabolism. Sphinx moths with a thoracic temperature of 25°C or less will initiate flight, as long as they are supported, from a flight mill, and their thoracic temperature rises with time, mimicking natural warm-up patterns (Heinrich, 1971a). The temperature increases in the natural warm-up pattern are more rapid, however, probably largely because the moving insects lose heat by forced convection. Remaining stationary during warm-up reduces the time and energy of warm-up by reducing convective cooling.

The flight muscles produce very nearly as much heat whether they work against each other or move the wings. For example, when *M. sexta* sphinx moths are held by their legs, they invariably execute wingbeats of large amplitude even at low thoracic temperatures. The rate of wing flapping increases directly with thoracic temperature but averages slightly lower than the rate of wing vibration during natural warm-up. The rate of thoracic temperature increase is nearly the same during wing vibration, when the upstroke and downstroke muscles contract synchronously working mostly against each other, as during flapping flight when they contract alternately and move the wings (Heinrich and Bartholomew, 1971).

An insect repeatedly trying to initiate flight by flapping its wings with a thoracic temperature too low for takeoff could damage its wings and attract the attention of potential predators. It would also lose heat by forced convection and thus delay the time until takeoff is possible. The evolutionary solution to these problems has been to stabilize or reduce the motion of the wings.

It is probable that warm-up evolved independently in all the major groups of insects, and possibly within them. In many instances among closely related species there are some that warm up by shivering and others that do not. The main variable affecting whether or not warm-up has evolved is the size of the animal, given that it flies for relatively long durations. In dragonflies, for example, "perchers" have not been observed to shiver, while "fliers" do so regularly prior to take-off (May, 1976). This difference is observed even within the same suborder (Anisoptera) in different species of similar size (*Aeshna multicolor* and *Libellula saturata*) that are active at the same time. The relatively advanced dragonfly *L. saturata* does not show warm-up behavior by shivering, but the more primitive *A. multicolor* does (Heinrich and Casey, 1978).

Kammer (1970b) argues from her neurophysiological data that warm-up evolved independently more than once from flight even within the family Sphingidae. Slight modifications of the motor output from a common mechanism that generates flight have resulted in different warm-up patterns. At a given temperature the frequency of activation of different muscle units of the main wing depressor and wing elevators is approximately the same during warm-up and flight, and certain phase relationships in the activation of different muscles seen during flight (where phase relationships may be variable) are also observed during warm-ups.

The differences in the phase relationships in activation of the flight muscles of sphingids concerned direct muscles (the subalar, basalar, and third axillary). These muscles, unlike the dorsoventral muscles, were usually activated out-of-phase with the dorsal longitudinal muscle. However, some of the other direct muscles were either activated in phase, or in antiphase with the dorsal longitudinal muscles, depending on the species (Kammer, 1970b).

A variety of behavioral patterns of warm-up are also observed in Lepidoptera (see Casey, this volume), suggesting that thermoregulation has evolved separately in each group. Many species, including the monarch butterfly, *D. plexippus,* the brush-footed butterflies such as the mourning cloak, *Nymphalis antiopa,* and most skippers, Hesperidae, warm up either by basking when sunshine is available or by shivering when they are in the shade (Kammer, 1968, 1970b). Other butterflies, including some tropical species (Heinrich, 1972), neither bask nor shiver. *Colias* butterflies of the Palearctic bask but do not shiver (Watt, 1968).

Basking postures also vary, suggesting independent evolutionary origins. Many pierid and satyrid butterflies bask by closing their wings and tipping their body and wing surfaces sideways. Hence they are called lateral baskers. Most other butterflies bask dorsally by opening their wings, facing away from the sun, and raising the anterior portion of the body so that the sun's rays strike the dorsal surface of the thorax and abdomen at right angles (Vielmetter, 1958). The wings are often partially closed, thus reducing convective heat loss. Skippers hold the hindwings out laterally, raising only the first pair (see Casey, this volume).

In summary, the variety of behavioral and physiological mechanisms of warm-up within closely related taxa is indicative of multiple independent evolution of thermoregulation. There are, however, numerous close parallels in morphology, physiology, and behavior even among widely separated orders, suggesting the operation of very similar evolutionary selective pressures, constraints, and paths of least resistance.

10.2.2 Other Shivering Behaviors

After modifications of flight behavior for preflight warm-up evolved, warm-up behavior probably evolved in turn to serve in other capacities. In foraging bees, for example, the interflight intervals are often only seconds or fractions of seconds in duration, and preflight warm-up during these intervals at low ambient temperature functions to maintain and stabilize an already elevated thoracic temperature (Heinrich, 1972b; Heinrich and Kammer, 1973). Furthermore, bumblebees may continuously maintain an elevated thoracic temperature by shivering for many hours at a time when incubating brood (Heinrich, 1972a). The mechanism of heat production to warm the nests in honeybees (Esch, 1964; Bastian and Esch, 1970), bumblebees (Kammer and Heinrich, 1974), and presumably other social insects that are vigorous flyers involves the same heat-generating mechanisms used for preflight warm-up. In bumblebees, however, heat is transferred by the circulatory system from the muscles of the thorax to the abdomen which is applied directly onto the brood (Heinrich, 1976), while in honeybees heat is lost to the air from the thorax and then heats the brood chamber.

Warm-up or flight behavior has been modified for several functions other than thermoregulation. Many bees vibrate flowers to shake loose pollen. The vibrations are produced by thoracic buzzing which is apparently similar to shivering, except that the power-producing muscles are only partially engaged with the wings, rather than being totally uncoupled as in warm-up. Buzzing sounds, serving for communication in honeybees (Wenner, 1964) and other social Hymenoptera that fly, and fanning for nest temperature regulation (Seeley and Heinrich, this volume), are also variations of flight or warm-up behavior.

10.2.3 Energetic Constraints

Having evolved the ability to be active at a high body temperature, an animal has the option of cooling down and becoming inactive and saving energy or of keeping warm continuously. The energy saving of keeping cool may be considerable. For example, the metabolic rate of a bumblebee at 5°C is <0.5 ml oxygen/g thorax per hour (Kammer and Heinrich, 1974) but, if the bee attempts to regulate its thoracic temperature (at 35°C) at 5°C, it increases its metabolic rate about 750 times (Heinrich, 1974b).

It is empirically evident that insects generally do not maintain a high body temperature continuously. Moths begin to cool down within a wingbeat of flight cessation, except when they are confined and prevented from flight and make repeated attempts at take-off. Saturniid moths, and some species of sphinx moths, do not feed as adults, and they have a limited amount of fuel stored over from the larval stage that must last them through their entire adult life. In the saturniid moth, *H. cecropia,* males, which do most of the flying prior to mating, contain 36% more lipid—their flight fuel—than females (Damrose and Gilbert, 1964). Animals prevented from flying by having their wings fastened together have a 13% decrease in body lipids in 5 days, while those that are allowed to beat their wings show a 30% decrease in body lipid over the same time (Damrose and Gilbert, 1964). Periods of flight and endothermy, however, are restricted (at least in restrained moths) to very short durations in the 24-hr cycle (Hanegan and Heath, 1970a).

Wax moths, *Galleria mellonella,* also do not feed as adults, and they die with only 33–35% of their original whole body caloric content remaining (Carefoot, 1973). Honeybees in small respirometers (captured leaving the hive) generally keep warm and engage in continuous escape behavior by buzzing and flight until they die (having used up their entire food reserves), generally within an hour (Heinrich, unpublished observations). The continuous endothermy (5–7°C above ambient temperature for up to several hours) observed in beetles being manipulated under laboratory conditions (Bartholomew and Casey, 1977) may similarly be related, at least in part, to escape behavior.

The strategy of keeping warm continuously is energetically prohibitive in animals as small as even the largest insect. The only feasible option that may bear further analysis is whether or not to keep warm continuously *within* the restricted activity periods, when continual flight is not necessary. For example, is it advantageous for a foraging bee to keep its body temperature high by shivering while it perches and probes for nectar on a flower before it flies to another flower, or should it allow its body temperature to decline so long as it is perched?

Bees generally spend only several seconds at each flower they visit, but they may spend several minutes if the flowers are artificially fortified with large amounts of viscous sugar syrup. Thoracic temperature remains high regardless of how long they stay to lap up this "nectar" (Heinrich, 1972b). It must be noted, however, that active ingestion of the nectar is generally only a small part of flower-handling time; bees seldom find large food bonanzas at individual flowers. They cannot see the nectar they imbibe and thus judge nectar amount by sight; they cannot *anticipate* when they will finish at a flower. However, it is to their advantage to be able to take off instantly after removing the nectar from a high-reward flower in order to be able to exploit the next one of the same kind that is likely to hold a similar reward. They should thus keep warm *continuously* while they are foraging, as long as the value of the nectar rewards gathered exceeds the energetic cost of foraging and endothermy.

While foraging on inflorescences such as those of goldenrod (*Solidago* spp.), with thousands of individual florets that have little nectar or that have been largely emptied by other bees, bumblebees often allow thoracic temperature to decline and they forage while walking about on an inflorescence (Heinrich, 1972c). They warm up again when ready to fly to another inflorescence. It is energetically advantageous to warm up repeatedly rather than to stay warm continuously, even if the durations of hypothermia are brief. For example, a bee weighing 0.6 g warms up to 35°C from 6°C in about 16 min, expending 15.7 cal in the process (Heinrich, 1974a). A cool-down from 35°C to within about 5°C of ambient temperature would take 7 min or less (Heinrich, 1972b). If the bee spends the 16 + 7 min of a warm-up and cool-down cycle shivering to maintain thoracic temperature at 35°C, it will expend 92 cal, rather than 16, given a metabolic rate of 80 ml oxygen/g body weight per hour (Heinrich, 1975a). Thus the temporary hypothermia encompassing less than a full heating and cooling cycle results in a sixfold energy saving. Similar calculations show, however, that the energy saving becomes less at higher ambient temperatures. For example, at 24°C, where warm-up to 35°C requires about 1 min, the saving is only twofold, rather than sixfold. The above savings presuppose, however, that the bees warm up, and cool down, at the maximum rates possible. [In the above calculation I have, for simplicity, disregarded resting metabolism because it is only an insignificant fraction of active metabolism (Kammer and Heinrich, 1974).]

It is advantageous to warm up as rapidly as possible because, the more time the animal spends in warm-up, the more heat is lost by convection and is wasted rather than being stored in the body to increase body temperature. For example, based on data on rates of heat loss and temperature increase, a *M. sexta* sphinx moth weighing 1.5 g would lose approximately 17 of the 26 cal of the energy expended during warm-up from 15°C to air temperature to convective cooling (Heinrich, 1975b). As the animal nears flight readiness, it expends energy at nearly the same rate as in flight (about 5 cal/min), and the benefit of this high energy expenditure in causing further increases in thoracic temperature becomes negligible when, at low air temperatures (about 15°C in *M. sexta*), rates of heat loss begin to equal rates of heat production. Because the rate of warm-up depends on the rate of heat production, which is limited by muscle temperature (Kammer, 1970a; Heinrich and Bartholomew, 1971), warm-up at low ambient temperatures is both energetically less efficient than at high ones, and it is also more costly in terms of time. It is perhaps therefore not surprising that most endothermic insects are much more easily induced to initiate warm-up behavior at high rather than at low T_A.

10.3 Mechanisms of Thermoregulation in Flight

Thermoregulation during flight potentially involves regulation of heat loss and/or regulation of heat production (see Kammer, this volume). As already mentioned, in insects (sphinx moths and bees) where detailed analyses of flight metabolism have been made, there is no evidence for regulation of heat production. Rate of heat production during flight was independent of ambient temperature. Endothermy is thus simply a consequence of flight metabolism, and no additional physiological mechanisms need be invoked to account for the heat produced. Nevertheless, it remains to be seen whether other insects, such as dragonflies and butterflies (Nachtigall, 1967), which can glide, minimize the endogenous heat production associated with flight activity specifically to reduce overheating at high air temperatures. May (1976, 1978) reports that *Tramea carolina* dragonflies "spent most of their time gliding, with only occasionally flaps" at high air temperatures. He indicated further that heat loss could also be augmented in *T. carolina* flying at high air temperatures by strongly depressing the abdomen, "sometimes as much as 45° or more below the horizontal." These dragonflies had on average a temperature excess of 13.5°C at 20°C, and only 6.5°C at 35°C, but at the present time it is not clear whether the halving of the temperature excess at high air temperature was due to decreased heat production by gliding or to increased cooling by vertically depressing the abdomen while gliding.

So far the major physiological mechanism for the regulation of body temperature in flight (examined in certain large cruising Odonata, Lepidoptera, and Hymenoptera) seems to be the transfer of excess heat by circulating hemolymph from the overheating thorax to the cooler abdomen. Development of the capacity to dissipate heat during flight by the circulatory system requires what are probably relatively minor evolutionary changes. All insects possess a heart used to circulate blood between thorax and abdomen. If the thorax is heated from flight metabolism and the abdomen remains cool, then heat is necessarily transported from thorax to abdomen. Second, since the Q_{10} effect alone would accelerate the pumping rate as abdominal temperature increases, the system could be in part self-regulatory. True regulation, however, requires neural control mechanisms and direct response to thoracic temperature.

The substrate for regulatory control of blood circulation for temperature regulation is already present in relatively primitive insects that show little or no temperature control in flight. Cockroaches, for example, have complexly neurally innervated hearts (Miller and Usherwood, 1971). For thoracic temperature control it is necessary to have the temperature of the thorax affect blood pumping from the abdomen, independent of abdominal temperature. This condition for thermoregulation is met in the sphinx moth, *M. sexta* (Heinrich, 1970a,b, 1971b), in bumblebees (Heinrich, 1976), and in cruising dragonflies (Heinrich and Casey, 1978), because in these animals abdominal heart activity increases dramatically in response to thoracic overheating. Additionally, it has been shown in various insects that control of the blood flow is also exerted during warm-up, when heat flow to the abdomen is minimized (Heath and Josephson, 1970; Heinrich and Bartholomew, 1971; Heinrich and Casey, 1978).

Most insects examined so far begin to dump heat into the abdomen long before their thoracic temperature becomes excessive. However, since the insects were pinned down during the experiments it is unlikely that they were attempting to keep warm. (They do not warm up and stay warm spontaneously under such conditions.) The slow heat loss observed under the confining experimental conditions may thus indicate a *passive* process. Indeed, moths (Bartholomew and Epting, 1975), katydids (Heath and Josephson, 1970), and bumblebees (Heinrich, 1972a) show an increase in abdominal temperature immediately after *ceasing* activity, indicating a physiologically facilitated heat loss when temperature control seems to be no longer necessary. Thus heat *retention* in the thorax may be an active process that is relaxed under certain conditions such as overheating. However, despite the above observations, facilitated heat transfer is in some cases probably also an active process rather than a mere corollary of heart activity. The dramatic jump in abdominal heart activity, and heat loss to the abdomen as thoracic temperature approaches lethal levels, as well as the qualitative differences in heart activity sometimes observed with heat transfer (Heinrich, 1976), cannot be accounted for on the basis of independent heart activity alone. Second, transection of the nerve cord abolishes the bulk of the heat transfer response (Heinrich, 1970b; Hanegan, 1973).

It is worth noting that the data so far indicate that all highly endothermic insects have also *simultaneously* evolved morphological adaptations to retain heat, as well as physiological mechanisms to dissipate it. For example, sphinx moths, bumblebees, and cruising dragonflies, which have evolved mechanisms for heat dissipation from the thorax, are also insects that have physiological mechanisms for warm-up and effective thoracic insulation in the form of dense scales, pile, or air sacs (Church, 1960). The insulation allows them to be active at lower air temperatures, and it is not a great liability at higher temperatures, as long as they also have the ability to dissipate heat.

11 BENEFITS OF THERMOREGULATION

Regulation of a high body temperature in large insects promotes high rates of sustained locomotor activity. As discussed previously, each insect has acquired a set of biochemical machinery that, in order to operate at peak efficiency, must be maintained within a relatively narrow range of temperatures by various physiological and behavioral means, and many potential benefits of thermoregulation are therefore obvious.

11.1 Broadening the Thermal Activity Niche

In the most general terms, a high body temperature permits insects to fly and thus to perform all vital functions that require flight. For example, sphinx moths feed, pursue mates, and oviposit while on the wing, and the costs and benefits of temperature regulation cannot, in these animals, be tied to *specific* functions.

In sphinx moths all flight involves hovering flight and, as far as has been determined, thoracic temperature of free moths during flight does not vary with the different functions performed, although thoracic temperature does decline when the moths are in tethered flight (Heinrich, 1971a). In other insects, however, different thoracic temperatures are maintained for different flight activities. For example, at ambient temperatures of 7–23°C, honeybees, *A. mellifera,* forage with a thoracic temperature near 31–32°C. However, while attacking, which requires much more rapid and agile flight, thoracic temperatures average 5°C higher (Heinrich, 1979a).

Numerous studies have shown the obvious—body temperature that governs flight also governs a large range of activities in both direct and indirect ways. *Colias* butterflies, for example, like numerous other insects must thermoregulate in order to fly over a wide range of ambient temperatures, and flight, though it is not continuous, is the basis of all

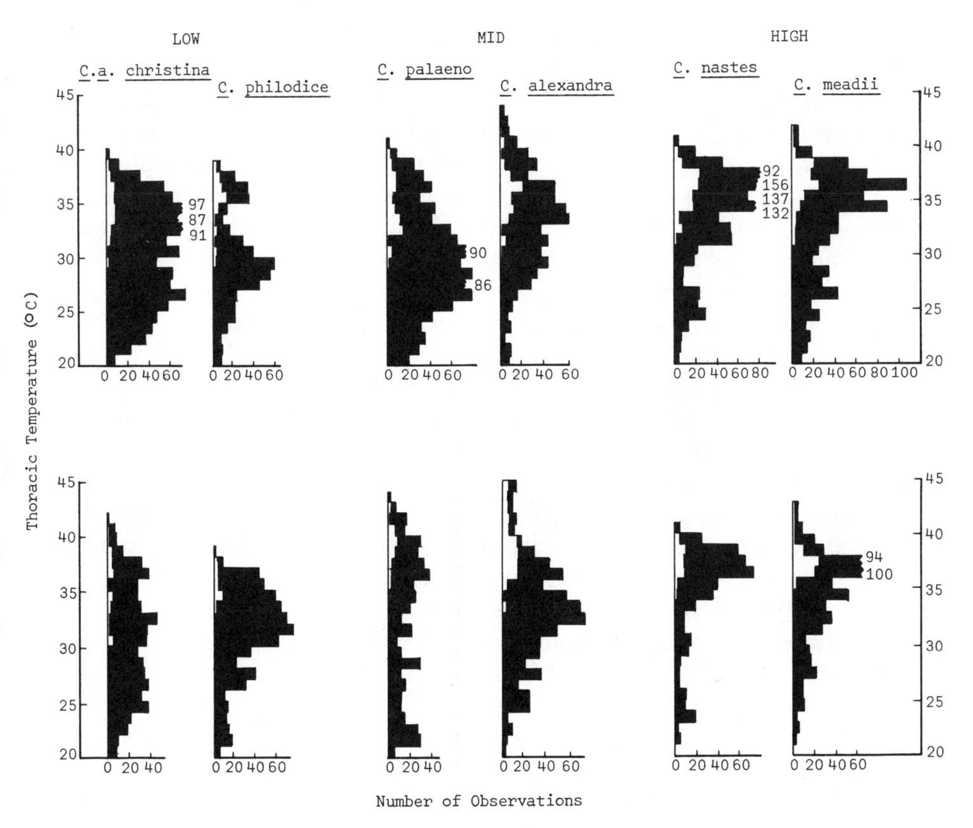

Fig. 8 Distribution of flight activity as a function of thoracic temperatures of Colias butterflies of low-middle- and high-altitude populations in western North America. White bars: flight; black bars: inactivity. The top graphs refer to males and those on the bottom refer to females. (From Watt, 1968.)

vital activities (Watt, 1968). Feeding (foraging) activity depends on flight, and feeding affects egg production; egg production declines in unfed females. Furthermore, females must fly and search vigorously for the widely dispersed food plants in order to oviposit, and males must fly to search for females. Butterflies fly at ambient temperatures as low as 7°C (A. M. Shapiro, personal communication), and they maintain a thoracic temperature near 37–38°C by basking at low temperatures and heat-avoiding orientation at high temperatures (Watt, 1968). If ambient thermal conditions become severe, individuals that can still maintain their thoracic temperature at appropriate levels should have enhanced reproductive success. As a consequence of their ability to thermoregulate, various butterflies can be active on cool mountaintops, as well as in the High Arctic (Kevan and Shorthouse, 1970).

The interrelationships between ambient thermal conditions and activity, and the role of behavioral thermoregulation, have been examined in detail for some of the tenebrionoid beetle fauna of the Namib Desert of southwest Africa (Hamilton, 1971; Edney, 1971; Holm and Edney, 1973; Henwood, 1975). The daily activity rhythms of the beetles correspond with fluctuations in ambient thermal conditions, and the ability to thermoregulate behaviorally allows them to broaden their temporal activity niches. Diurnal forms maintain their body temperatures near upper tolerable levels by shuttling between shade and sunshine, by stilting above the hot substrate or climbing above it, and/or by burrowing and basking. Nocturnal forms, which inhabit a cooler, more thermally uniform environment, are active at body temperatures 8–10°C lower. Despite the ability of the diurnal forms to thermoregulate behaviorally, which broadens their temperature niche, their activity is nevertheless sometimes limited by temperature. *Cardiosus* spp. are active throughout the middle of the day on cool days, but on hot days activity becomes bimodal, being restricted to the morning and evening (Hamilton, 1971). Similarly, *O. rugatipennis* and *O. plana* have unimodal activity patterns in the winter, when it is warm at midday, and bimodal patterns in the summer when it is too hot at midday (Holm and Edney, 1973). Usually, however, a completely new fauna is present in summer and winter, and the fauna varies diurnally as well. Holm and Edney (1973) speculate that the functional significance of the beetles' seasonal and diurnal partitioning is to avoid interference in courting among the different species. There is no evidence of lack of food and, in any case, the food source does not vary temporally in either quantity or quality. Once committed to be active during a particular season or time of day, the beetles thermoregulate behaviorally and enhance activity within that time niche.

Dragonflies of various species in any one habitat are also active over a relatively wide range of ambient temperature, and their ability to remain active is related to temperature regulation. Species that thermoregulate well extend their activity periods early as well as late in the day, while those less able to thermoregulate are often restricted to midday activity (May, 1976, 1977). In general, large species that are continuous flyers produce their own body heat, and they dissipate the excess when threatened with thoracic overheating (Heinrich and Casey, 1978). Large endothermic flyers are commonly active early in the morning, as well as during overcast and at or shortly after sundown (Corbet, 1963). Perchers, in contrast, are generally active only during sunshine, and they spend considerable time basking early in the day when it is cool. Furthermore, at

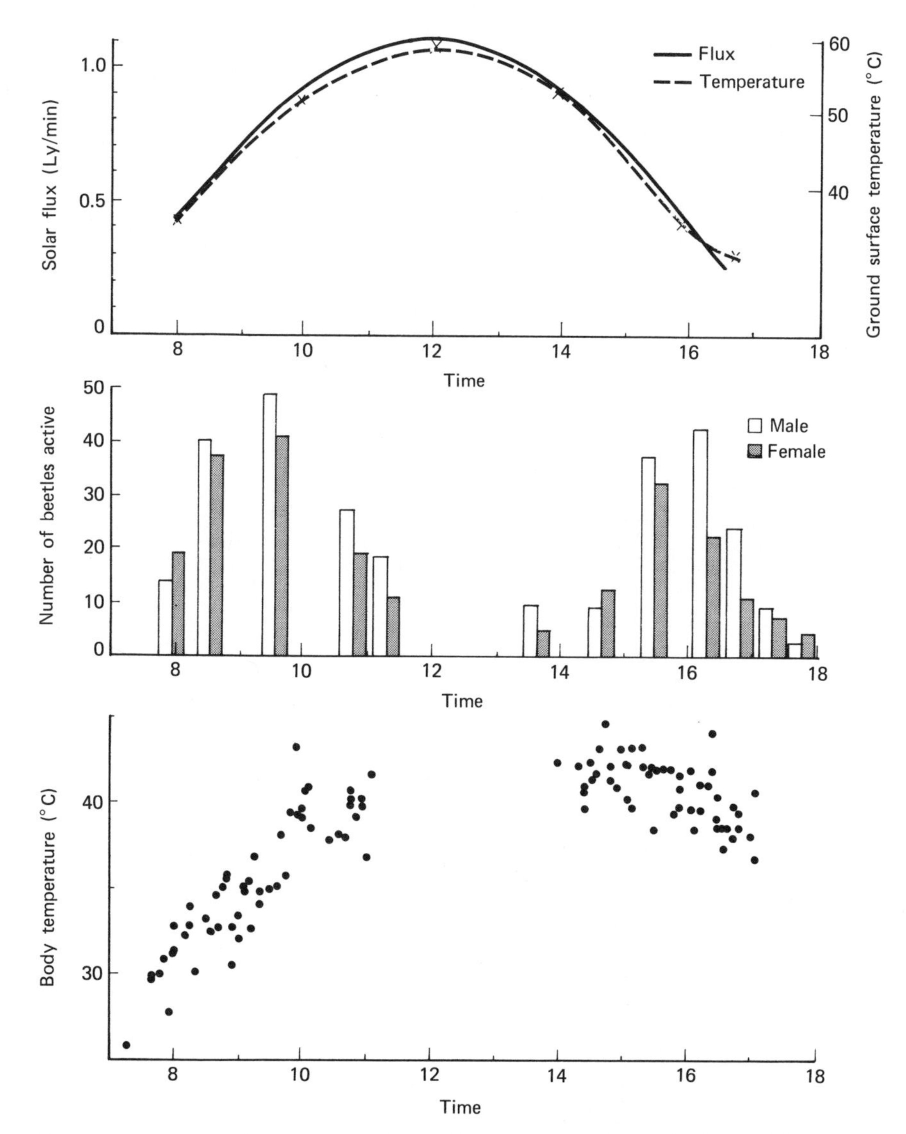

Fig. 9 Thermal conditions at ground level throughout the day on April 26, 1973, in the Namib Desert, and corresponding activity and thoracic temperatures of a population of the tenebrionid beetle O. plana. (From Henwood, 1975.)

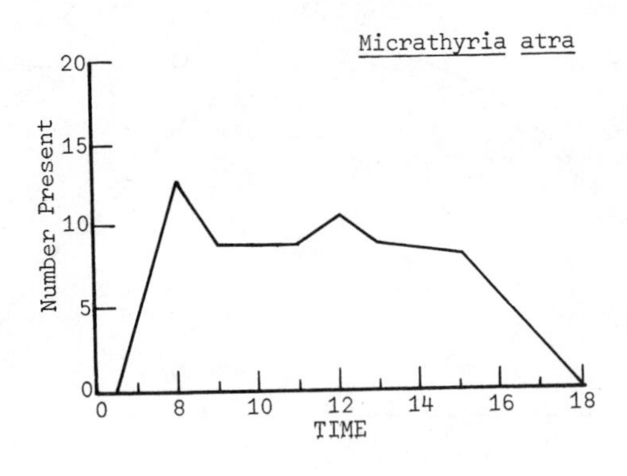

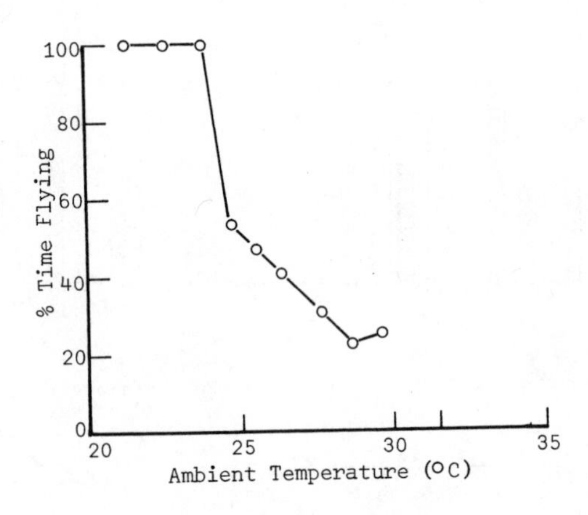

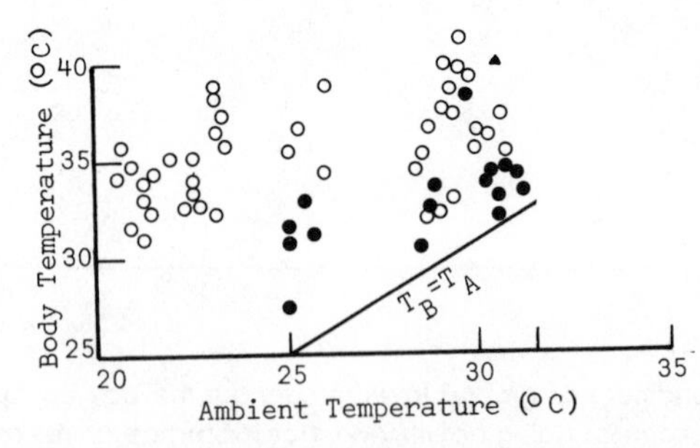

Fig. 10 Activity and thoracic temperature relationship in **Micrathyria atra**. The dragonflies are "active" throughout the day, possibly by reducing flight activity at high ambient temperature to keep thoracic temperature from rising too high. ○, in flight; ●, perching. (Adapted from May, 1977.)

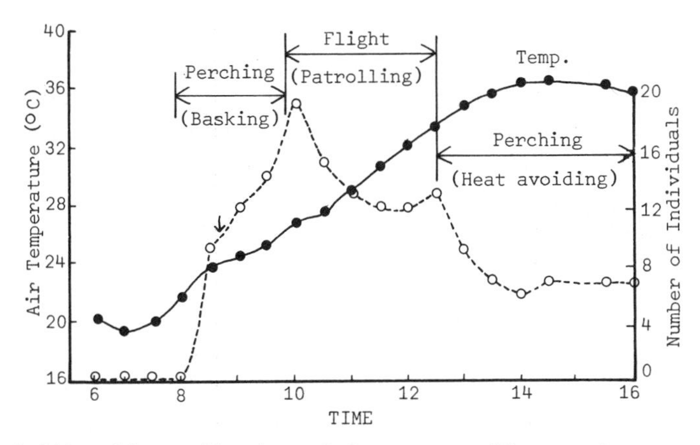

Fig. 11 Activities of the perching dragonfly L. satyrata, at different ambient temperatures (in sunshine) throughout the day. (Adapted from Heinrich and Casey, 1978.)

high air temperatures (>35°C) some perchers such as *Libellula saturata* reduce their flight activity (mating, territorial defense, feeding), possibly because they cannot dissipate excess heat from the thorax (Heinrich and Casey, 1978).

The above examples were chosen to indicate that thermoregulation may affect all aspects of flight activity of insects that ultimately are fed back to affect reproductive success. In many insects, however, thermoregulation is restricted to specific behaviors. For example, large grasshoppers, although they cannot avoid being endothermic while in flight, may be poikilothermic while feeding, mating, or ovipositing, while they are on the ground. But they could presumably have either a high or a low body temperature and accomplish the same tasks that do not require flight. In a number of insects the maintenance of a high thoracic temperature has been shown to have particular significance in specific contexts.

11.2 Reproductive Activity

In the katydid, *Neoconocephalus robustus,* thermoregulation by males is necessary in order for them to sing and attract females. Singing is generated by rubbing the wings together, and the wing movements are driven by the thoracic flight muscles. During singing katydids not only generate enough heat to raise thoracic temperature well above ambient, but a high thoracic temperature is also required in order for the motions of the wings to be rapid and strong enough to produce the required sound frequency and amplitude (Heath and Josephson, 1970). The song is preceded by a warm-up, and thoracic temperature during singing is maintained at 30°C or above. Stridulation costs energetically about the same as flight (Stevens and Josephson, 1977)—about 3.6 cal/min·g^{-1} thorax—but the endothermy as such is not costly, for it is a by-product of the singing activity. Heath and Josephson (1970) suggest that the high cost of the singing may be met in mating advantage because of competitive interaction with another species, *Neoconocephalus ensiger,* which is active in the same habitat at the same time. *Neoconocephalus ensiger* has a soft, intermittent song powered by much slower thoracic muscle contractions (10–15 Hz, rather than 150–200 Hz for *N. robustus*).

Counter and Henke (1977), on the other hand, suggest that *N. ensiger* females may be attracted from a distance by the *N. robustus* calls and that the *N. ensiger* males position themselves to intercept these females by attracting them with their own singing at close range. Apparently at a distance the *N. robustus* calls mask those of *N. ensiger.*

In the tree crickets, *Oecanthus niveus,* in contrast to *N. robustus,* chirping frequency is a direct function of air temperature, and these "thermometer" crickets do not regulate their thoracic temperature (Bessey and Bessey, 1898). Mole crickets (*Gryllotalpa*), on the other hand, also warm up prior to singing (Bennet-Clark, 1970), but nothing is known about possible thermoregulation or its significance in these animals.

Body temperature may also more directly affect reproduction. Hocking and Sharplin (1965) and Kevan (1975), who have observed mosquitoes and other small diptera basking in the parabolic flowers of *Dryas integrifolia* and *Papaver radicatum* (which face the sun throughout the 24-hr cycle in the High Arctic) suggest that the corollas of the flowers act as solar reflectors focusing heat that may hasten the development of the plant's reproductive tissues as well as those of the flower-hopping insects. Both thoracic and abdominal temperatures were elevated in the insects perched in the flowers.

The extent to which abdominal temperature affects the maturation of developing eggs is not known. Most endothermic insects so far examined appear to minimize heat flow to the abdomen at low temperatures when they shiver to keep warm, although accelerated heat flow to the abdomen to heat brood is, in bumblebees, a mechanism for speeding up the developmental rate of the already laid eggs, the larvae, and the pupae. Nest temperature regulation in social insects directly and indirectly appears to be primarily related to accelerating the developmental rate of the immatures (Seeley and Heinrich, this volume).

11.3 Larval Development

Larvae, particularly those of nonsocial insects, may themselves regulate their temperature and thus affect their own rate of development. The desert-dwelling larvae of the sphinx moth, *H. lineata,* for instance, orient in the morning perpendicular to the rays of the sun and, by this basking behavior as well as by heat avoidance at high air temperatures, they regulate body temperatures. Thermoregulation allows them to be active and to feed at their maximum rates for long periods of the day (Casey, 1976b). Temperature regulation thus conveys a selective advantage to *H. lineata* because it decreases the duration of the larval stage and therefore the amount of time the caterpillars are exposed to predators. It also helps to ensure that they finish the larval stage of the life cycle during the short period that the annual desert plants upon which they feed are available.

Another sphinx moth caterpillar, *M. sexta,* which is also abundant in the Mojave Desert where Casey studied *H. lineata,* feeds on the long-lived perennial herb *Solanum metalloides* which, because of its long tap roots, remains lush when all annuals upon which *H. lineata* subsists have died. *Manduca sexta* caterpillars do not regulate their body temperature (Casey, 1977), even though their feeding rates and growth rates are also direct functions of body temperature (Casey, 1976, 1977). At 10°C *M. sexta* caterpillars fed only sporadically and showed almost no growth. Feeding and growth rates increased to a maximum at 30°C. At 35°C, fifth-instar caterpillars gained 0.1 g/day, and at 30°C they gained 2.3 g/day (Casey, 1977).

Manduca sexta caterpillars, unlike *H. lineata,* do not have the defensive behavior of biting, thrashing, and regurgitation. Possibly they are partially protected from predators by the glycosides they ingest from their food plant, and their food supply may be available longer, so they could have less need to feed nearly continuously, as does *H. lineata.*

Some caterpillars (particularly those that do not need to remain hidden to escape avian predators) aggregate into groups that permit mutual heating and increased rates of development. Mosebach-Pukowski (1938) observed that colonies of approximately 100 individuals of *Vanessa io* and *V. urtica* were 1.5–2.0°C warmer than isolated siblings. Temperature excess was a function of group size, and groups of 25 *V. urtica* caterpillars pupated 24 hr sooner than isolated siblings, while groups of over 200 pupated 67 hr sooner. Both isolated and grouped caterpillars reached the same size. The faster developmental rates of the grouped larvae should reduce the time the animals are subjected to predators and disease, as well as allow them to use up food before it becomes limiting. *Vanessa* larvae

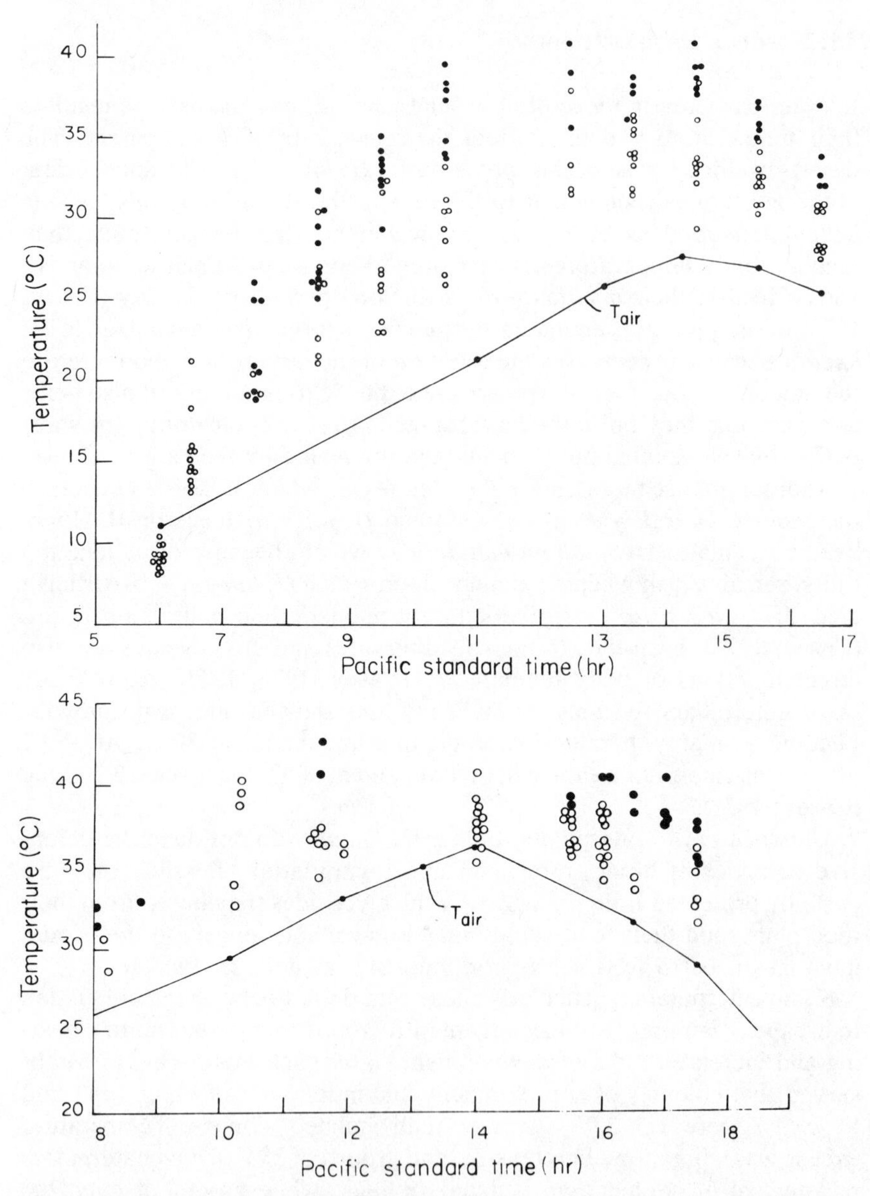

Fig. 12 Body temperatures of H. lineata caterpillars in the Mojave Desert of California on a warm day (top) and a cool day (bottom). ●, On ground; ○, on vegetation. (From Casey, 1976.)

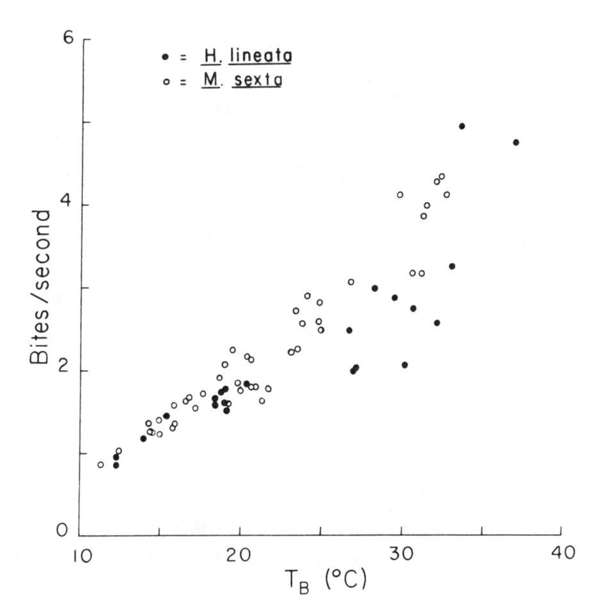

Fig. 13 Biting rate during feeding of **H. lineata** (●) and **Manduca sexta** (○) caterpillars as a function of body temperature. (From Casey, 1976a.)

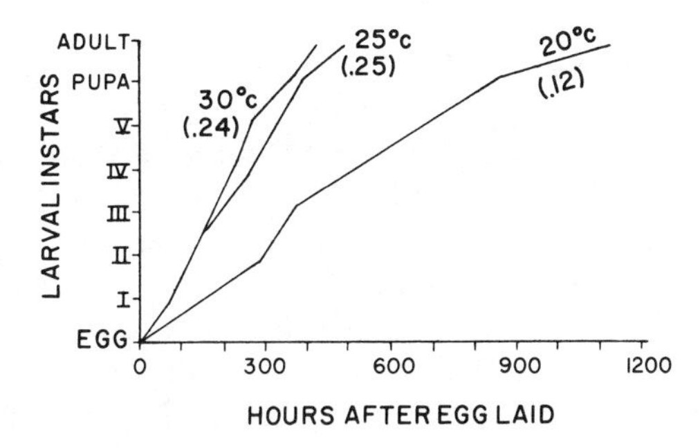

Fig. 14 Developmental rates of **Colias eurytheme** caterpillars as a function of rearing (approximate body) temperature, and feeding rates (square millimeters of leaf surface eaten per second, in parentheses). (From Sherman and Watt, 1973.)

fed on nettles and presumably incur risks if they must leave one herb after stripping it of its leaves and search for another.

The wax moth, *G. mellonella,* typically builds up its populations rapidly in abandoned hives, eating all the available wax, pollen, honey, and dead bees. It would be highly advantageous, because food supplies become exhausted in a few weeks, to be able to grow rapidly and complete development while food is still available. In these larvae, as in others, the developmental rate is presumably highly temperature-dependent, and these caterpillars maintain higher body temperatures than any other lepidopterous larvae. Hase (1926) observed temperatures up to 42°C in caterpillar colonies feeding on dry bee combs; the caterpillars were generating their own heat. Smith (1941) also observed *Galleria* colonies heating to 39–41°C. The larvae thermoregulate, in part, by clumping and scattering. If 500–1000 larvae are in combs filling a volume of 3 ft^2 in an abandoned hive, for example, they aggregate into a volume of approximately 35 in^2 near the top strata of the combs (Smith, 1941). When the temperature exceeds 40°C, the larvae scatter over the surface of the combs.

Tent caterpillars (*Malocosoma*) actively seek sunlit leaves or twigs when they are cool (<30°C) but retreat to their shady tents if overheated (Sullivan and Wellington, 1953). However, the strategy of feeding while exposed to sunlight to increase body temperature and growth rate may not be feasible to many of those caterpillars that must also remain hidden from predators (see Heinrich, 1979b).

11.4 Scramble and Contest Competition

The advantages and costs of thermoregulation have been examined in detail in the scramble competition for nectar resources among bumblebees. These bees of arctic and temperate regions must maintain a minimum thoracic temperature of at least 30°C in order to fly from flower to flower. When flowers provide ample nectar, their thoracic temperature increases to about 35°C or above, and they forage at a faster rate. Heat is produced by shivering while perched on flowers at low air temperatures, and the bees—particularly large-bodied (400–600 mg) queens—are able to forage at air temperatures in shade at 5°C or less where other bees are unable to be active (Heinrich, 1972d).

Even small-bodied honeybees maintain thoracic temperatures near 30°C while foraging, and the African honeybee, *A. m. adansonii* (50–80 mg), is active even at 10°C or less (Fletcher, 1978) where the European bee, *A. m. mellifera,* generally does not forage and is unable to remain in continuous free flight (Heinrich, 1979a). The ability of the African bee to be active at low air temperatures may give it a competitive advantage over European and other bees, since it allows it to forage during the cooler parts of the day, as early as 5 AM and as late as 10:30 PM, as well as by moonlight at night (Fletcher, 1978). The bees warm up not only when on the flowers but also before leaving the hive (Heinrich, 1979a).

The metabolic costs of maintaining a high thoracic temperature in order to be flight-ready at all times are relatively great in terms of the energy contents of some of the flowers. Low-reward flowers are avoided at low air temperatures or, if they occur in tight inflorescences, bees may allow thoracic temperatures to decline for the duration of time they spend on the inflorescence (Heinrich, 1972b,c).

An endogenously elevated thoracic temperature has also been shown to favor dung beetles, particularly the ball roller, *Scarabaeus laevistriatus,* during competition for dung (Bartholomew and Heinrich, 1978; Heinrich and Bartholomew, 1978). These African beetles arrive at the dung of a variety of large herbivores at dusk and after dark, a time when beetles of many species can be so abundant that one elephant dropping attracts thousands of beetles per minute and a dung pile is sometimes consumed, carried away, and buried, or rendered useless for ball building, in $\frac{1}{2}$ hr. There is a large advantage for *S. laevistriatus* in arriving at a fresh dung pile quickly and making a ball and rolling it away before it is consumed by other beetles. If a male beetle is too slow, it may not make or secure a dung ball, which is used as a nuptial gift in mating. If a female does not finish a dung ball and roll it away in time, she has nothing into which to deposit her eggs.

Beetles of both sexes appear to work frenziedly, and their speed of motion is a direct function of thoracic temperature. During ball building the beetles can maintain a thoracic temperature near 40°C for at least $\frac{1}{2}$ hr, and in these "hot" beetles the patting motions of the front legs on the dung balls during ball construction are a rapid blur. Numerous factors affect the speed of ball construction, so that no clear correlation between rate of ball construction and thoracic temperature is apparent (see Heinrich and Bartholomew, 1979). However, the rate at which completed balls are rolled away is clearly a function of thoracic temperature. On the average beetles with a thoracic temperature of 42°C (close to that in flight) rolled their balls at 11.4 m/min, while those with a thoracic temperature of 32°C rolled them at only 4.8 m/min. All the *S. laevistriatus* observed (Heinrich and Bartholomew, 1978) had elevated thoracic temperature during ball rolling but, if they did not and instead allowed their T_{Th} to remain at ambient temperature (25°C), then the regression of ball-rolling velocity on thoracic temperature indicates that they should have been moving their balls at a velocity of less than 0.3 m/min. Thus endothermy allowed the beetles to speed their getaway by at least 38 times. The balls of slow beetles were often infested and eaten by small invading beetles. Diurnal ball rollers, on the other hand, which were subjected to relatively little competition, moved in apparent slow motion by comparison to *S. laevistriatus*, and their thoracic temperatures were never as elevated as those of most of the *S. laevistriatus*.

Rather than attempting to make their own balls, the beetles (*S. laevistriatus*) often tried to steal already formed balls from others. Vigorous fights always ensued, and of 112 observed contests the probability of a warmer animal winning was 0.9, while the probability of a larger animal winning was 0.6. There was no significant difference in body temperature with mass. Thus thoracic temperature was more important than mass in determining the outcome of contests (Heinrich and Bartholomew, 1979).

11.5 Predator Escape

The speed of flight in any one insect, like that of other movements, is related to thoracic temperature. In honeybees, for example, the most rapid flight is observed in bees that attack, and these insects have thoracic temperatures approximately 5°C higher than bees returning to the hive at the same ambient temperature (Heinrich, 1979b). Recently Fletcher (1978) reviewed evidence indicating that the rapid flight of African honeybees may make them less vulnerable than European bees to attack from various predators and parasites; several species of wasps and birds, as well as parasitic conopid flies, specialize in honeybees in Africa, capturing those that fly the most slowly. European bees, which fly more slowly than African bees, are more vulnerable to attack by all these enemies which sometimes decimate honeybee colonies. Also, the ability to thermoregulate allows the bees to forage even at night (Fletcher, 1978) when there are relatively fewer parasites and avian and wasp predators.

In the examples so far I have stressed the adaptive significance of being active at low ambient temperatures by elevating body temperature either by exogenous or endogenous heat input. It has also been suggested that tolerance of high temperature may be a means of escaping predators (Heath and Wilkin, 1970). In the Arizona deserts the cicada, *D. apache*, is active during the warmest part of the summer, where it exploits the hot midday period by locating perches, such as the underside of twigs, that are microclimatically milder than the surroundings that potential bird predators would have to tolerate.

Activity in *D. apache* males consists in singing to attract females. Since singing involves the tymbal muscles, and not bulky flight muscles as in katydids, it does not result in as large increases in body temperature (see, however, Josephson, this volume). The animals lose motor control at a body temperature of 45.6°C, and singing occurs at air temperatures of 38–44°C (Heath and Wilkin, 1970). During flight at air temperatures above 39°C the animals are unlikely to fly for more than 3 sec because of the added heat load from endogenous heat production. However, since they remain in place while singing as well as while feeding (by sucking plant juices), they generally do not heat up to lethal temperatures, and they are able to exploit the hot desert environment when most of their predators have taken cover.

Rather than escaping to high temperatures to avoid predators, Tozer (1979) has evidence indicating that the stonefly, *Zapada cinctipes,* active during the winter in the Sierra Nevada at an elevation of near 2000 m, survives severe nighttime low temperatures (mean minimum night temperature −6.2°C, mean minimum water temperature 8.6°C) by moving underwater at night. They also move underwater when air temperature is experimentally decreased to 5°C or less. Stoneflies normally resume activity during the day when the ice melts.

11.6 Combating Disease

Another functional significance of maintaining or tolerating a high body temperature, particularly in larvae, could be in combating disease. Kluger (1979) reviews evidence for "fever"—the maintenance of an elevated body temperature following infection by various species of bacteria—in fish, reptiles, amphibians as well as in birds and mammals. Poikilothermic animals develop "fever" by basking or other behavioral means, whereas endotherms elevate their metabolic rate. The elevation of body temperature in bacterially infected lizards, *Dipsosaurus dorsalis,* has been reported to promote their survival (Kluger et al., 1975), but few data are available for nonvertebrate animals.

Casterlin and Reynolds (1977) report that crayfish, *Cambams bartoni,* select a 2°C higher body temperature when killed bacteria, *Aeromonas hydrophilia,* are injected into the gill cavity. However, there are so far no reports of fever in insects. Nevertheless, numerous data indicate that a high body temperature is also effective in disease control in this group of organisms.

Many species of insects resist virus infection when they are reared at high temperatures (Tanada, 1967). High body temperatures (37°C) prevent lethal infection of nuclear polyhedrosis virus-exposed lavae in the armyworm, *Pseudaletia unipunctata,* by reducing viral penetration into the host cell as well as by suppressing the production of viral toxins (Watanabe and Tanada, 1972). As reviewed by Watanabe and Tanada

(1972), failure of insect viruses to cause lethal infection has also been reported for granulosis virus in *Pieris rapae* reared at 36°C, a nuclear polyhedrosis virus in *Diprion hercyniae* at 29.4°C, *Trichoplusia ni* and *Heliothis zea* at 39°C, and in *B. mori* at 36°C. Noninclusion viruses protect *Sericesthis pruinosa* at 28°C in *G. mellonella* at 30°C.

The above data on various caterpillars were obtained under laboratory conditions. However, as already indicated, many caterpillars in the field regularly achieve body temperatures through behavioral thermoregulation that should be high enough to retard or prevent lethal viral infections. *Galleria mellonella,* for example, when invading an abandoned beehive in large numbers, increase their body temperature to 40°C or more (Smith, 1941), well above the 30°C required to reduce virus infectivity in this species. Casey (1976a) found that sphinx moth caterpillars, *H. lineata,* in the field in the Mojave Desert of California, maintained a body temperature between 32 and 39°C at ambient temperatures from 20 to 35°C (Fig. 12). On warm days in the desert the caterpillars had a body temperature above 30°C until midmorning. In July, in Minnesota, forest tent caterpillars, *Malacosoma disstria,* had body temperatures averaging 30°C while basking in sunshine on the leaves at 11 AM at an ambient temperature of 22°C, while those in shade were only 0.7°C above ambient temperature. Similarly, the larvae of *D. plexippus* at the same place and near the same time averaged 7.4°C above ambient temperature while they were in sunshine (Heinrich, unpublished observations). Australian sawfly larvae, *Perga dorsalis,* on the other hand, initiated cooling behaviors when body temperatures reached and exceeded 30°C (Seymour, 1974).

Caterpillars can gain considerable protection from viral infection even if they do not maintain a high body temperature continuously. In the silkworm, *B. mori,* there is no synthesis of Flacherie virus antigen at 37°C, and very little synthesis at 32°C, but when larvae are transferred to 27°C, toxin synthesis reappears. Generally, the longer infected animals can maintain a high body temperature, the greater the probability of their survival and the less their infectivity (Inoue et al. 1972). Even though high temperature affords complete protection from the virus, the larvae cannot be maintained indefinitely at 37°C because it is harmful to their growth. *Manduca sexta* (Casey, 1977), as well as *Colias eurytheme* and *C. eriphyle,* larvae achieve maximum growth rates at 25–35°C (Sherman and Watt, 1973). But they do not survive constant temperatures of 35°C. However, in *B. mori,* infected larvae produce normal cocoons if they are subjected to 37°C for only a short period (Inoue and Tanada, 1977). Most of the above caterpillars can be heated at least to 40°C for short durations without damage.

Why do not all caterpillars bask? As elaborated elsewhere (Heinrich,

1979c), feeding on the surface of leaves to take advantage of direct solar radiation may involve heavy tradeoffs with bird predation, possibly explaining why most caterpillars remain hidden. Non-cryptic caterpillars that bask have evolved defensive hairs, spines, and chemical and other defenses.

12 SUMMARY

Some insect species have evolved the ability to live in areas of the earth's greatest temperature extremes. Long-range responses in temperate and arctic areas involve seasonal adjustments of life cycles. The anticipation of seasonal extremes is primarily through photoperiodic cues and responses of the endocrine system for diapause induction. Animals prepare physiologically to overwinter in the egg, larval, pupal, or adult stage, depending on the species. Winter-hardened stages of the life cycle lower their freezing and supercooling points by building up large concentrations of antifreeze, principally glycerol, in the blood. Glycerol also functions in some species to aid in survival of freezing. Freezing tolerance is due primarily to the prevention, or the delay, in intracellular ice crystal formation. Many winter-hardy species tolerate extracellular ice crystals.

Some poikilothermic insects acclimate and remain active over moderate ranges of temperature. But in large flying insects the largest heat loads may be metabolically rather than environmentally induced, and these show little or no acclimation.

Endogenous heat production in insects is a necessary by-product of flight metabolism. As large insects evolved rapid flight, they probably evolved biochemical machinery specialized for rapid rates of catalysis, which required specialization of muscles to operate over a relatively high and narrow range of temperatures. A necessary cost of rapid flight was therefore the time and energy costs of preflight shivering. Warm-up behavior, which is a modification of flight behavior, probably evolved numerous times, even within given insect families. Preflight warm-up has been modified further for heat production in thoracic temperature stabilization used in foraging, for heat production in temperature regulation in the nests of some Hymenoptera, for sound production, and possibly predator avoidance behaviors. Temperature regulation in flight became possible from relatively minor evolutionary changes or modifications in anatomy and functioning of the circulatory system.

The advantages of temperature regulation have been great. In some insects all the major activities—feeding, mating, dispersal, ovipositing—are associated with flight, and a regulated body temperature is necessary for them to remain active. In others, temperature regulation is restricted to specific activities, where it confers reproductive advantage by speeding up growth rate, increasing foraging yield, promoting success in the outcome of scramble and contest competition, in predator avoidance behaviors, and in resistance to some diseases.

ACKNOWLEDGMENTS

I thank George A. Bartholomew, Timothy M. Casey, John H. Crowe, and Ann E. Kammer for valuable comments and criticisms on a draft of the manuscript.

REFERENCES

Ahearn, G. A. (1970). Changes in hemolymph accompanying heat death in the desert tenebrionid beetle *Centrioptera muricata*. *Comp. Biochem. Physiol.* **33,** 845–857.

Anderson, R. L. and Mutchmor, J. A. (1968). Temperature acclimation and its influence on the electrical activity of the nervous system in three species of cockroaches. *J. Insect Physiol.* **14,** 243–251.

Asahina, E. (1969). Frost resistance in insects. *Adv. Insect Physiol.* **6,** 1–49.

Asahina, E. and Tanno, K. (1966). Freezing resistance in the diapausing pupa of the *Cecropia* silkworm at liquid nitrogen temperatures. *Low Temp. Sci.* **B24,** 25–34.

Asahina, E. and Tanno, K. (1968). A frost resistant adult insect, *Pterocoranus molitorius* (Hymenoptera, Ichneumonidae). *Low Temp. Sci.* **B26,** 85–89.

Barber, S. B. and Pringle, J. W. S. (1966). Functional aspects of flight in belastomatid bugs (Heteroptera). *Proc. R. Soc. Lond.* **B164,** 21–39.

Bartholomew, G. A. and Casey, T. M. (1977). Endothermy during terrestrial activity in large beetles. *Science* **195,** 882–883.

Bartholomew, G. A. and Casey, T. M. (1978). Oxygen consumption of moths during rest, pre-flight warm-up, and flight in relation to body size and wing morphology. *J. Exp. Biol.* **76,** 11–25.

Bartholomew, G. A. and Epting, R. J. (1975). Allometry of post-flight cooling rates in moths: A comparison with vertebrate homeotherms. *J. Exp. Biol.* **63,** 603–613.

Bartholomew, G. A. and Heinrich, B. (1973). A field study of flight temperatures in moths in relation to body weight and wing loading. *J. Exp. Biol.* **58,** 123–135.

Bartholomew, G. A. and Heinrich, B. (1978). Endothermy in African dung beetles during flight, ball making, and ball rolling. *J. Exp. Biol.* **73,** 65–83.

Bastian, J. and Esch, H. (1970). The nervous control of the indirect flight muscles of the honey bee. *Z. Vergl. Physiol.* **67,** 307–324.

Beck, S. D., Clouter, E. J. and McLeod, D. G. R. (1962). Photoperiod and insect development. *Proc. 23rd Biol. Colloq. Oreg. State Univ.*, pp. 46–64.

Bennet-Clark, H. C. (1970). The mechanism and efficiency of sound production in male crickets. *J. Exp. Biol.* **52,** 619–652.

Bessey, C. A. and Bessey, E. A. (1898). Further notes on thermometer crickets. *Am. Nat.* **32,** 263–264.

Brock, T. D. (1967). Life at high temperatures. *Science* **158,** 1012–1019.

Buchthal, F., Weis-Fogh, T. and Rosenflack, P. (1957). Twitch contractions of isolated flight muscles of locusts. *Acta Physiol. Scand.* **39,** 246–276.

Buffington, J. D. (1969). Temperature acclimation of respiration in *Culex pipiens pipiens* (Diptera: Culicidae) and the influence of seasonal selection. *Comp. Biochem. Physiol.* **30,** 865–878.

Bullock, T. H. (1955). Compensation for temperature in the metabolism and activity of poikilotherms. *Biol. Rev.* **30,** 311–342.

Bünning, E. and Joerrens, G. (1960). Tagesperiodische antagonistische Schwankungen der Blauviolett- und Gelbrot-Empfindlichkeit als Grundlage der photoperiodischen Diapause-Induction bei *Pieris brassicae*. *Z. Naturforsch.* **15,** 205–213.

Buxton, P. A. (1935). Changes in the composition of adult *Culex pipiens* during hibernation. *Parasitology* **27,** 263–265.

Calhoun, E. H. (1954). Temperature acclimation in insects. *Nature* **173,** 582.

Calhoun, E. H. (1960). Acclimation to cold in insects. *Entomol. Exp. Appl.* **3,** 27–32.

Carefoot, T. H. (1973). The energy budget of the adult stage of the greater wax moth, *Galleria mellonella* (L.). *Can. J. Zool.* **51,** 1035–1039.

Carpenter, F. M. (1971). Adaptations among Paleozoic insects. *Proc. N. Am. Paleontol. Conv.* **2,** 1236–1251.

Casey, T. M. (1976a). Activity patterns, body temperature and thermal ecology in two desert caterpillars (Lepidoptera: Sphingidae). *Ecology* **57,** 485–497.

Casey, T. M. (1976b). Flight energetics in sphinx moths: Heat production and heat loss in *Hyles lineata* during free flight. *J. Exp. Biol.* **64,** 545–560.

Casey, T. M. (1977). Physiological responses to temperature of caterpillars of a desert population of *Manduca sexta* (Lepidoptera: Sphingidae). *Comp. Biochem. Physiol.* **57A,** 53–58.

Casterlin, M. E. and Reynolds. W. W. (1977). Behavioral fever in crayfish. *Hydrobiology* **56,** 99–101.

Chino, H. (1957). Conversions of glycogen to sorbitol and glycerol in the diapause egg of the bombyx silkworm. *Nature* **180,** 606–607.

Chippendale, G. M. (1977). Hormonal regulation of larval diapause. *Ann. Rev. Entomol.* **22,** 121–138.

Church, N. S. (1960). Heat loss and the body temperature of flying insects. II. Heat conduction within the body and its loss by radiation and convection. *J. Exp. Biol.* **37,** 187–212.

Cloudsley-Thompson, J. L. (1962). Lethal temperatures of some desert arthropods and the mechanism of heat death. *Entomol. Exp. Appl.* **5,** 270–280.

Corbet, P. S. (1963). A biology of dragonflies. Quadrangle: Chicago.

Counter, S. A., Jr., and Henke, W. (1977). Commensal auditory communication in two species of *Neoconosephalis* (Orthoptera). *J. Insect Physiol.* **23,** 817–824.

Crabtree, B. and Newsholme, E. A. (1972). The activities of phosphorylase, hexokinase, phosphofructokinase, lactate dehydrogenase and the glycerol 3-phosphate dehydrogenases in muscles from vertebrates and invertebrates. *Biochem. J.* **126,** 49–58.

Crampton, G. C. (1916). The phylogenetic origin and the nature of the wings of insects according to the paranotal theory. *J. N.Y. Entomol. Soc.* **24,** 1–38.

Crowe, J. H. and Clegg, J. S., Eds. (1978). *Dry Biological Systems.* Academic: New York.

Damrose, K. A. and Gilbert, L. I. (1964). The role of lipid in adult development and flight muscle metabolism in *Hyalophora cecropia. J. Exp. Biol.* **41,** 573–590.

Danileviskii, A. S. (1965). *Photoperiodism and Seasonal Development in Insects.* Oliver and Boyd: London.

David, M. H., Mills, R. B., and White, G. D. (1977). Effect of low temperature acclimation on developmental stages of stored product insects. *Environ. Entomol.* **6**(1), 181–184.

deWilde, J. (1969). Diapause and seasonal synchronization in the adult Colorado beetle (*Leptinotarsa decemlineata* Say). *Symp. Soc. Exp. Biol.* **23,** 263–284.

DeVries, A. L. (1970). Freezing resistance in Antarctic fishes. In *Antarctic Ecology*, vol. 1, M. Holgate, Ed., pp. 320–328. Academic: New York.

Digby, P. S. B. (1955). Factors affecting the temperature excess of insects in sunshine. *J. Exp. Biol.* **32,** 279–298.

Dorsett, D. A. (1962). Preparation for flight by hawk moths. *J. Exp. Biol.* **39,** 379–388.

Downer, R. G. H. and Mathews, J. R. (1976). Patterns of lipid distribution and utilization in insects. *Am. Zool.* **16,** 733–745.

Downes, J. A. (1965). Adaptations of insects in the Arctic. *Ann. Rev. Entomol.* **10,** 257–274.

Duman, J. G. (1977). Environmental effects on antifreeze levels in larvae of the darkling beetle, *Meracantha contracta. J. Exp. Zool.* **201,** 333–337.

Duman, J. G. (1979). Thermal-hysteresis factors in overwintering insects. *J. Insect Physiol.* **25,** 805–810.

Edney, E. B. (1971). The body temperature of tenebrionid beetles in the Namib Desert of southern Africa. *J. Exp. Biol.* **55,** 253–272.

Elder, H. Y. (1971). High frequency muscles used in sound production by a katydid. II. Ultrastructure of the singing muscles. *Biol. Bull.* **141,** 434–448.

Engelmann, F. (1970). *The Physiology of Insect Reproduction*, pp. 221–222. Pergamon: Oxford, New York.

Esch, H. (1964). Über den Zusammenhang swischen Temperatur, Aktionpotentialen und Thoraxbewegungen bei der Honigbiene (*Apis mellifica* L.). *Z. Vergl. Physiol.* **48,** 547–551.

Esch, H. (1976). Body temperature and flight performance of honey bees in a servo-mechanically controlled wind tunnel. *J. Comp. Physiol.* **109,** 265–277.

Fletcher, D. J. C. (1978). The African bee, *Apis mellifera adansonii*, in Africa. *Ann. Rev. Entomol.* **23,** 151–171.

Fukuda, S. (1952). Function of the pupal brain and subesophageal ganglion in the production of non-diapause and diapause eggs in the silkworm. *Ann. Zool. Japan* **25,** 149–155.

Gollnick, P. D. and Ianuzzo, C. D. (1968). Colonic temperature responses of rats during exercise. *J. Appl. Physiol.* **24,** 747–750.

Hamilton, W. J., III. (1971). Competition and thermoregulatory behavior of the Namib Desert tenebrionid beetle genus *Cardiosis. Ecology* **52,** 810–822.

Hanegan, J. L. (1973). Control of heart rate in cecropia moths: Response to thermal stimulation. *J. Exp. Biol.* **59,** 67–76.

Hanegan, J. L. and Heath, J. E. (1970a). Activity patterns and energetics of the moth, *Hyalophora cecropia. J. Exp. Biol.* **53,** 611–627.

Hanegan, J. L. and Heath, J. E. (1970b). Temperature dependence of the neural control of the moth flight system. *J. Exp. Biol.* **53,** 629–639.

Harwood, R. F. and Takata, N. (1965). Effect of photoperiod and temperature on fatty acid composition of the mosquito *Culex tarsalis*. *J. Inst. Physiol.* **11,** 711–716.

Hase, A. (1926). Über Wärmeentwicklung im Kolonien von Wachsmottenraupen. *Naturwissenschaften* **14,** 995–997.

Hazel, J. R. and Prosser, C. L. (1974). Molecular mechanisms of temperature compensation in poikilotherms. *Physiol. Rev.* **54,** 620–677.

Heath, J. E. and Josephson, R. K. (1970). Body temperature and singing in the katydid, *Neoconocephalus robustus* (Orthoptera, Tettigoniidae). *Biol. Bull.* **138,** 272–285.

Heath, J. E. and Wilkin, P. J. (1970). Temperature responses of the deset cicada, *Diceroprocta apache* (Homoptera, Cicadidae). *Physiol. Zool.* **43,** 145–154.

Heath, J. E., Hanegan, J. L., Wilkin, P. J., and Heath, M. S. (1971). Adaptations of the thermal responses of insects. *Am. Zool.* **11,** 147–158.

Heath, J. E., Wilkin, P. J. and Heath, M. S. (1972). Temperature responses of the cactus dodger *Cacama valvata* (Homoptera, Cicadidae). *Physiol. Zool.* **45,** 238–246.

Heinrich, B. (1970a). Thoracic temperature stabilization in a free-flying moth. *Science* **168,** 580–582.

Heinrich, B. (1970b). Nervous control of the heart during thoracic temperature regulation in a sphinx moth. *Science* **169,** 606–607.

Heinrich, B. (1971a). Temperature regulation of the sphinx moth, *Manduca sexta*. I. Flight energetics and body temperature during free and tethered flight. *J. Exp. Biol.* **54,** 141–157.

Heinrich, B. (1971b). Temperature regulation of the sphinx moth, *Manduca sexta*. II. Regulation of heat loss by control of blood circulation. *J. Exp. Biol.* **54,** 153–166.

Heinrich, B. (1972a). Physiology of brood incubation in the bumblebee queen *Bombus vosnesenskii*. *Nature* **239,** 223–225.

Heinrich, B. (1972b). Temperature regulation in bumblebees, *Bombus vagans*: A field study. *Science* **175,** 185–187.

Heinrich, B. (1972c). Energetics of temperature regulation and foraging in a bumblebee, *Bombus terricola* Kirby. *J. Comp. Phys.* **77,** 49–64.

Heinrich, B. (1972d). Patterns of endothermy in bumblebee queens, drones and workers. *J. Comp. Phys.* **77,** 65–79.

Heinrich, B. (1972e). Thoracic temperatures of butterflies in the field near the equator. *Comp. Biochem. Physiol.* **43A,** 459–467.

Heinrich, B. (1974a). Thermoregulation in endothermic insects. *Science* **185,** 747–756.

Heinrich, B. (1974b). Thermoregulation in bumblebees: I. Brood incubation by *Bombus vosnesenskii* queens. *J. Comp. Physiol.* **88,** 129–140.

Heinrich, B. (1975a). Thermoregulation in bumblebees. II. Energetics of warm-up and free flight. *J. Comp. Physiol.* **96,** 155–166.

Heinrich, B. (1975b). Thermoregulation and flight energetics of desert insects. In *Environmental Physiology of Desert Organisms*, W. F. Hadley, Ed. Dowden, Hutchinson and Ross: Stroudsburg, Pa.

Heinrich, B. (1976). Heat exchange in relation to blood flow between thorax and abdomen in bumblebees. *J. Exp. Biol.* **64,** 561–585.

Heinrich, B. (1977). Why have some animals evolved to regulate a high body temperature? *Am. Nat.* **111,** 623–640.

Heinrich, B. (1979a). Thermoregulation of African and European honeybees during foraging, attack, and hive exits and returns. *J. Exp. Biol.* **80,** 217–229.

Heinrich, B. (1979b). Keeping a cool head: Honeybee thermoregulation. *Science* **205,** 1269–1271.

Heinrich, B. (1979c). Foraging strategies of caterpillars: Leaf damage and possible predator avoidance strategies. *Oecologia* **42,** 325–337.

Heinrich, B. and Bartholomew, G. A. (1971). An analysis of pre-flight warm-up in the sphinx moth, *Manduca sexta*. *J. Exp. Biol.* **55,** 223–239.

Heinrich, B. and Bartholomew, G. A. (1979). Roles of endothermy and size in inter- and intra-specific competition for elephant dung in an African dung beetle, *Scarabaeus laevistriatus*. *Physiol. Zool.* **52,** 484–496.

Heinrich, B. and Casey, T. M. (1973). Metabolic rate and endothermy in sphinx moths. *J. Comp. Physiol.* **82,** 195–206.

Heinrich, B. and Casey, T. M. (1978). Heat transfer in dragonflies: "Fliers" and "perchers." *J. Exp. Biol.* **74,** 17–36.

Heinrich, B. and Kammer, A. E. (1973). Activation of the fibrillar muscles in the bumblebee during warm-up, stabilization of thoracic temperature and flight. *J. Exp. Biol.* **58,** 677–688.

Heinrich, B. and Pantle, C. (1975). Thermoregulation in small flies (*Syrphus*. sp.): Basking and shivering. *J. Exp. Biol.* **62,** 599–610.

Henwood, K. (1975). A field-tested thermoregulation model for two diurnal Namib Desert tenebrionid beetles. *Ecology* **56,** 1329–1342.

Hinton, H. E. (1960a). Cryptobiosis in the larva of *Polypedilum vanderplanki* Hinton (Chironomidae). *J. Insect Physiol.* **5,** 286–300.

Hinton, H. E. (1960b). A fly larva that tolerates dehydration and temperatures of −270 to +102°C. *Nature* **188,** 336–337.

Hochachka, P. and Somero, G. N. (1973). *Strategies of Biochemical Adaptation*, pp. 179–279. Saunders: Philadelphia.

Hocking, B. and Sharplin, C. D. (1965). Flower basking by Arctic insects. *Nature*, **206,** 213.

Hoffmann, K. H. (1976). Catalytic efficiency and structural properties of invertebrate muscle pyruvate kinases: Correlation with body temperature and oxygen consumption rates. *J. Comp. Physiol.* **110,** 185–195.

Hoffmann, K. H. and Marstatt, H. (1977). The influence of temperature on catalytic efficiency of pyruvate kinase of crickets (Orthoptera: Gryllidae). *J. Therm. Biol.* **2,** 203–207.

Holeton, G. F. (1974). Metabolic cold adaptation of polar fish: Fact or artefact? *Physiol. Zool.* **47**(3), 137–152.

Holm, E. and Edney, E. B. (1973). Daily activity of Namib Desert arthropods in relation to climate. *Ecology* **54,** 45–56.

Hoy, M. A. (1978). Variability in diapause attributes of insects and mites: Some evolutionary and practical implications. In *Evolution of Insect Migration and Diapause*, H. Dingle, Ed., pp. 101–128. Springer-Verlag: New York.

Inoue, H., Ayuzawa, C., and Kawamura, A., Jr. (1972). Effect of high temperature on the multiplication of infectious flacherie virus in the silkworm, *Bombyx mori*. *Appl. Entomol. Zool.* **7,** 155–160.

Inoue, H. and Tanada, Y. (1977). Thermal therapy of the flacherie virus disease in the silkworm. *Bombyx mori*. *J. Invert. Pathol.* **29,** 63–68.

Jungries, A. M. (1978). Insect dormancy. In *Dormancy and Developmental Arrest: Experimental Analysis in Plants and Animals*, M. E. Culter, Ed. Academic: New York.

Kammer, A. E. (1968). Motor patterns during flight and warm-up in Lepidoptera. *J. Exp. Biol.* **48,** 89–109.

Kammer, A. E. (1970a). Thoracic temperature, shivering, and flight in the monarch, *Danaus plexippus* (L.). *Z. Vergl. Physiol.* **68,** 334–344.

Kammer, A. E. (1970b). A comparative study of motor patterns during pre-flight warm-up in hawkmoths. *Z. Vergl. Physiol.* **70,** 45–56.

Kammer, A. E. (1971a). Influence of acclimation temperature on the shivering behavior of the butterfly *Danaus plexippus* (L.). *Z. Vergl. Physiol.* **72,** 364–369.

Kammer, A. E. (1971b). The motor output during turning flight in a hawkmoth, *Manduca sexta. J. Insect Physiol.* **17,** 1073–1086.

Kammer, A. E. and Heinrich, B. (1974). Metabolic rates related to muscle activity in bumblebees. *J. Exp. Biol.* **61,** 219–227.

Kammer, A. E. and Heinrich, B. (1978). Insect flight metabolism. *Adv. Insect Physiol.* **13,** 133–228.

Kevan, P. G. (1975). Sun-tracking solar furnaces in High Arctic flowers: Significance for pollination and insects. Science **189,** 723–726.

Kevan, P. G. and Shorthouse, J. D. (1970). Behavioral thermoregulation by High Arctic butterflies. *Arctic* **23,** 268–279.

Kluger, M. J. (1979). Phylogeny of fever. *Fed. Proc.* **38,** 30–34.

Kluger, M. J., Ringler, D. H., and Anver, M. R. (1975). Fever and survival. *Science* **188,** 166–168.

Kohayashi, M. and Ishitoya, Y. (1964). Hormonal system on the control of the egg diapause in the silkworm, *Bombyx mori* L. *J. Seric. Sci. Japan* **33,** 111–114.

Kukalova, J. (1968). Permian mayfly nymphs. *Psyche* **75**(4), 310–327.

Kukalova-Peck, J. (1978). Origin and evolution of insect wings and their relation to metamorphosis, as documented by the fossil record. *J. Morphol.* **156,** 53–126.

Lange, C. A. (1963). The effect of temperature on the growth and chemical composition of the mosquito. *J. Insect Physiol.* **9,** 279–286.

Levitt, J. (1978). Role of SH and SS groups in damage to biological systems at low water activities. In *Dry Biological Systems*, J. H. Crowe and J. S. Clegg, Eds., pp. 243–256. Academic: New York.

Low, P. S. and Somero, G. N. (1976). Adaptation of muscle pyruvate kinases to environmental temperature and pressures. *J. Exp. Zool.* **198,** 1–12.

MacDougall, J. D., Reddan, W. G., Layton, C. R., and Dempsey, J. A. (1974). Effects of metabolic hyperthermia on performance during heavy and prolonged exercise. *J. Appl. Physiol.* **36,** 538–544.

Mansingh, A. (1971). Physiological classification of dormancies in insects. *Can. Entomol.* **103,** 783–1009.

May, M. L. (1976). Thermoregulation and adaptation to temperature in dragonflies (Odonata: Anisoptera). *Ecol. Monogr.* **46,** 1–32.

May, M. L. (1977). Thermoregulation and reproductive activity in tropical dragonflies of the genus *Micrathyria. Ecology* **58,** 787–798.

May, M. L. (1978). Thermal adaptations of dragonflies. *Odontologica* **7,** 27–47.

Meats, A. (1973). Rapid acclimation to low temperature in the Queensland fruit fly, *Dacus tryoni. J. Insect Physiol.* **19,** 1903–1911.

Meats, A. (1976). Developmental and long-term acclimation to cold by the Queensland fruit-fly (*Dacus tryoni*) at constant and fluctuating temperatures. *J. Insect Physiol.* **22**(7), 1013–1019.

Meats, A. and Fay, H. A. C. (1976). The effect of acclimation on mating frequencies and mating competitiveness in the Queensland fruit fly, *Dacus tryoni*, in optimal and cool mating regimes. *Physiol. Entomol.* **1**(3), 207–212.

Mellanby, K. (1940). The activity of certain arctic insects at low temperature. *J. Anim. Ecol.* **9**, 296–301.

Mellanby, K. (1954). Acclimation and the thermal death points in insects. *Nature* **173**, 582–583.

Miller, K. L. (1969). Freezing tolerance in an adult insect. *Science* **166**, 105–106.

Miller, K. L. (1978). Freezing tolerance in relation to cooling rate in an adult insect. *Cryobiology* **15**, 345–349.

Miller, P. L. (1974). Respiration-aerial gas transport. In *Physiology of Insecta* vol. 6, M. Rockstein, Ed., pp. 345–402. Academic: New York.

Miller, T. and Usherwood, P. N. R. (1971). Studies of cardioregulation in the cockroach, *Periplaneta americana*. *J. Exp. Biol.* **54**, 329–348.

Morrissey, R. E. and Baust, J. G. (1976). The ontogeny of cold tolerance in the gall fly, *Eurasta solidagensis*. *J. Insect Physiol.* **22**(3), 431–437.

Mosebach-Pukowski, E. (1938). Über die Raupengesellschaften von *Vanessa io* und *Vanessa urticae*. *Z. Morphol. Tiere* **33**, 358–380.

Mutchmor, J. A. and Richards, A. G. (1961). Low temperature tolerance of insects in relation to the influence of temperature on muscle apyrase activity. *J. Insect Physiol.* **7**, 141–158.

Nachtigall, W. (1967). Aerodynamische Messungen am Tragflügelsystem segelnder Schmetterlingen. *Z. Vergl. Physiol.* **54**, 210–231.

Neville, A. C. (1965). Energy economy in insect flight. *Sci. Prog.* **53**, 203–219.

Newall, R. C. (1969). Effects of fluctuations in temperature on the metabolism of intertidal invertebrates. *Am. Zool.* **9**, 293–307.

Newall, R. C. (1973). Environmental factors affecting the acclimation responses of ectotherms. In *Effects of Temperature on Ectothermic Organisms*, W. Wieser, Ed., pp. 151–164. Springer-Verlag, Heidelberg.

Odesser, D. B., Hayes, D. K., and Schecther, M. S. (1972). Phosphodiesterase activity in pupae and diapausing and non-diapausing larvae of the European corn borer, *Ostrinia nubilalis*. *J. Insect Physiol.* **18**(6), 1097–1105.

Ohyama, Y. and Asahina, E. (1972). Frost resistance in adult insects. *J. Insect Physiol.* **18**, 267–282.

Patterson, J. L. and Duman, J. G. (1979). Composition of a protein antifreeze from larvae of the beetle, *Tenelerio molitor*. *J. Exp. Biol.* **210**, 361–367.

Perutz, M. F. (1978). Electrostatic effects in proteins. *Science* **201**, 1187–1191.

Portier, P. (1930). Respiration pendent le vol chez les Lépidoptères. *C. R. Soc. Biol.* **105**, 760–764.

Prebble, M. L. (1941). The diapause and related phenomena in *Gilpinia polytoma* (Harty). V. Diapause in relation to epidemiology. *Can. J. Res.* **D19**, 437–454.

Rankin, M. A. (1978). Hormonal control of insect migratory behavior. In *Evolution of Insect Migration and Diapause*, H. Dingle, Ed., pp. 5–32. Springer-Verlag: New York.

Sacktor, B. (1974). Biological oxidations and energetics in insect mitochondria. In *The*

Physiology of Insecta, 2nd ed., vol. 4, M. Rockstein, Ed., pp. 271–353. Academic: New York.

Salt, R. W. (1959). Role of glycerol in the cold-hardiness of *Bracon cephi* (Gehan.). *Can. J. Zool.* **37,** 59–69.

Salt, R. W. (1961). Principles of insect cold-hardiness. *Ann. Rev. Entomol.* **6,** 55–74.

Salt, R. W. (1966). Factors influencing nucleation in supercooled insects. *Can. J. Zool.* **44,** 117–133.

Salt, R. W. (1969). The survival of insects at low temperatures. *Symp. Soc. Exp. Biol.* **23,** 331–350.

Schechter, M. S., Hayes, D. K., and Sullivan, W. N. (1971). Manipulation of photoperiod to control insects. *Israel J. Entomol.* **6**(2), 143–166.

Scholander, P. F., Flagg, W., Irving, R. J., and Irving, L. (1953). Studies on the physiology of frozen plants and animals in the Arctic. *J. Cell Comp. Physiol.* **42,** suppl. 1, 1–56.

Seymour, R. S. (1974). Convective and evaporative cooling in sawfly larvae. *J. Insect Physiol.* **20,** 2447–2457.

Shapley, H. (1924). Note on the thermokinetics of Dolichodrine ants. *Proc. Nat. Acad. Sci. U.S.* **10,** 436–439.

Sherman, P. W. and Watt, W. B. (1973). The thermal ecology of some *Colias* butterfly larvae. *J. Comp. Physiol.* **83,** 25–40.

Singh, S. P. and Das, A. B. (1977). Thermal acclimation in respiratory metabolism of the cockroach *Periplaneta americana* (Linn.). *Ind. J. Exp. Biol.* **15**(2), 108–112.

Smith, D. S. (1961). The organization of the flight muscle in a dragonfly, *Aeshna* sp. (Odonata). *J. Biophys. Biochem. Cytol.* **11,** 119–146.

Smith, T. L. (1941). Some notes on the development and regulation of heat among *Galleria* larvae. *Arkansas Acad. Sci.* **1,** 29–33.

Somero, G. N. (1975). Temperature as a selective factor in protein evolution: The adaptive strategy of "compromise." *J. Exp. Zool.* **194,** 175–188.

Sømme, L. (1964). Effects of glycerol on cold-hardiness in insects. *Can. J. Zool.* **42,** 87–101.

Sømme, L. (1965). Further observations on glycerol and cold-hardiness in insects. *Can. J. Zool.* **43,** 765–770.

Sotavalta, O. (1947). The flight-tone (wing-stroke frequency) of insects. *Acta Entomol. Fenn.* **4,** 1–117.

Stegwee, D. (1964). Respiratory chain metabolism in the Colorado potato beetle. II. Respiration and oxidative phosphorylation in "sarcosomes" from diapausing beetles. *J. Insect Physiol.* **10,** 97–102.

Stevens, E. D. and Josephson, R. K. (1977). Metabolic rate and body temperature in singing katydids. *Physiol. Zool.* **50,** 31–42.

Stokes, J. L. (1967). Heat sensitive enzyme and enzyme synthesis in psychrophilic microorganisms. In *Molecular Mechanisms of Temperature Adaptation*, C. L. Prosser, Ed., pp. 311–323. Publ. No. 84. American Association for the Advancement of Science: Washington, D.C.

Sullivan, C. R. and Wellington, W. G. (1953). The light reactions of larvae of the tent caterpillars, *Malacosoma disstria* Hbn., *M. americanum* (Fab.), and *M. pluviale* (Dyar) (Lepidoptera: Lasiocampidae). *Can. Entomol.* **85,** 297–310.

Takehara, I. and Asahena, E. (1960). Frost resistance and glycerol content in overwintering insects. *Low Temp. Sci.* **B18,** 57–65.

Tanada, Y. (1967). Effect of high temperatures on the resistance of insects to infectious diseases. *J. Sericult. Sci. Japan* **36,** 333–339.

Tanno, K. (1964). High sugar levels in the solitary bee, *Ceratina. Low Temp. Sci.* **B22,** 51–57.

Tanno, K. (1967). Freezing injury in fat-body cells of the poplar sawfly. In *Cellular Injury and Resistance in Freezing Organisms*, E. Asahina, Ed., pp. 245–257. Institute of Low Temperature Science: Sapporo, Japan.

Tanno, K. (1968). Frost resistance in the poplar sawfly, *Trichiocampus populi* Okamoto. IV. Intracellular freezing in fat-cells and injury occurring upon metamorphosis. *Low Temp. Sci.* **B26,** 71–78.

Taylor, C. R. (1974). Exercise and thermoregulation. In *Environmental Physiology,* D. Robertshaw, Ed. MTP International Review of Science, Ser. 1. Butterworth: London.

Taylor, C. R. and Lyman, C. P. (1972). Heat storage in running antelopes: Independence of brain and body temperatures. *Am. J. Physiol.* **222,** 112–117.

Tekle, A. (1960). The physiology of hibernation and its role in the geographical distribution of populations of the *Culex pipiens* complex. *Am. J. Trop. Med. Hyg.* **9,** 321–330.

Thiessen, C. I. and Mutchmor, J. A. (1967). Some effects of thermal acclimation on muscle apyrase activity and mitochondrial number in *Periplaneta americana* and *Musca domestica. J. Insect Physiol.* **13,** 1837–1842.

Tiegs, O. W. (1955). The flight muscles of insects—Their anatomy and histology; with some observations on the structure of striated muscle in general. *Phil. Trans.* **B238,** 221–347.

Tozer, W. (1979). Underwater behavioral thermoregulation in the adult stonefly, *Zapada cinctipes. Nature* **281,** 566–567.

Vielmetter, W. (1958). Physiologie des Verhaltens zur Sonnenstrahlung bei den Tagfalter *Argynnis paphia* L. I. Untersuchungen im Freiland. *J. Insect Physiol.* **2,** 13–37.

Watanabe, H. and Tanada, Y. (1972). Infection of nuclear-polyhedrosis virus in armyworm, *Pseudoletia unipuncta* Haworth (Lepidoptera: Noctuidae), reared at high temperatures. *Appl. Entomol. Zool.* **7,** 43–51.

Watt, W. B. (1968). Adaptive significance of pigment polymorphisms in *Colias* butterflies. I. Variation of melanin pigment in relation to thermoregulation. *Evolution* **22,** 437–458.

Wenner, A. M. (1964). Sound communication in honeybees. *Sci. Am.* 1–10.

Wieser, W. (1973). Effects of temperature on ectothermic organisms. Springer-Verlag: New York.

Wigglesworth, V. B. (1976). The evolution of insect flight. In *Insect Flight*, R. C. Rainey, Ed., pp. 255–269. Wiley: New York.

Williams, C. M. (1969). Photoperiodism and endocrine aspects of insect diapause. *Symp. Soc. Exp. Biol.* **23,** 285–300.

Wilson, D. M. (1962). Bifunctional muscles in the thorax of grasshoppers. *J. Exp. Biol.* **39,** 669–677.

Wunder, B. A. (1974). The effect of activity on body temperature of Ord's kangaroo rat (*Dipodomys ordii*). *Physiol. Zool.* **47,** 26–36.

Wyatt, G. R. and Kalf, G. F. (1958). Organic compounds of insect hemolymph. *Proc. 10th Intern. Congr. Entomol. Montreal* **2,** 333.

Zebe, E. (1960). Condensing enzyme und β-keto-acyl-thiolase in verschiedenen Muskeln. *Biochem. Z.* **322,** 328–332.

CONCLUDING DISCUSSION

The following discussion with the audience and among the participants was recorded immediately after the presentation of all the papers. It has here been shortened and edited. The initials refer to the names of the symposium speakers.

Q Dave Byman (Colorado State University)

Dr. Heinrich, you showed that there is a difference in the activity of solitary bees versus bumblebees in relation to ambient temperature. Do you find that solitary bees operate over a smaller portion of the day and that their activity is more limited by cold weather?

BH

Both. Bumblebees (*Bombus* spp.) are often active at much lower temperatures than solitary bees (principally andrenids) in Maine. They can start earlier when it is still cold. Bumblebees can also be active at high temperatures but, as I indicated, there were few bumblebees and many solitary bees active at the flowers (raspberry) during the warm later parts of the day. Possibly it was easier for the small solitary bees to thermoregulate at high rather than at low T_A. However, later in the day there was less nectar available, and this nectar was perhaps not sufficient for the bumblebees but still ample for the smaller solitary bees with more modest energy budgets.

Q (Anonymous)

I am curious, Dr. Casey, why you excluded orientation in temperature gradients from your consideration of temperature regulation. It seems likely to be a component of the situation, and the controlling mechanisms are likely to be similar to those you discussed. I perceive that what you were talking about is really more *posturing* mechanisms, and as such just a part of the larger scheme of behavior.

TC

I really didn't want to get into taxis-kinesis arguments, which are based primarily on temperature gradients, because they are extremely difficult to evaluate in relation to the natural situation that interested me. It becomes exceedingly difficult to evaluate behavior under natural conditions when there is good evidence to suggest that the animals are orienting with respect to infrared radiation on the ground, with respect to wind velocity, and with respect to direct solar radiation.

Q Bill Reynolds (Penn State University)

Have there been any studies on aquatic insects, particularly diving beetles, that are immersed in water where we need not consider radiation and evaporation?

GB

Large beetles (Dytiscidae) can crawl out of the water at 20°C, warm up, and fly away. Some years ago D. Leston, J. W. S. Pringle, and D. C. S. White (*J. Exp. Biol.* **42,** 409, 1965) made measurements on water beetles that warmed up to 38°C and then flew. Water beetles function in water down to almost freezing, but they probably use completely different sets of muscular machinery for swimming and flying.

Q Richard Bellamy (University of Massachusetts)

Dr. Heinrich, I wonder if small insects can fly at a variety of ambient temperatures? Has there been any biochemical work done on temperature acclimation in these insects?

BH

The two examples I showed, *Culex tarsalis* and *Musca domestica,* happened to have been from laboratory culture. Both fly over an extremely wide range of body temperatures. *Culex tarsalis* is a mosquito that lives throughout the western United States, down into Mexico, and up through Canada, and the interesting thing is that some individuals of this species could fly in a temperature-controlled lab room at 9°C, quite close to the lower limit of temperature at which mosquitoes in the arctic fly. But almost nothing is known of thermal adaptation or functioning at the underlying biochemical level.

Q Albert Bennett (University of California, Irvine)

Dr. Bartholomew, I find your apostasy in regard to vertebrates rather disturbing. I want to ask you to comment on the insect versus mammal and bird condition with respect to limitations on oxygen transport and utilization. It may be that the sorts of systems vertebrates use cannot be miniaturized to achieve the high levels of weight-specific oxygen consumption reached by some insects. Do you believe that insect systems reach an upper size limit that prevents insects from becoming much larger?

GB

I have often thought of this and I do have ideas that I can share with you. I think probably that the system that depends on cardiovascular transport of oxygen and carbon dioxide can go to a very, very small size. But the actual rates of metabolism are very low in these creatures. I doubt if it is specifically the inability of the transport system that says 2 g is small. I think that it simply gets so difficult that other factors become limiting. That, plus other things. I think the more interesting question is: Why is it that insects don't get any bigger than they do with their tracheal system? And one of the things that has impressed me is that the very largest insects, like the very large dung beetles that Heinrich and I have studied that get up to 25 g, and the rhinoceros beetles of tropical America which get up to 30 g, are all strong and willing flyers. They fly long distances at high velocities. I won't repeat the arguments here—but Heinrich and I convinced ourselves that they are unloading an enormous amount of heat (*J. Exp. Biol.* **73**, 65, 1978, and summarized by A. Kammer, this volume) and are in fact having to regulate their body temperature as a consequence. It seems to me fairly clear that, even in the very largest of insects, the oxygen transport system by the trachea is quite adequate. The very largest birds, however, cannot fly. In fact, the cutoff size for flight for birds is far short of their maximum size. The same is true of animals; the biggest flying mammals—the pteropid bats—weigh only a kilo. It looks to me as if the oxygen supply system is not the limiting factor on insect size. It's going to be something else. It looks to me that the tracheal system could go on and get bigger. If this were not the case, the very biggest insects would be nonflying insects crawling around with very low energy rates. While I don't think the oxygen supply system of insects is limiting to the largest size presently extant, I do think it possible that the cardiovascular system and lungs of vertebrates cannot supply enough oxygen to let the birds and mammals get much smaller. However, I have not thought this through far enough to take a very strong stand. It really is terribly complicated whenever you come to study limits. One of the traps that biologists have been falling into ever since the first biologist made a generalization was looking for simple answers to complex questions, and usually single answers to questions that require multiple answers.

Q Richard Cobb (State University of Buffalo)
Dr. Casey, you pointed out that in butterflies the wings act as very important heat exchangers. Did you establish whether or not that was a conductive process, or does it involve some type of convection?

TC
The information from L. T. Wasserthal's study (*J. Insect Physiol.* **221**, 1921, 1975) is by no means clear. Basically, he finds that there is selective absorption as a result of the darker wing base. This tends to increase the

ambient temperature beneath the surface of the wings. I think Dr. Kammer has data showing that cutting off the wings and moving them close together reestablishes heat transfer. Certainly Dr. Douglas (Boston University) does.

AK

Yes. I think that the rate of blood flow through the wings quite clearly is not sufficient to be responsible for any kind of heat transfer. Apparently there is warming and changes in the airflow around the animal, and the warm air then warms the body—rather than any conduction in the wings. Experimentally you get some warming just by having the wings near the animal, as opposed to removing them entirely, even though you sever them and displace them slightly. It then makes sense that the basal part of the wing is important. This is also true in lateral baskers because there is an air space between the wings that is warmed and thus warms the body. However, nobody has looked carefully at the cuticle to see how good it is at conducting heat.

TC

Matt Douglas, who cut off the wings and moved them within approximately a millimeter away from the body, reported virtually the same heating effect by the wings, which would tend to negate the conductive process.

Q Hal Dewitt (University of Wisconsin)

I wondered if anyone has given any thought to the thinness of the waist connecting the thorax and the abdomen in relation to the need and use of a countercurrent heat exchange system. In other words, insects requiring a high thoracic temperature would benefit from isolation of the thorax from the abdomen. Is that related to the thinness of the waist?

BH

Sphinx moths, for example, have a fairly wide waist, but they still achieve isolation or insulation of the thorax from the abdomen by having an intervening air space. The anatomy required for countercurrent exchange is probably more likely given a narrow waist, but it need not be a prerequisite. A narrow waist is observed in the Hymenoptera whether they are large and endothermic or very tiny and poikilothermic, like small parasitic wasps. I think in this group the narrow waist may play a more primary role in maintaining maneuverability of the abdomen for stinging and possibly egg laying.

Q Steve Troutleg (Southeastern Missouri State University)
Today we've heard a lot about the temperature of the thorax and the temperature of the abdomen. There's a glaring emptiness here around the temperature of the head. I'm wondering about this absence.

GB
Tim Casey and I took the mouth and prothoracic temperatures of some big beetles. The prothorax is cool and the metathorax and head is still cooler, but we did find that head temperatures were 5–6°C above ambient in some big beetles.

BH
Another comment to your question. Bees in flight at high air temperatures will extrude a droplet of fluid from their tongue, wag it in the breeze, and that causes evaporative cooling. This reduces head temperature and might prevent the head from overheating when the hot blood from the hot flight muscles rushed to the head. (This has subsequently been confirmed; see B. Heinrich, *Science* **205,** 1269, 1979.)

AK
A related point of course is that if head temperature is regulated then my comments about the thorax being the main site of temperature regulation have to be reevaluated and we have to begin to deal also with neural mechanisms in the head in relation to body temperature regulation. As I mentioned earlier, in my talk, there is one experiment by Hanegan and Heath (*J. Exp. Biol.* **59,** 67, 1973) using thermodes to heat the thoracic ganglia of a large moth, *H. cecropia,* giving reason to think there is thermal sensitivity in the thoracic ganglia. But no one, as far as I know, has tried selective heating of the head to see if it has thermal sensitivity (This has subsequently been examined, see B. Heinrich, *J. Exp. Biol.* **80,** 217, 1980). However, there is a nice study on locusts that I didn't have time to talk about, showing that there is a temperature threshold for firing in certain motor neurons. But no one has looked at this problem in relation to thermoregulation.

Q Harvey Lilliwhite (University of Kansas)
Are evaporative heat losses from the tracheal system significant in relation to heat exchange in warm-bodied insects?

BH
I calculated it at one time for locusts, honeybees, and moths (B. Heinrich, in *Environmental Physiology of Desert Organisms,* N. F. Hadley, Ed.,

1975) based on the assumption that the air would become saturated with water vapor as it goes through the tracheal system, and on the amount of water that would be taken up at different temperatures and relative humidities. There was a substantial amount of heat that would be lost by evaporation, but it would make relatively little difference in overall temperature excess.

RJ

Church, N. S. (*J. Exp. Biol.* **37,** 171, 1960) looked at this directly in locusts. He compared the amount by which they warmed during flight in dry and moist air, in moist air there being little evaporative cooling. The thoracic temperature excess during flight was about 6°C, and the difference between the temperature excess in moist and dry air was less than 1°C, indicating little evaporative cooling.

AK

E. B. Edney and R. Barrass (*J. Insect Physiol.* **8,** 488, 1963) showed a significant effect of evaporative cooling in tse-tse flies, *Glossina morsitans*, feeding on the hot hide of a mammal in sunshine. The flies opened their spiracles and cooled about 1½°C. But that's an unusual situation in which water is temporarily available to the insect in superabundance, and it's more of an emergency measure during feeding than any kind of regulated state.

RJ

It's probably a consequence of evaporative cooling and a reflection of the low resting metabolic rate of insects that when you first insert a temperature probe you find that the inside of most insects is slightly below ambient temperature.

BH

Yes, this is true even in insects such as cockroaches that do not regulate their body temperature, at the higher T_A where there is a higher saturation deficit. Their body temperature is several tenths of a degree Celsius above T_A at low T_A, and several tenths below T_A at high T_A, provided the air is dry. I should point out, however, that in metabolically very active flying insects it would make a difference on the distribution of heat on whether the air was flowing in from the abdomen to cause evaporative cooling (and to become saturated) before entering the heated thorax, or whether it was entering the thorax causing cooling, and condensed to give off heat in the abdomen. This possibility has been discussed in greater detail (see B. Heinrich, in *Environmental Physiology of Desert Organisms* in N. S.

Hadley, Ed., 1975) but there is no experimental evidence so far that such a reflux condenser is used as part of the cooling system.

Q (DeWitt)
In insects that are active flyers, is there, in fact, a unidirectional airflow in the tracheal system?

AK
There is in flying locusts, that is, air comes in the anterior spiracles and goes out the posterior (Weiss-Fogh, T. *J. Exp. Biol.* **47,** 561, 1967). In the honeybee in flight most of the air enters the thoracic spiracles and leaves by the abdominal ones (Bailey, L. *J. Exp. Biol.* **31,** 589, 1954).

Q (Reynolds)
In the discussion on the abdomen as a heat exchanger and on the upper lethal temperatures, I didn't perceive mention of the fact that it's difficult to lose heat to an environment warmer than onself other than by water loss.

BH
This is certainly true. The point has been made repeatedly for vertebrate homeotherms relying on evaporative cooling that the higher the body temperature they can tolerate, the more readily they lose heat by convection and the more they save water. A high body temperature in vertebrates may thus be in part related to water economy. But insects, except for the exceptions just mentioned, don't rely on evaporative cooling. They are for the most part highly adapted for water conservation. The higher their temperature set point, the easier it is for them to get rid of heat by convection, but the more costly it is in terms of time and energy for warm-up.

Q Richard L. Marsh (University of Michigan)
Dr. Heinrich, it bothers me that in your consideration of the evolution of high body temperatures in insects, you have left out advantages due to what I call molecular efficiency, which is the rate of activity per molecule of enzyme. There seems to be an advantage accruing to an animal operating at high temperatures in terms of being able to operate at high rates with a given amount of protein packaged in the cell. You implied that insects were endothermic because they had to be; because they were producing heat, they were obligatorily endothermic. You implied that there was no *inherent* advantage to endothermy such that smaller insects might stay warm despite the fact that they could keep cool under all environmental conditions they would be exposed to.

BH

It seems to me that enzymes can be adapted to operate at different temperatures. But that does not exclude the idea that enzymes are also inherently more efficient at a higher temperature, say about 39–42°C. There is a recent article by N. O. Calloway (*J. Theor. Biol.* **57**, 331, 1976) in which he claims that water has certain properties that would lend themselves to more efficient functioning in biological systems at near 40°C, thus setting severe constraints on the evolution of temperature adaptation in enzymes.

AK

I think that when you consider the range of temperatures over which animals function, and Bernd mentioned beetles and other insects that operate at near-freezing temperatures, then it seems that evolution can build an enzyme that will function at any temperature within, say, the biological range. There isn't an inherent advantage to working at a high temperature. There is, however, a Q_{10} effect that does operate over a much smaller range. Over a small range of temperatures, maybe about 10°C in bees, you have a Q_{10} effect on muscles designed to work at those high temperatures. But that Q_{10} effect operates to advantage then within the larger context of enzymes that have evolved to work at whatever temperature the animal is in fact encountering when active. You may not "buy" that.

Q (Marsh)

Yes. If your assumption were true, that evolution could make enzymes work just as efficiently at low temperatures—and remember that I've defined efficiency as activity per amount of enzyme protein—I would buy your argument totally, but there is no evidence that that's the case. The evidence is to the contrary. Although some compensation is possible, the best data that's always quoted from G. Somero indicate that there is still a huge temperature effect when you consider animals adapted to operate at high temperature and those at low temperature. Additionally, in a study of growth rates in different species of algae from different temperature environments, there was a Q_{10} of approximately 1.86. So what I pointed out is that I think you're perhaps dismissing a little too easily an inherent advantage of being warm in terms of molecular efficiency.

BH

Your point is well taken. I don't mean to dismiss it. There are probably more reasons than the one I discussed.

Author Index

Numbers in *italics* indicate the pages on which full references are listed.

Subject Index